ENERGY BAND THEORY

PURE AND APPLIED PHYSICS
A SERIES OF MONOGRAPHS AND TEXTBOOKS

CONSULTING EDITOR

H. S. W. MASSEY

University College, London, England

Volume 1. F. H. FIELD and J. L. FRANKLIN, Electron Impact Phenomena and the Properties of Gaseous Ions. 1957

Volume 2. H. KOPFERMANN, Nuclear Moments. English Version Prepared from the Second German Edition by E. E. SCHNEIDER. 1958

Volume 3. WALTER E. THIRRING, Principles of Quantum Electrodynamics. Translated from the German by J. BERNSTEIN. With Corrections and Additions by WALTER E. THIRRING. 1958

Volume 4. U. FANO and G. RACAH, Irreducible Tensorial Sets. 1959

Volume 5. E. P. WIGNER, Group Theory and Its Application to the Quantum Mechanics of Atomic Spectra. Expanded and Improved Edition. Translated from the German by J. J. GRIFFIN. 1959

Volume 6. J. IRVING and N. MULLINEUX, Mathematics in Physics and Engineering. 1959

Volume 7. KARL F. HERZFELD and THEODORE A. LITOVITZ, Absorption and Dispersion of Ultrasonic Waves. 1959

Volume 8. LEON BRILLOUIN, Wave Propagation and Group Velocity. 1960

Volume 9. FAY AJZENBERG-SELOVE (ed.), Nuclear Spectroscopy. Parts A and B. 1960

Volume 10. D. R. BATES (ed.), Quantum Theory. In three volumes. 1961

Volume 11. D. J. THOULESS, The Quantum Mechanics of Many-Body Systems. 1961

Volume 12. W. S. C. WILLIAMS, An Introduction to Elementary Particles. 1961

Volume 13. D. R. BATES (ed.), Atomic and Molecular Processes. 1962

Volume 14. AMOS DE-SHALIT and IGAL TALMI, Nuclear Shell Theory. 1963

Volume 15. WALTER H. BARKAS. Nuclear Research Emulsions. Part I, 1963
Nuclear Research Emulsions. Part II. In preparation

Volume 16. JOSEPH CALLAWAY, Energy Band Theory. 1964

Volume 17. JOHN M. BLATT, Theory of Superconductivity. 1964

In preparation

R. E. BURGESS, Fluctuation Phenomena in Solids
F. A. KAEMPFFER, Concepts in Quantum Mechanics

ENERGY BAND THEORY

Joseph Callaway
Department of Physics
University of California
Riverside, California

1964

ACADEMIC PRESS New York and London

COPYRIGHT © 1964, BY ACADEMIC PRESS INC.
ALL RIGHTS RESERVED
NO PART OF THIS BOOK MAY BE REPRODUCED IN ANY
FORM, BY PHOTOSTAT, MICROFILM, OR ANY OTHER MEANS,
WITHOUT WRITTEN PERMISSION FROM THE PUBLISHERS.

ACADEMIC PRESS INC.
111 Fifth Avenue, New York 3, New York

United Kingdom Edition published by
ACADEMIC PRESS INC. (LONDON) LTD.
Berkeley Square House, London W.1

LIBRARY OF CONGRESS CATALOG CARD NUMBER: 63–16720

LIBRARY
LOS ANGELES COUNTY MUSEUM
EXPOSITION PARK

PRINTED IN THE UNITED STATES OF AMERICA

To Mary

Preface

Solid state physics has grown rapidly in the past two decades. Unprecedented developments in technology have been accompanied by a substantial refinement and extension of the fundamental theory. New phenomena have been discovered and interpreted, and old puzzles clarified. The one-electron theory has come to maturity with the development of powerful methods of calculation and successful applications to real crystals. It is now possible to describe the results of numerous experiments concerning both semiconductors and metals in the language of band theory in a consistent fashion with the use of a small number of parameters. Calculations from first principles are able to reproduce most of the important qualitative features and some quantitative characteristics of the energy levels of actual materials.

This book contains a discussion of the principles and methods of the calculation of the energy levels of electrons in crystals. In Chapter 1, the language of band theory is developed with attention to the general features of band structures which may be deduced from considerations of crystal symmetry. Chapter 2 includes a description of the principal methods for the solution of the Schrödinger equation in a periodic potential together with a survey of the problems encountered in the construction of the potential function in the Hartree-Fock approximation. Some of the results of experimental investigations of the band structures of materials, including the alkali metals, the noble metals, and common semiconductors, are surveyed in Chapter 3, and compared with theoretical calculations. The effects of external perturbations, including electric and magnetic fields, on band structures are considered in Chapter 4, together with a discussion of the energy levels of electrons bound to point impurities. A calculation of optical constants is also included.

Some major omissions should be noted. Transport theory is not discussed. Alloys and disordered materials are not considered. Many-body theory, the electron lattice interaction, and superconductivity are not included. The neglect of these topics is not in any respect an attempt to deny their fundamental place in the theory of solids, but rather seemed to be required if the size of this book and the labor of preparation were to be restricted to reasonable limits.

I am indebted to Dr. Thomas Wolfram for many valuable discussions, and to Miss Sara Cecil for the preparation of the manuscript.

JOSEPH CALLAWAY

Riverside, California
June 1963

Contents

PREFACE vii

Chapter 1
The Language of Band Theory

    1.1 Bloch's Theorem 2
    1.2 The Reciprocal Lattice 4
    1.3 The Brillouin Zone 5
    1.4 Energy Bands in the Free Electron Limit 9
    1.5 Space Groups 11
    1.6 Irreducible Representations 16
    1.7 The Effective Mass 29
    1.8 The Density of States 37
    1.9 The Fermi Surface in the Free Electron Approximation 43
    1.10 Spin Orbit Coupling 46
    1.11 Time Reversal Symmetry 52

Chapter 2
Methods of Calculation

    2.1 General Discussion 55
    2.2 Plane Wave Expansions 56
    2.3 Plane Wave Expansions: An Example 64
    2.4 Orthogonalized Plane Waves 68
    2.5 The Pseudopotential 74
    2.6 The Cellular Method 77
    2.7 The Effective Mass 80
    2.8 Variational Methods 86
    2.9 The Augmented Plane Wave Method 95
    2.10 The Tight Binding Approximation 102
    2.11 Wannier Functions 108
    2.12 The Hartree-Fock Equations 112
    2.13 Determination of the Crystal Potential 120
    2.14 The Quantum Defect Method 124
    2.15 The Pseudopotential in Relation to the Quantum Defect Method . 130

Chapter 3
Band Structure of Materials

3.1 The Alkali Metals: Cohesive Energy 133
3.2 Cohesive Energies of the Alkali Metals: Results: Lattice Constant and Compressibility 140
3.3 Band Calculations in the Alkali Metals 146
3.4 Wave Functions for the Alkali Metals 154
3.5 Valence Crystals: Diamond, Germanium, and Silicon 157
3.6 Valence Crystals: Results of Calculations 167
3.7 Zinc Blende Structures 174
3.8 Aluminum 177
3.9 The Noble Metals 181
3.10 d Bands and Transition Metals 191
3.11 Bismuth 203
3.12 Graphite 209
3.13 Summary 212

Chapter 4
Point Impurities and External Fields

4.1 General Discussion 215
4.2 Point Impurities 221
4.3 The Effective Mass Equation 233
4.4 The Steady Magnetic Field 244
4.5 The Magnetic Susceptibility of Free Electrons 257
4.6 The Cyclotron Frequency for an Arbitrary Fermi Surface 265
4.7 The Steady Diamagnetic Susceptibility: Arbitrary Band Structure . 269
4.8 The Steady Electric Field 276
4.9 Optical Properties of Semiconductors 283
4.10 Optical Absorption by Free Carriers 289
4.11 Optical Properties in a Magnetic Field 302

Appendix 1. Some Symmetrized Linear Combinations of Plane Waves . . 307
Appendix 2. Summation Relations 311
Appendix 3. The Effective Mass Equation in the Many Body Problem . . 315
Appendix 4. Evaluation of a Tunneling Integral 320
Appendix 5. Spin Density Waves 324

Bibliography 327

AUTHOR INDEX 349
SUBJECT INDEX 355

Chapter 1

The Language of Band Theory

One of the principal objectives of physics in this century is to account for the observed properties of macroscopic solid bodies on the basis of the quantum mechanical theory of the behavior of atomic particles. Success will have been achieved when it becomes possible to calculate the quantities which describe the constitution of materials and their response to alterations of environment from knowledge of the component elements and their proportions. It is already possible, in principle, to perform these computations, since the forces between charged particles are known and the basic principles of quantum mechanics and quantum statistics are firmly established. Practically, rigorous procedures tend to produce mathematical problems of unmanageable complexity except in idealized cases.

It is necessary to concentrate study on those aspects of the structure of materials which seem amenable to approximate analysis. Much can be inferred from a description of the energy levels of electrons in solids. It can be determined immediately whether a material should be a metal or an insulator; calculations of binding energies are possible, and some of the essential ingredients are available for determination of transport coefficients. This book contains a discussion of the general nature of the energy levels, the principal methods of calculation, the effect of external electric and magnetic fields on a given set of levels, and the principles underlying some of the experimental procedures for testing our ideas.

It is already implied, in speaking of energy levels of electrons in solids, that it is possible to separate the descriptions of the electronic and nuclear motions, and that the influence of the nuclear motions on the electrons is small. This is a famous and fundamental approximation which, however, occasionally fails, as in the phenomena of superconductivity. The problem of determining the effects of the interaction of the electrons with each other is also both serious and very difficult. A fundamental idealization which is often made is to restrict attention to the states of a single electron in a rigid, infinite, periodic lattice. The assembly of many electrons is then regarded as a collection of noninteracting particles in the periodic potential, occupying states in accord with the rules of the Fermi statistics Fortunately, it turns out that many of the properties of real materials can be described successfully in the language developed for this simple model. Quantitative results can even be obtained in some interesting cases. The present discussion is largely based on this approximation, but there will be occasional comments on its accuracy, and the means by which it may be improved.

1.1 Bloch's Theorem

The periodicity of a crystal is described by specifying a set of vectors $\mathbf{R}_i$, such that if $V(\mathbf{r})$ is the potential energy of an electron at position $\mathbf{r}$ in the crystal, this potential is unchanged if the point of observation is displaced by any vector $\mathbf{R}_i$ which belongs to a set of translation vectors.

$$V(\mathbf{r} + \mathbf{R}_i) = V(\mathbf{r}) \qquad (1.1)$$

All the vectors $\mathbf{R}_i$ may be expressed in terms of three independent primitive translation vectors $\mathbf{a}_1$, $\mathbf{a}_2$, $\mathbf{a}_3$ in the following way:

$$\mathbf{R}_i = n_{i1}\,\mathbf{a}_1 + n_{i2}\,\mathbf{a}_2 + n_{i3}\,\mathbf{a}_3 \qquad (1.2)$$

The n_{ij} are integers. It is convenient to introduce a set of translation operators $T(\mathbf{R}_i)$ which have the property that, if $f(\mathbf{r})$ is any function of position,

$$T(\mathbf{R}_i)f(\mathbf{r}) = f(\mathbf{r} + \mathbf{R}_i) \qquad (1.3)$$

These operators evidently form a group; they also commute with each other. As a result of (1.1), the translation operators commute with the

Hamiltonian operator for one electron in the periodic potential. This Hamiltonian is the sum of the kinetic energy operator and the potential energy mentioned previously:

$$H = \frac{-\hbar^2}{2m} \nabla^2 + V(\mathbf{r}) \tag{1.4}$$

thus, we have:

$$[T(\mathbf{R}_i), H] = 0 \tag{1.5}$$

Consequently, the wave function of the electron may be chosen to be simultaneously an eigenfunction of the energy and of all the translations. Let ψ be such an eigenfunction:

$$T(\mathbf{R}_i)\psi(\mathbf{r}) = \psi(\mathbf{r} + \mathbf{R}_i) = \lambda_i \psi(\mathbf{r}) \tag{1.6}$$

If we multiply this equation by its complex conjugate, we have

$$|\psi(\mathbf{r} + \mathbf{R}_i)|^2 = |\lambda_i|^2 |\psi(\mathbf{r})|^2$$

Since the electron distribution determines in part the potential in which the electrons move, it is necessary for consistency that the electron distribution have the same periodicity as the potential. Consequently,

$$|\psi(\mathbf{r} + \mathbf{R}_i)|^2 = |\psi(\mathbf{r})|^2 \tag{1.7}$$

It follows that λ_i must be a complex number of modulus unity:

$$\lambda_i = e^{i\theta_i}$$

Now let two translation operators, say $T(\mathbf{R}_j)$ and $T(\mathbf{R}_i)$, act in succession. This translation is equivalent to that produced by the single operator $T(\mathbf{R}_j + \mathbf{R}_i)$.

$$T(\mathbf{R}_j)T(\mathbf{R}_i)\psi(\mathbf{r}) = \psi(\mathbf{r} + \mathbf{R}_i + \mathbf{R}_j) = \lambda_j \lambda_i \psi(\mathbf{r}) = \lambda_{j+i} \psi(\mathbf{r})$$

Evidently, the product of the eigenvalues corresponding to different displacements must be equal to the eigenvalue of the combined translation. This condition will be satisfied if

$$\theta_i = \mathbf{k} \cdot \mathbf{R}_i$$

where $\mathbf{k}$ is an arbitrary vector that is the same for each of the operations.

The wave function of an electron in a periodic potential is characterized by the particular vector **k** which appears in the eigenvalue of each translation operation, and the vector **k** is generally written as a subscript on ψ. The allowed values of **k** will be determined in Section 1.3.

We have now established Bloch's theorem (Bloch, 1928):

$$\psi_\mathbf{k}(\mathbf{r} + \mathbf{R}_i) = e^{i\mathbf{k}\cdot\mathbf{R}_i} \psi_\mathbf{k}(\mathbf{r}) \tag{1.8}$$

This theorem may be interpreted as a boundary condition on the solution of the Schrödinger equation in the periodic potential.

It is desirable to define a function $u_\mathbf{k}(\mathbf{r})$ by the relation

$$\psi_\mathbf{k}(\mathbf{r}) = e^{i\mathbf{k}\cdot\mathbf{r}} u_\mathbf{k}(\mathbf{r}) \tag{1.9}$$

Then from Bloch's theorem

$$\psi_\mathbf{k}(\mathbf{r} + \mathbf{R}_i) = e^{i\mathbf{k}\cdot(\mathbf{r}+\mathbf{R}_i)} u_\mathbf{k}(\mathbf{r} + \mathbf{R}_i) = e^{i\mathbf{k}\cdot\mathbf{R}_i} [e^{i\mathbf{k}\cdot\mathbf{r}} u_\mathbf{k}(\mathbf{r})]$$

Thus $u_\mathbf{k}$ must be unchanged by a translation through any lattice vector: It has the full periodicity of the potential.

1.2 The Reciprocal Lattice

It is convenient to define a set of vectors $\mathbf{K}_j$ through the relation

$$\mathbf{K}_j \cdot \mathbf{R}_i = 2\pi n_{ij} \tag{1.10}$$

The quantity n_{ij} is to be an integer (negative and zero values included). A relation of this form is required to hold for all translation vectors $\mathbf{R}_i$. The end points of all vectors $\mathbf{K}_j$ define a lattice of points, which is called the reciprocal lattice.

In (1.2) the lattice translation vectors $\mathbf{R}_i$ were expressed as linear combinations of the basic translation vectors $\mathbf{a}_j$ with integer coefficients. A similar relation holds for the reciprocal lattice vectors $\mathbf{K}_j$.

$$\mathbf{K}_j = g_{j1}\mathbf{b}_1 + g_{j2}\mathbf{b}_2 + g_{j3}\mathbf{b}_3 \tag{1.11}$$

The numbers g_{ji} are integers and the vectors $\mathbf{b}_i$ are fundamental translation vectors of the reciprocal lattice. The defining relation for the $\mathbf{K}_j$, Eq. (1.10), always holds provided that vectors $\mathbf{b}_j$ satisfy

$$\mathbf{b}_j \cdot \mathbf{a}_i = 2\pi \delta_{ij} \tag{1.12}$$

Vectors $\mathbf{b}_j$ that satisfy this relation can always be found. In particular,

$$\mathbf{b}_1 = \frac{2\pi(\mathbf{a}_2 \times \mathbf{a}_3)}{\mathbf{a}_1 \cdot (\mathbf{a}_2 \times \mathbf{a}_3)} ; \quad \mathbf{b}_2 = \frac{2\pi(\mathbf{a}_3 \times \mathbf{a}_1)}{\mathbf{a}_1 \cdot (\mathbf{a}_2 \times \mathbf{a}_3)} ; \quad \mathbf{b}_3 = \frac{2\pi(\mathbf{a}_1 \times \mathbf{a}_2)}{\mathbf{a}_1 \cdot (\mathbf{a}_2 \times \mathbf{a}_3)} \quad (1.13)$$

We now have a prescription for constructing the reciprocal lattice from the direct lattice vectors $\mathbf{a}_i$.

Now consider plane waves of the type

$$e^{i\mathbf{K}_s \cdot \mathbf{r}}$$

in which $\mathbf{K}_s$ is a reciprocal lattice vector. Such functions have the full translational periodicity of the potential. This is easily seen since, by (1.10),

$$e^{i\mathbf{K}_s \cdot (\mathbf{r}+\mathbf{R}_i)} = e^{2\pi i n_{is}} e^{i\mathbf{K}_s \cdot \mathbf{r}} = e^{i\mathbf{K}_s \cdot \mathbf{r}}$$

This is the same periodic property possessed by the function $u_\mathbf{k}$ defined in (1.9). Hence $u_\mathbf{k}$ may be expanded as a Fourier series in these plane waves:

$$u_\mathbf{k} = \sum_s a_{\mathbf{k},s}\, e^{i\mathbf{K}_s \cdot \mathbf{r}} \quad (1.14)$$

The wave function $\psi_\mathbf{k}$ can then be expressed as

$$\psi_\mathbf{k} = \sum_s a_{\mathbf{k},s}\, e^{i(\mathbf{k}+\mathbf{K}_s) \cdot \mathbf{r}} \quad (1.15)$$

This expression for $\psi_\mathbf{k}$ obviously satisfies the boundary conditions imposed by Bloch's theorem. It is still necessary to choose the coefficients $a_{\mathbf{k},s}$ so that the one-particle Schrödinger equation is satisfied. This expansion is the basis of some of the techniques for the calculation of wave functions in solids which will be explored in Chapter 2.

1.3 The Brillouin Zone

It is desirable to construct a unit cell in the reciprocal lattice that has the full symmetry of this lattice. This may be done as follows. One particular lattice point is chosen as the origin. The vectors connecting this point with other lattice points are drawn. The planes that are perpendicular bisectors of these vectors are then constructed. The smallest solid figure containing the origin bounded by these planes is the unit cell of the reciprocal lattice, usually called the Brillouin zone (Brillouin, 1931). It is usually necessary to consider only a small number of sets of bisecting planes.[1]

Points **k** on the surface of the Brillouin zone must satisfy the condition $\mathbf{k}^2 = (\mathbf{k} - \mathbf{K}_n)^2$ or $\mathbf{K}_n^2 - 2\mathbf{k} \cdot \mathbf{K}_n = 0$ (in which $\mathbf{K}_n$ is some reciprocal lattice vector).

Consider two position vectors in **k**-space, **k**′, **k**″, which satisfy the relation

$$\mathbf{k}'' = \mathbf{k}' + \mathbf{K}_s \tag{1.16}$$

The vectors **k**′ and **k**″ are said to be equivalent. Evidently

$$e^{i\mathbf{k}' \cdot \mathbf{R}_i} = e^{i\mathbf{k}'' \cdot \mathbf{R}_i} \tag{1.17}$$

for all lattice vectors $\mathbf{R}_i$. Consequently, wave functions $\psi_{\mathbf{k}'}$ and $\psi_{\mathbf{k}''}$ satisfy the same boundary conditions and may then describe the same state. From the definition of the Brillouin zone, it follows that no two points in the interior of the zone satisfy (1.16), whereas each exterior point is related to an interior point by a relation of this type. A point on the surface of the zone will be equivalent to at least one other point on the surface.

It is possible to characterize all the electron states in the periodic potential by **k** vectors lying in the interior or on the surface of the Brillouin zone. The energy of these states may be regarded as a function of the **k** vector. This function will be multivalued: there will be many different energies for a single **k**. As long as we only consider values of **k** lying inside the zone, the energy will be a continuous function of **k** (this statement will be proved in Section 1.7); a single such continuous manifold is referred to as an energy band. Discontinuities in the energy may occur only on the surface of the zone.

The restriction of the definition of $E(\mathbf{k})$ to **k** values inside the Brillouin zone is only a convention. Since **k**-space may be filled by a set of Brillouin

---

[1] If this procedure is applied to construct a unit cell in the direct lattice, the result is a figure known as the Wigner-Seitz cell.

zones, one centered on each lattice point of the reciprocal lattice, an alternative convention is to define the energy as a function of **k** throughout the entire **k**-space by requiring it to be a periodic (and multivalued) function of **k**, repeating its values in each zone.[2]

As examples of the foregoing considerations, let us examine the body-centered cubic and face-centered cubic lattices. These are of particular importance since they are the most symmetric structures assumed by single elements. In the former case, possible choices for the three primitive translation vectors are $(a/2)$ $(\mathbf{i} + \mathbf{j} + \mathbf{k})$, $(a/2)$ $(\mathbf{i} + \mathbf{j} - \mathbf{k})$, and $(a/2)$ $(\mathbf{i} - \mathbf{j} + \mathbf{k})$, where a is the lattice parameter. For the face-centered cubic lattice we have vectors $(a/2)$ $(\mathbf{j} + \mathbf{k})$, $(a/2)$ $(\mathbf{i} - \mathbf{k})$, and

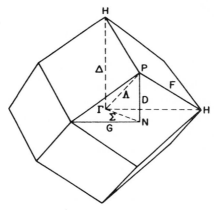

FIG. 1. Brillouin zone for the body-centered cubic lattice. Points and lines of symmetry are indicated.

$(a/2)$ $(\mathbf{i} - \mathbf{j})$. If the basis vectors of the reciprocal lattice are now constructed according to Eq. (1.13), it is immediately seen that the lattice reciprocal to the body-centered cubic structure is face-centered cubic. The basic reciprocal lattice vectors are those given for the face-centered cubic structure with $a/2$ replaced by $2\pi/a$, and the lattice

---

[2] In the literature one occasionally encounters so-called higher Brillouin zones. These are figures formed in the construction procedure previously discussed by planes bisecting vectors to more distant neighbors. In the present work, only a single Brillouin zone will be considered for each lattice.

reciprocal to the face-centered cubic is body-centered cubic. The Brillouin zones for these structures may be constructed according to the procedures given. The zones are shown in Figs. 1 and 2. The symmetry points of the zone have been labeled according to the notation introduced by Bouckaert, Smoluchowski, and Wigner (Bouckaert et al., 1936).

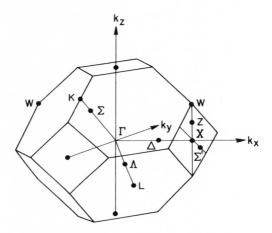

FIG. 2. Brillouin zone for the face-centered cubic lattice. Points and lines of symmetry are indicated.

In both cases the center of the zone is designated Γ. The 100, 111, and 110 axes are labeled Δ, Λ, and Σ, respectively. In the zone for the body-centered cubic lattice, the principal "symmetry points" of the zone are Γ, H, P, and N. The last three points are the intersections of the Δ, Λ, and Σ axes (respectively) with the faces of the zone. A particular point H has coordinates $(2\pi/a)$ $(1, 0, 0)$. All six points H can be obtained from the original one by adding reciprocal lattice vectors of the type $(2\pi/a)$ $(-1, 1, 0)$ or $(2\pi/a)(-2, 0, 0)$, etc. All these points are equivalent in the sense of (1.16). A particular point P has a **k** vector $(2\pi/a)$ $(\frac{1}{2}, \frac{1}{2}, \frac{1}{2})$. It is easily seen that three other points P are equivalent to it: $(2\pi/a)$ $(-\frac{1}{2}, -\frac{1}{2}, \frac{1}{2})$, $(2\pi/a)$ $(-\frac{1}{2}, \frac{1}{2}, -\frac{1}{2})$, and $(2\pi/a)$ $(\frac{1}{2}, -\frac{1}{2}, -\frac{1}{2})$. However, the points with coordinates $(2\pi/a)$ $(-\frac{1}{2}, -\frac{1}{2}, -\frac{1}{2})$, $(2\pi/a)$ $(-\frac{1}{2}, \frac{1}{2}, \frac{1}{2})$, $(2\pi/a)$ $(\frac{1}{2}, -\frac{1}{2}, \frac{1}{2})$, and $(2\pi/a)$ $(\frac{1}{2}, \frac{1}{2}, -\frac{1}{2})$, while equivalent to each other, are not equivalent to any of the points in the first group. There are two inequivalent points P. The twelve points N have $\mathbf{k}$ vectors of the type $(2\pi/a)$ $(\frac{1}{2}, \frac{1}{2}, 0)$. These points are equivalent in pairs, the point defined being equivalent to the point $(2\pi/a)$ $(-\frac{1}{2}, -\frac{1}{2}, 0)$. There are six inequivalent points N.

In the Brillouin zone for the face-centered cubic lattice, the points X have the same coordinates as the points H discussed previously, but since the reciprocal lattice vectors are now of the type $(2\pi/a)$ $(1, 1, 1)$, $(2\pi/a)$ $(2, 0, 0)$, etc., each point X is equivalent to only one other such point. There are three inequivalent points X. Similarly, the point L with coordinates $(2\pi/a)$ $(\frac{1}{2}, \frac{1}{2}, \frac{1}{2})$ is equivalent to point $(2\pi/a)$ $(-\frac{1}{2}, -\frac{1}{2}, -\frac{1}{2})$ but not to any others; thus, there are four inequivalent points L. One also finds there are twenty-four corner points W with coordinates $2\pi(1, \frac{1}{2}, 0)$ but these are equivalent in groups of four. For instance, the points $(2\pi/a)$ $(1, \frac{1}{2}, 0)$, $(2\pi/a)$ $(-1, \frac{1}{2}, 0)$, $(2\pi/a)$ $(0, -\frac{1}{2}, 1)$, and $(2\pi/a)$ $(0, -\frac{1}{2}, -1)$ are equivalent. There are six inequivalent points. Finally, each of the twelve points K of the type $(2\pi/a)$ $(\frac{3}{4}, \frac{3}{4}, 0)$ is equivalent to two of the twenty-four points U, whose coordinates are (for the particular K given) $(2\pi/a)$ $(-\frac{1}{4}, -\frac{1}{4}, 1)$ and $(2\pi/a)$ $(-\frac{1}{4}, -\frac{1}{4}, -1)$.

1.4 Energy Bands in the Free Electron Limit

We consider now the limiting case in which the periodic potential of the lattice structure becomes arbitrarily weak while the symmetry properties of the wave functions are preserved. Any function of the form $e^{i\mathbf{K}_n \cdot \mathbf{r}}$ is then an acceptable $u_\mathbf{k}$, so the energy of a state of wave vector $\mathbf{k}$ is just

$$E = \frac{\hbar^2}{2m} |\mathbf{k} + \mathbf{K}_n|^2. \tag{1.18}$$

Here $\mathbf{K}_n$ is any reciprocal lattice vector. It turns out, somewhat surprisingly, that the energy bands of electrons in many metals can be rather well approximated by this simple expression.

To illustrate the method and results, we shall consider the body-centered cubic lattice. We shall write the vector $\mathbf{k}$ in the form $(2\pi/a)$ (x, y, z) and the reciprocal lattice vector $\mathbf{K}_n$ as $(2\pi/a)$ (n_1, n_2, n_3). Then if we define $\lambda = ma^2 E/2\hbar^2 \pi^2$ we have

$$\lambda = (x+n_1)^2 + (y+n_2)^2 + (z+n_3)^2 \tag{1.19a}$$

Consider first the 100 axis, Δ; $y = z = 0$. The lowest band is characterized by $n_1, n_2, n_3 = 0$, and is just the parabola $\lambda_1 = x^2$. The next highest bands are those found by taking n_1, n_2, n_3 from the first reciprocal lattice vector type 1, 1, 0. These give:

$$\lambda_2 = (x-1)^2 + 1; \qquad \lambda_3 = x^2 + 2; \qquad \lambda_4 = (1+x)^2 + 1 \tag{1.19b}$$

Each of these bands is fourfold degenerate: there are four possible choices of reciprocal lattice vectors which lead to the above expressions. Turning to reciprocal lattice vectors of the (200) type, we have

$$\lambda_5 = (x-2)^2; \qquad \lambda_6 = x^2 + 4; \qquad \lambda_7 = (x+2)^2 \tag{1.19c}$$

The second of these is fourfold degenerate; the other bands are not degenerate.

On the 111 axis, $x = y = z \leqslant \frac{1}{2}$. We have, considering the same reciprocal lattice vectors as above, the following bands:

$$\lambda_1 = 3x^2; \qquad \lambda_2 = 2(x-1)^2 + x^2; \qquad \lambda_3 = 3x^2 + 2; \tag{1.19d}$$

$$\lambda_4 = 2(x+1)^2 + x^2; \qquad \lambda_5 = (x-2)^2 + 2x^2; \qquad \lambda_6 = (x+2)^2 + 2x^2$$

λ_1 is a nondegenerate band; λ_3 is sixfold degenerate; the others given are triply degenerate. Finally, for the 110 axis, we have $x = y \leqslant \frac{1}{2}, z = 0$. The lowest levels are

$$\lambda_1 = 2x^2; \quad \lambda_2 = 2(x-1)^2; \quad \lambda_3 = (x-1)^2 + x^2 + 1; \quad \lambda_4 = 2(x^2+1);$$

$$\lambda_5 = (x+1)^2 + x^2 + 1; \quad \lambda_6 = 2(x+1)^2; \quad \lambda_7 = (x-2)^2 + x^2; \tag{1.19e}$$

$$\lambda_8 = 2x^2 + 4; \qquad \lambda_9 = (x+2)^2 + x^2$$

The degeneracies are: λ_1, λ_2, and λ_6 are not degenerate; $\lambda_4, \lambda_7, \lambda_8$, and λ_9 are doubly degenerate; and λ_3 and λ_5 are fourfold degenerate. These bands are illustrated in Fig. 3. These procedures may be extended to any point in the zone, and may be applied to any lattice structure.

We shall see in Section 1.6 that the principal effect of including a periodic potential in the energy band calculation is the removal of much of the degeneracy present in the free electron limit. For instance, a twofold degeneracy is the maximum permitted along any symmetry axis.

The discussion of the free electron approximation will be continued after we have considered the symmetry properties of electron wave functions in more detail.

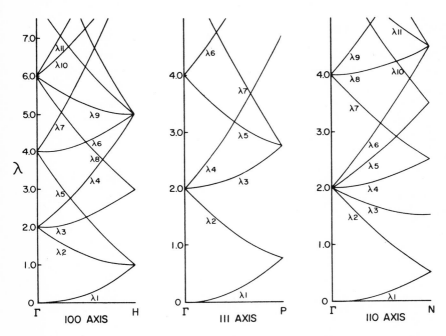

FIG. 3. Energy bands in the free electron limit. The dimensionless quantity $\lambda = ma^2 E/2\hbar^2 \pi^2$ is plotted as a function of wave vector for points along the 100, 110, and 111 axes in the Brillouin zone of the body-centered cubic lattice.

1.5 Space Groups

Most interesting crystals possess more symmetry than the translational invariance discussed in Section 1.1. Other operations exist which carry the crystal into itself. In the simplest cases these are rotations and reflections. The operators representing these symmetry transformations will commute with the crystal Hamiltonian and are important in classifying the electron states. In this section, the symmetry operators will be considered in a general way; in the next, the classification of the electron states will be developed. For a more detailed analysis, the reader is referred to the book by Jones (1960) and the review article by Koster (1957) which contains references to the original literature. An excellent account of the basic principles of the group theory applied here has been given by Wigner (1959).

The operations which carry a crystal into itself form a group of a rather specialized nature called space group. We shall indicate the essential notation and results briefly, after which the formal definitions will be given.

An operator of a space group contains a part which is either a proper or improper rotation, α, and a translation part, t, and is denoted by the symbol $\{\alpha|t\}$. This operator corresponds to the coordinate transformation

$$\mathbf{x}' = \alpha \mathbf{x} + \mathbf{t} \tag{1.20}$$

[In (1.20), α will be represented by a 3 × 3 orthogonal matrix.] Two such operators: $\{\alpha|t\}$ and $\{\beta|t'\}$ multiply in the following way:

$$\{\beta|t'\}\{\alpha|t\} = \{\beta\alpha|\beta t + t'\} \tag{1.21}$$

The unit operator is represented by $\{\epsilon|0\}$. From (1.21) it may be verified that the inverse of the operator $\{\alpha|t\}$ is

$$\{\alpha|t\}^{-1} = \{\alpha^{-1}|-\alpha^{-1}t\} \tag{1.22}$$

The pure translation operations, which in Section 1.1 were denoted as $T(\mathbf{R}_i)$ are now expressed as $\{\epsilon|\mathbf{R}_i\}$.

The operators $\{\alpha|t\}$ may be expressed in matrix form on the basis of position vectors $\mathbf{x}$. Denote each $\mathbf{x}$ as

$$\begin{pmatrix} 1 \\ x_1 \\ x_2 \\ x_3 \end{pmatrix}$$

Then (1.20) becomes

$$\begin{pmatrix} 1 \\ x_1' \\ x_2' \\ x_3' \end{pmatrix} = \begin{pmatrix} 1 & 0 & 0 & 0 \\ t_1 & \alpha_{11} & \alpha_{12} & \alpha_{13} \\ t_2 & \alpha_{21} & \alpha_{22} & \alpha_{23} \\ t_3 & \alpha_{31} & \alpha_{32} & \alpha_{33} \end{pmatrix} \begin{pmatrix} 1 \\ x_1 \\ x_2 \\ x_3 \end{pmatrix} \tag{1.23}$$

or

$$\begin{pmatrix} 1 \\ x' \end{pmatrix} = \begin{pmatrix} 1 & 0 \\ t & \alpha \end{pmatrix} \begin{pmatrix} 1 \\ x \end{pmatrix}$$

The multiplication equation (1.21) becomes

$$\begin{pmatrix} 1 & 0 \\ t' & \beta \end{pmatrix} \begin{pmatrix} 1 & 0 \\ t & \alpha \end{pmatrix} = \begin{pmatrix} 1 & 0 \\ t' + \beta t & \beta \alpha \end{pmatrix} \quad (1.21a)$$

and the inverse is

$$\begin{pmatrix} 1 & 0 \\ t & \alpha \end{pmatrix}^{-1} = \begin{pmatrix} 1 & 0 \\ -\alpha^{-1}t & \alpha^{-1} \end{pmatrix} \quad (1.22a)$$

In order to give a formal definition of a space group, it is necessary to recall some of the basic definitions of group theory (Wigner, 1959):

(1) *Conjugate elements.* If X and A are members of a group, the element $B = XAX^{-1}$ is said to be conjugate to A. If two elements A, C are conjugate to a third element D, they are conjugate to each other.

(2) *Class.* Those elements of a group which are conjugate to each other form a class. Stated in less abstract terminology, two transformations (operations) A, B are in the same class if it is possible to find a new coordinate system in which transformation B has the same effect that transformation A had in the previous system.

(3) *Invariant subgroup.* A subgroup that consists entirely of whole classes is an invariant subgroup. If A is an element of the subgroup, then every element $B = XAX^{-1}$ is also a member of the subgroup when X is any member of the full group.

A space group may now be defined as a group of operators of the form $\{\alpha|t\}$ which possesses an invariant subgroup of pure translations. There are only a finite number of possible space groups in a space of finite dimensions: 230 in three-dimensional space.

It is useful to consider the rotational parts of the operators alone. These are of the type $\{\alpha|0\}$. These operators form a group known as the point group. It can be shown that the only operations which can occur in a point group are proper rotations through integral multiples of 60° and 90° about specified axes, and improper rotations consisting of combinations of the proper rotations with the inversions. There are

14   CHAPTER 1. THE LANGUAGE OF BAND THEORY

only 32 possible point groups in a three-dimensional space. It can also be shown that if, for a given space group, we form the factor group[3] of that space group with respect to the invariant subgroup of primitive translations, this factor group is isomorphic with the point group composed of the rotational parts of the operators of the space group.

Next consider the translation parts of the operators. Observe first that if $\{\epsilon|R_m\}$ is a translation operation, so is $\{\epsilon|\alpha R_m\}$ if $\{\alpha|t\}$ is any member of the space group. This follows from the rule of multiplication and the construction of the inverse, since

$$\{\epsilon|\alpha R_n\} = \{\alpha|t\}\{\epsilon|R_n\}\{\alpha|t\}^{-1} \qquad (1.24)$$

The translations parts of operators for which the rotational part is the identity operation ϵ are necessarily lattice translations R_n. This is not always true for other operations. However, it can be shown that all operators belonging to a given space group which have a common rotational part α may be expressed as $\{\alpha|v(\alpha) + R_n\}$ where R_n is a lattice translation and $v(\alpha)$ is either zero or a translation which is not a lattice translation ($\{\varepsilon|v(\alpha)\}$ does not belong to the space group). Certain space groups, called *"symmorphic"* (or simple) have $v(\alpha) = 0$ for all α. The symmorphic space groups contain the entire point group as a subgroup.

Cubic structures (simple cubic, body-centered cubic, face-centered cubic lattices) have the full cubic group as a point group. In several cases, this point group is a subgroup of the space group; in other structures, such as the diamond lattice, some operations of the space group involve nonprimitive translations. We will now describe the cubic group in some detail in order to have a specific example to which the analysis of the next section may be applied.

The group consists of the 48 operations which leave a cube invariant. These operations may be described as follows: (operations are grouped according to classes):

    I. The identity, E.

    II. Rotation by $\pm 90°$ about a coordinate (fourfold) axis: Class C_4 — six operations.

    III. Rotations by 180° about the same axis: Class C_4^2 — three operations.

---

[3] See Wigner (1959, p. 68) for definition of the factor group.

1.5 SPACE GROUPS

IV. Rotations by π about a twofold axis. There are six such axes: Class C_2 — six operations.

V. Rotations by $\pm 2\pi/3$ about a threefold axis (a body diagonal). There are four such axes: Class C_3 — eight operations.

VI. The inversion with respect to the origin: Class J — one operation.

TABLE I

THE CUBIC GROUP

Class	Operation			Class	Operation		
E	x	y	z	J	$-x$	$-y$	$-z$
$C_4{}^2$	$-x$	$-y$	z	$JC_4{}^2$	x	y	$-z$
	x	$-y$	$-z$		$-x$	y	z
	$-x$	y	$-z$		x	$-y$	z
C_4	$-y$	x	z	JC_4	y	$-x$	$-z$
	y	$-x$	z		$-y$	x	$-z$
	x	$-z$	y		$-x$	z	$-y$
	x	z	$-y$		$-x$	$-z$	y
	z	y	$-x$		$-z$	$-y$	x
	$-z$	y	x		z	$-y$	$-x$
C_2	y	x	$-z$	JC_2	$-y$	$-x$	z
	z	$-y$	x		$-z$	y	$-x$
	$-x$	z	y		x	$-z$	$-y$
	$-y$	$-x$	$-z$		y	x	z
	$-z$	$-y$	$-x$		z	y	x
	$-x$	$-z$	$-y$		x	z	y
C_3	z	x	y	JC_3	$-z$	$-x$	$-y$
	y	z	x		$-y$	$-z$	$-x$
	z	$-x$	$-y$		$-z$	x	y
	$-y$	$-z$	x		y	z	$-x$
	$-z$	$-x$	y		z	x	$-y$
	$-y$	z	$-x$		y	$-z$	x
	$-z$	x	$-y$		z	$-x$	y
	y	$-z$	$-x$		$-y$	z	x

VII. Classes JC_4, $JC_4{}^2$, JC_2, JC_3, which are combinations of the listed operations with the inversion J: twenty-three operations.

Evidently there are forty-eight operations in all, divided into ten classes.

The operations of the cubic group may be conveniently described in another manner. If we consider a position vector **R** with components x, y, z, the operations of the cubic group are then specified as the possible rearrangements or permutations of x, y, z, including changes of sign. The classifications are given in Table I.

The proper rotations which send a cube into itself form a subgroup of the cubic group (E, $C_4{}^2$, C_4, C_2, C_3). Also, the twenty-four operations in classes E, $C_4{}^2$, JC_4, JC_2, and C_3 form a different subgroup: the tetrahedral group. These are the operations which send a regular tetrahedron into itself.

1.6 Irreducible Representations

The significance of the discussion of group theory results from a general principle of quantum mechanics: The wave functions of a quantum system must form bases for irreducible representations of the group of operators which commute with the Hamiltonian of the system (see Wigner, 1959, Chapter 11).

A group is said to be *represented* by a set of matrices B_i, if to each element in the group there corresponds a matrix such that products correspond to products, etc.

A matrix M is said to be the *direct sum* of matrices $m_1, m_2, \ldots$ if every element of M is zero except for square blocks (the submatrices $m_1, m_2, \ldots$) along the diagonal. If each matrix of a representation can be expressed as a direct sum in this way, and if the dimensions of the corresponding submatrices are the same in every case, then the submatrices themselves are a *representation* of the group, and the original representation has been *reduced*. In order to carry out the reduction, it is necessary to find a unitary transformation U which will bring the matrices to the required form, $M_i = U^{-1} B_i U$ (the same matrix U for all the B_i).

m_1	0	0
0	m_2	0
0	0	m_3

In terms of the vectors of the space on which the B_i operate, the reduction separates out the subspaces which are carried into themselves by all the B_i (*invariant subspaces*). If no transformation exists by means of which the matrices of a representation may each be expressed as direct sums, then the representation is said to be *irreducible*.

The sum of the diagonal elements (trace) of a matrix representing an operation of the group is called its *character*. The character is the same for the matrices representing all the operations of the same class. The *dimension* of a representation is the number of rows or columns of any matrix, and is given by the character of the identity operation in that representation.

There are two general theorems pertaining to the irreducible representations of a finite group which are important in our applications. These are stated without proof:

(1) The number of irreducible representations equals the number of classes.

(2) The sum of the squares of the dimensions of the representations equals the number of elements of the group.

There are ten irreducible representations of the full cubic group: Four are one-dimensional, two are two-dimensional, and four are three-dimensional representations. [Observe that $4 \times (1)^2 + 2 \times (2)^2 + 4 \times (3)^2 = 48$.]

Before considering the representations of space groups, we consider the subgroup of lattice translations. Since this is an Abelian group (the operators commute with each other), all the irreducible representations are one-dimensional. It will be recognized that Bloch's theorem is a consequence of the general statement at the beginning of this section. It follows from (1.8) that the translation operators $T(\mathbf{R}_i) = \{\boldsymbol{\epsilon}|\mathbf{R}_i\}$ are represented by (one-dimensional matrices) $e^{i\mathbf{k}\cdot\mathbf{R}_i}$, where $\mathbf{k}$ lies inside or on the surface of the Brillouin zone.

We now must consider the symmetries of the reciprocal lattice. If $\boldsymbol{\alpha}$ and $\boldsymbol{\alpha}^{-1}$ are operations in the point group of a lattice, then $\boldsymbol{\alpha}^{-1}\mathbf{R}_n$ is a lattice translation if $\mathbf{R}_n$ is, and $\boldsymbol{\alpha}^{-1}\mathbf{R}_n \cdot \mathbf{K}_j = 2\pi N$, where N is some integer, for all $\mathbf{K}_j$. Since $\boldsymbol{\alpha}^{-1}$ is an orthogonal transformation, it also follows that

$$\mathbf{R}_n \cdot \boldsymbol{\alpha}\mathbf{K}_j = 2\pi N \qquad (1.25)$$

An equation of the form of (1.25) must hold for all operations in the point group. Then $\boldsymbol{\alpha}\mathbf{K}_j$ must be a reciprocal lattice vector if $\mathbf{K}_j$ is. This means that if a lattice is invariant with respect to the operations of a given point group, the reciprocal lattice is also invariant under the same point group. (But not space group: the translations may be different, as we have already seen in two cases.) Now let $\{\beta|b\}$ be an operator belonging to the space group (of the direct lattice) which has the property that

$$\beta k = k + K_j \tag{1.26}$$

The operator β sends the vector k into itself or into an equivalent vector. It can easily be verified that these operations form a group, which is called the group of the wave vector. This group is a subgroup of the space group of the lattice concerned; it also contains the entire subgroup of lattice translations, and is therefore a space group. For a general point of the Brillouin zone, the group of the wave vector contains only the translations, since no operation of the point group except the identity will leave the wave vector unaltered. For points on symmetry axes, for the center, or for points on the surface of the zone, the group of the wave vector will be larger.

Our detailed considerations of the representations of space groups will be limited to the symmorphic groups. For a more detailed analysis, the reader should consult the article by Koster (1957), previously cited. We discuss the representations with respect to points in the zone. At a general point, it is sufficient that the wave function have the Bloch form (1.9) in order that it be an acceptable basis function for an irreducible representation of the space group. An operator $\{\alpha|t\}$ acting on such a function will send it into a function characterized by a different wave vector. If there are n operators in the point group, $(n - 1)$ other functions may be obtained in this way, which are characterized by wave vectors αk. The figure of these k vectors is referred to as a "star." It exhibits all of the rotational and reflectional symmetry of the lattice. The Bloch functions characterized by k vectors in the star are basis functions for an n-dimensional irreducible representation of the space group. In this representation, the matrices representing lattice translations are diagonal, with elements $e^{i k' \cdot R_j}$ where k' runs over all the vectors of the star.

When the point group of the wave vector contains operations in addition to the identity, the irreducible representations of the space group are not n-dimensional but rather of lower dimension. There will be as many distinct irreducible representations of the space group G

1.6 IRREDUCIBLE REPRESENTATIONS

as there are irreducible representations of the point group of the wave vector. If there are S such irreducible representations of dimension d_i (small representations), and there are g operations in the point group of the wave vector, the S irreducible representations of G have dimension D_i, where

$$D_i = \frac{n}{g} d_i \qquad (1.27)$$

There are d_i orthogonal functions $\phi_{\mathbf{k}}^j$ (where j runs from 1 to d_i) which are multiplied by $e^{i\mathbf{k}\cdot\mathbf{R}_n}$ under translation through $\mathbf{R}_n$ (and thus belong to the same representation of the translation subgroup) and which are also the basis functions for an irreducible representation of the point group of the wave vector. The representations of all point groups are known, and have been listed by Koster (1957). If $\Gamma(\boldsymbol{\beta})$ is a representation of the operation $\{\boldsymbol{\beta}|0\}$ in the point group of the wave vector, the representation of the operator $\{\boldsymbol{\beta}|\mathbf{b}\}$ in the space group (κ) of the wave vector $\mathbf{k}$ is $e^{i\mathbf{k}\cdot\mathbf{b}}\Gamma(\boldsymbol{\beta})$. There will exist $q = n/g$ operators in the space group $\{\boldsymbol{\alpha}_i|\mathbf{a}_i\}$ where $i = 1 \ldots q$, and $\{\boldsymbol{\alpha}_1|\mathbf{a}_1\} = \{\boldsymbol{\epsilon}|0\}$, such that the full space group G may be expressed as the sum of its left cosets[4] with respect to κ.

$$G = \kappa + \{\boldsymbol{\alpha}_2|\mathbf{a}_2\}\kappa + \ldots + \{\boldsymbol{\alpha}_q|\mathbf{a}_q\}\kappa \qquad (1.28)$$

The $D_i = q d_i$ functions

$$\phi_{\mathbf{k}'}^j = \{\boldsymbol{\alpha}_l|\mathbf{a}_l\}\phi_{\mathbf{k}}^j \qquad (1.29)$$

where $l = 1 \ldots q$ and $j = 1 \ldots d_i$ form bases for an irreducible representation of the space group, G.

States which belong to the same irreducible representation of the space group must have the same energy, since the wave functions can be transformed into each other by the operators of the group. For a general point in $\mathbf{k}$-space, the vectors of the star are all different. States characterized by wave vectors in the star will all belong to the same representation of the space group but are located at different points of the Brillouin zone. It follows that the energy will have the complete symmetry of the reciprocal lattice. For points in the zone for which the

---

[4] See Wigner (1959, p. 60) for a definition of "coset."

group of the wave vector contains more than the lattice translations, further analysis is required. There will then be only $n/g = q$ distinct points in the zone where the corresponding states will have the same energy. It is now necessary to classify states according to the irreducible representation of the point group of the wave vector. If this point group permits degeneracy, some energy bands will stick together at that point.

TABLE II

GROUP OF Δ

Class	Operation			Designation	E	α	β	γ	δ	ε	ζ	η
E	x	y	z	E	E	α	β	γ	δ	ε	ζ	η
$C_4{}^2$	x	$-y$	$-z$	α	α	E	γ	β	ε	δ	η	ζ
C_4	x	$-z$	y	β	β	γ	α	E	ζ	η	ε	δ
	x	z	$-y$	γ	γ	β	E	α	η	ζ	δ	ε
$JC_4{}^2$	x	$-y$	z	δ	δ	ε	η	ζ	E	α	γ	β
	x	y	$-z$	ε	ε	δ	ζ	η	α	E	β	γ
JC_2	x	$-z$	$-y$	ζ	ζ	η	δ	ε	β	γ	E	α
	x	z	y	η	η	ζ	ε	δ	γ	β	α	E

For the purpose of calculating wave functions in solids, and in order to make calculations with them, it is useful to determine the irreducible representations of the point groups of the wave vectors considered, and in particular to find basis functions for a particular representation. If degeneracy is permitted, it is also useful to determine an orthogonal set of basis functions for the degenerate representation. We will illustrate the determination of the representations by an example which is simple enough to be analyzed in a very elementary fashion.

Consider the point group of the (100) axis, Δ, in a cubic crystal. This is the group of the square. Table II lists the operations and

1.6 IRREDUCIBLE REPRESENTATIONS

gives the multiplication table for the group. The entries in the multiplication table are worked out in the following manner:

$$\zeta\delta = (x, -z, -y)(x, -y, z) = (x, -z, y) = \beta$$

The table is constructed so that the operator which designates a row appears on the left in multiplication. The multiplication is not commutative: for instance $\delta\zeta = \gamma$ while $\zeta\delta = \beta$. According to the rules given at the beginning of this section, there are five irreducible representations, four of which are one-dimensional and the other two-dimensional. This is all, since $[4 \times (1)^2 + 1 \times (2)^2 = 8]$. The one-dimensional representations have as bases functions which are carried into themselves (or their negatives) by the operations of the group. The group operations are here represented by numbers, either $+1$ or -1. The two-dimensional representation involves 2×2 square matrices and is based on a two-dimensional vector space.

The simplest procedure in the construction of basis functions for the representations is to consider elements of the form $x^n y^m z^l$ and their linear combinations. In this example, a possible set of basis functions can be chosen very simply; there are, however, an infinite number of basis functions of this type for each representation:

1. There is a symmetric representation, denoted Δ_1, according to which each operation is represented by the number $+1$. (Throughout this book, representations are designated according to the fundamental paper of Bouckaert *et al*, 1936, in which a more complete discussion of representation theory may be found. The notation is designed to facilitate understanding of the connections between bands that occur at points of symmetry.) Acceptable basis functions for this representation are $1, x, 2x^2 - y^2 - z^2$, etc.

2. In the representation Δ_2, the operators E, α, δ, and ε are represented by $+1$; the operators β, γ, ζ, and η are represented by -1. A basis function for this representation is $y^2 - z^2$. The reader can easily verify that this assignment is consistent with the multiplication table.

3. In the representation $\Delta_2{}'$, the operators E, α, ζ, and η are represented by $+1$; the operators β, γ, δ, and ε are represented by -1. A basis function for this representation is yz.

4. In the representation Δ_1', the operators E, α, β, and γ are represented by $+1$; the operators δ, ε, ζ, and η by -1. A basis function for this representation is $yz(y^2 - z^2)$.

5. For the two-dimensional representation Δ_5, the following assignment of matrices to operations may be made:

$$E = \begin{pmatrix} 1 & 0 \\ 0 & 1 \end{pmatrix}; \quad \alpha = \begin{pmatrix} -1 & 0 \\ 0 & -1 \end{pmatrix}; \quad \beta = \begin{pmatrix} 0 & -1 \\ 1 & 0 \end{pmatrix}; \quad \gamma = \begin{pmatrix} 0 & 1 \\ -1 & 0 \end{pmatrix};$$

$$\delta = \begin{pmatrix} -1 & 0 \\ 0 & 1 \end{pmatrix}; \quad \varepsilon = \begin{pmatrix} 1 & 0 \\ 0 & -1 \end{pmatrix}; \quad \zeta = \begin{pmatrix} 0 & -1 \\ -1 & 0 \end{pmatrix}; \quad \eta = \begin{pmatrix} 0 & 1 \\ 1 & 0 \end{pmatrix}$$

(1.30)

The functions y, z are suitable basis functions, as are also the functions xy, xz.

The essential properties of a representation can be presented in the character table. From the theorem of the invariance of the trace of a matrix under unitary transformations it follows that the character is independent of the basis functions employed and depends only on the representation and the class. In Table III, the character table for the group of Δ is presented.

TABLE III

CHARACTER TABLE: GROUP OF Δ

Representation	Basis	E	C_4^2	C_4	JC_4^2	JC_2
Δ_1	$1, x, 2x^2 - y^2 - z^2$	1	1	1	1	1
Δ_2	$y^2 - z^2$	1	1	-1	1	-1
Δ_2'	yz	1	1	-1	-1	1
Δ_1'	$yz(y^2 - z^2)$	1	1	1	-1	-1
Δ_5	$y, z; xy, xz$	2	-2	0	0	0

The cubic group whose operations were listed in Table I is the point group of the wave vector at the zone center and the corner R in the simple cubic lattice, at the zone center and the corner H for the body-centered cubic lattice, and the zone center in the face-centered cubic

1.6 IRREDUCIBLE REPRESENTATIONS

structure. There are ten irreducible representations: four one-dimensional, two two-dimensional, and four three-dimensional ones. When we add the squares of the dimensions we find: $4 \times (1)^2 + 2 \times (2)^2 + 4 \times (3)^2 = 48$ as required. Many of the representations are of no importance in the level structures of real materials since they have very high energies. The character table for this group is presented in Table IV which also contains a listing of polynomial basis functions according to Bouckaert et al. (1936) and Von der Lage and Bethe (1947). In the case of a representation of dimension greater than one, only one basis function is given in most cases; others may be determined by interchange of x, y, and z.

The determination of basis functions makes possible an approximate

TABLE IV

CHARACTER TABLE FOR THE CUBIC GROUP

Representation	Basis	E	C_4^2	C_4	C_2	C_3	J	JC_4^2	JC_4	JC_2	JC_3
Γ_1	1	1	1	1	1	1	1	1	1	1	1
Γ_2	$x^4(y^2 - z^2)$ $+ y^4(z^2 - x^2) +$ $z^4(x^2 - y^2)$	1	1	-1	-1	1	1	1	-1	-1	1
Γ_{12}	$x^2 - y^2$, $2z^2 - x^2 - y^2$	2	2	0	0	-1	2	2	0	0	-1
Γ_{15}	x, y, z	3	-1	1	-1	0	-3	1	-1	1	0
Γ_{25}	$z(x^2 - y^2)$	3	-1	-1	1	0	-3	1	1	-1	0
Γ_1'	$xyz[x^4(y^2 - z^2)$ $+ y^4(z^2 - x^2)$ $+ z^4(x^2 - y^2)]$	1	1	1	1	1	-1	-1	-1	-1	-1
Γ_2'	xyz	1	1	-1	-1	1	-1	-1	1	1	-1
Γ_{12}'	$xyz(x^2 - y^2)$	2	2	0	0	-1	-2	-2	0	0	1
Γ_{15}'	$xy(x^2 - y^2)$	3	-1	1	-1	0	3	-1	1	-1	0
Γ_{25}'	xy, yz, zx	3	-1	-1	1	0	3	-1	-1	1	0

correspondence with atomic central field wave functions which are conventionally designated s, p, d, f .... This can be accomplished if one observes that a term $x^m y^n z^p$ is proportional to r^{m+n+p} times a linear combination of spherical harmonics whose total angular momentum quantum number, $l = m + n + p$. The correspondence is useful in considering the states likely to be occupied by electrons deriving from atomic levels of a given symmetry. The correspondence is, for states at the center of the zone, $\Gamma_1 \to s$; $\Gamma_{15} \to p$; Γ_{12}, $\Gamma_{25}' \to d$ Γ_2', Γ_{25}, $\Gamma_{15} \to f$, ... This correspondence is only approximate in that if a solid-state wave function, belonging to Γ_{25}' for instance, is expanded in spherical harmonics, terms involving $l = 4$ and $l = 6$, etc., may be present in addition to those of $l = 2$; and is useful only at points of high symmetry where functions of small l belong to different representations. At a general point of the zone, all spherical harmonics may be present in the expansion of any wave function. In Tables V–XII, the character tables are presented for the points of high symmetry in the Brillouin zones of the body-centered cubic and face-centered cubic lattices shown in Figs. 1 and 2.

Wave functions are eigenfunctions of definite parity only when the group of the wave vector contains the inversion. When it does not, basis functions of even and odd parity occur together, as in the case of the group of P.

The group of P is the tetrahedral group previously described. There are only five irreducible representations: half the number found for the cubic group. It is interesting to see the way in which the representations at Γ combine to give those at P: P_1 contains functions belonging to Γ_1 and Γ_2'; P_2 contains those belonging to Γ_2 and Γ_1'; P_3 contains Γ_{12} and Γ_{12}'; P_4, Γ_{15} and Γ_{25}'; P_5, Γ_{25} and Γ_{15}'.

Basis functions for N have been chosen which are appropriate to the point $(2\pi/a)(\frac{1}{2}, \frac{1}{2}, 0)$. The operation $C_2\|$ is a rotation about an axis parallel to the wave vector, while $C_2 \perp$ is a rotation about an axis perpendicular to the wave vector. These operations are not equivalent. The representations are all nondegenerate. The representation N_1 is of particular interest since it contains, in addition to an s-like function, two different d-like functions.

The points on the surface of the Brillouin zone of the face-centered cubic lattice do not have as high symmetry as to the corresponding points

1.6 IRREDUCIBLE REPRESENTATIONS

TABLE V
Character Table, Group of $P = (2\pi/a)\ (\frac{1}{2}, \frac{1}{2}, \frac{1}{2})$

Representation	Basis	E	$3C_4{}^2$	$8C_3$	$6JC_4$	$6JC_2$
P_1	$1, xyz$	1	1	1	1	1
P_2	$x^4(y^2 - z^2) + y^4(z^2 - x^2) + z^4(x^2 - y^2)$	1	1	1	-1	-1
P_3	$x^2 - y^2, xyz(x^2 - y^2)$	2	2	-1	0	0
P_4	$x, y, z; xy; yz; zx$	3	-1	0	-1	1
P_5	$z(x^2 - y^2)$	3	-1	0	1	-1

TABLE VI
Character Table, Group of $N = (2\pi/a)\ (\frac{1}{2}, \frac{1}{2}, 0)$

Representation	Basis	E	$C_4{}^2$	$C_2\|\|$	$C_2\perp$	J	$JC_4{}^2$	$JC_2\perp$	$JC_2\|\|$
N_1	$1, xy, 3z^2 - r^2$	1	1	1	1	1	1	1	1
N_2	$z(x - y)$	1	-1	1	-1	1	-1	-1	1
N_3	$z(x + y)$	1	-1	-1	1	1	-1	1	-1
N_4	$x^2 - y^2$	1	1	-1	-1	1	1	-1	-1
N_1'	$x + y$	1	-1	1	-1	-1	1	1	-1
N_2'	$z(x^2 - y^2)$	1	1	1	1	-1	-1	-1	-1
N_3'	z	1	1	-1	-1	-1	-1	1	1
N_4'	$x - y$	1	-1	-1	1	-1	1	-1	1

TABLE VII
Character Table, Group of Λ, F[a]

Representation	Basis	E	$2C_3$	$3JC_2$
Λ_1	$1, x + y + z$	1	1	1
Λ_2	$x(y^2 - z^2) + y(z^2 - x^2) + z(x^2 - y^2)$	1	1	-1
Λ_3	$2x - y - z, y - z$	2	-1	0

[a] $F = (2\pi/a)(\frac{1}{2} + x, \frac{1}{2} - x, \frac{1}{2} - x);\quad 0 \leqslant x \leqslant \frac{1}{2}$

TABLE VIII

Character Table, Group of $X = (2\pi/a)(1, 0, 0)$

Representation	Basis	E	$2C_4^2 \perp$	$C_4^2 \parallel$	$2C_4 \parallel$	$2C_2$	J	$2JC_4^2 \perp$	$JC_4^2 \parallel$	$2JC_4 \parallel$	$2JC_2$
X_1	$1, 2x^2 - y^2 - z^2$	1	1	1	1	1	1	1	1	1	1
X_2	$y^2 - z^2$	1	1	1	-1	-1	1	1	1	-1	-1
X_3	yz	1	-1	1	-1	1	1	-1	1	-1	1
X_4	$yz(y^2 - z^2)$	1	-1	1	1	-1	1	-1	1	1	-1
X_5	xy, xz	2	0	-2	0	0	2	0	-2	0	0
X_1'	$xyz(y^2 - z^2)$	1	1	1	1	1	-1	-1	-1	-1	-1
X_2'	xyz	1	1	1	-1	-1	-1	-1	-1	1	1
X_3'	$x(y^2 - z^2)$	1	-1	1	-1	1	-1	1	-1	1	-1
X_4'	x	1	-1	1	1	-1	-1	1	-1	-1	1
X_5'	y, z	2	0	-2	0	0	-2	0	2	0	0

pertaining to the body-centered cubic lattice. There are three inequivalent points X (instead of the single H) so that the point group of X contains sixteen operations. It is necessary to distinguish rotations about axes parallel and perpendicular to the wave vector. Notice that the representation X_1 contains a d function as well as an s function. There are four inequivalent points L compared to two inequivalent points P. In addition to the basis functions listed for the doubly degenerate representation L_3, there are two additional linear combinations of spherical harmonics with $l = 2$ which belong to L_3; thus all the d-like functions are contained in L_1 and L_3. There are six inequivalent points W. Since the group of W does not contain the inversion, the representations may contain both even and odd basis functions. For instance, the representation W_3 contains both p- and d-like functions.

TABLE IX

Character Table, Group of $L = (2\pi/a)\,(\tfrac{1}{2}, \tfrac{1}{2}, \tfrac{1}{2})$

Representation	Basis	E	$2C_3$	$3C_2$	J	$2JC_3$	$3JC_2$
L_1	$1, xy + yz + xz$	1	1	1	1	1	1
L_2	$yz(y^2 - z^2) + xy(x^2 - y^2) + xz(z^2 - x^2)$	1	1	-1	1	1	-1
L_3	$2x^2 - y^2 - z^2;\ y^2 - z^2$	2	-1	0	2	-1	0
L_1'	$x(y^2 - z^2) + y(z^2 - x^2) + z(x^2 - y^2)$	1	1	1	-1	-1	-1
L_2'	$x + y + z$	1	1	-1	-1	-1	1
L_3'	$y - z;\ 2x - y - z$	2	-1	0	-2	1	0

TABLE X

Character Table, Group of $W_1 = (2\pi/a)\,(1, \tfrac{1}{2}, 0)$

Representation	Basis	E	$C_4{}^2$	$2C_2$	$2JC_4$	$2JC_4{}^2$
W_1	$1, 2y^2 - x^2 - z^2$	1	1	1	1	1
W_1'	xz	1	1	1	-1	-1
W_2	xyz	1	1	-1	1	-1
W_2'	$y, z^2 - x^2$	1	1	-1	-1	1
W_3	$xy, yz;\ x, z$	2	-2	0	0	0

TABLE XI

CHARACTER TABLE, GROUP OF $\Sigma = (2\pi/a)\,(x, x, 0)$

Representation	Basis	E	C_2	$JC_4{}^2$	JC_2
Σ_1	$1, x+y$	1	1	1	1
Σ_2	$z(x-y);\ z(x^2-y^2)$	1	1	-1	-1
Σ_3	$z;\ z(x+y)$	1	-1	-1	1
Σ_4	$x-y;\ x^2-y^2$	1	-1	1	-1

TABLE XII

CHARACTER TABLES OF G, K, U, D, Z, S

Representation	Z	E	$C_4{}^2$	$JC_4{}^2$	$JC_4{}^2 \perp$
	G, K, U, S	E	C_2	$JC_4{}^2$	JC_2
	D	E	$C_4{}^2$	JC_2	$JC_2 \perp$
K_1	$1, x+y$	1	1	1	1
K_2	$z(x-y),\ z(x^2-y^2)$	1	1	-1	-1
K_3	$z,\ z(x+y)$	1	-1	-1	1
K_4	$x-y;\ x^2-y^2$	1	-1	1	-1

$$G = \frac{2\pi}{a}\,(\tfrac{1}{2}+x, \tfrac{1}{2}-x, 0) \quad \text{(bcc)}; \qquad K = \frac{2\pi}{a}\,(\tfrac{3}{4}, \tfrac{3}{4}, 0) \quad \text{(fcc)}$$

$$U = \frac{2\pi}{a}\,(1, \tfrac{1}{4}, \tfrac{1}{4}) \quad \text{(fcc)}; \qquad D = \frac{2\pi}{a}\,(\tfrac{1}{2}, \tfrac{1}{2}, x) \quad \text{(bcc)}$$

$$Z = \frac{2\pi}{a}\,(1, x, 0) \quad \text{(fcc)}; \qquad S = \frac{2\pi}{a}\,(1, x, x) \quad \text{(fcc)}$$

1.7 The Effective Mass

We now wish to study in a general way the behavior of the energy as a function of wave vector inside the Brillouin zone. We have already observed that the energy has the full symmetry of the reciprocal lattice. In addition, the invariance of the Hamiltonian under time reversal (which changes $\mathbf{k}$ into $-\mathbf{k}$) requires that $E(\mathbf{k}) = E(-\mathbf{k})$ regardless of crystallographic symmetry. (Time reversal symmetry will be discussed in detail in Section 1.11.) Our principal concern here is with the perrbutation theory which enables the determination of the energy as a function of $\mathbf{k}$ for values of $\mathbf{k}$ in the neighborhood of some $\mathbf{k_0}$ for which the energy levels are assumed to be known. The perturbation procedure can be developed in the following way:

Let $\psi_n(\mathbf{k}, \mathbf{r})$ be the wave function for a state in the nth band at position $\mathbf{k}$ in the Brillouin zone, and let $\psi_j(\mathbf{k_0}, \mathbf{r})$ similarly pertain to the jth band at $\mathbf{k_0}$. For the present, all states will be assumed to be nondegenerate. The functions

$$\chi_j(\mathbf{k}, \mathbf{r}) = e^{i(\mathbf{k} - \mathbf{k_0}) \cdot \mathbf{r}} \psi_j(\mathbf{k_0}, \mathbf{r}) = e^{i\mathbf{s} \cdot \mathbf{r}} \psi_j(\mathbf{k_0}, \mathbf{r}) \tag{1.31}$$

(where we have set $\mathbf{k} = \mathbf{k_0} + \mathbf{s}$) are satisfactory basis functions for the expansion of the wave function for a state of wave vector $\mathbf{k}$. This may easily be seen by writing $\psi_j(\mathbf{k_0}, \mathbf{r})$ in the Bloch form: $e^{i\mathbf{k_0} \cdot \mathbf{r}} u_j(\mathbf{k_0}, \mathbf{r})$. It is shown in Section 4.1 that the $\chi_j(\mathbf{k}, \mathbf{r})$ are a complete, orthonormal set if the $\psi_n(\mathbf{k}, \mathbf{r})$ are.

$$\psi_n(\mathbf{k}, \mathbf{r}) = \sum_j A_{nj}(\mathbf{k}) \chi_j(\mathbf{k}, \mathbf{r}) = e^{i\mathbf{s} \cdot \mathbf{r}} \sum_j A_{nj} \psi_j(\mathbf{k_0}, \mathbf{r}) \tag{1.32}$$

The $\psi_j(\mathbf{k_0}, \mathbf{r})$ are solutions of the Schrödinger equation for an energy $E_j(\mathbf{k_0})$. This expansion is substituted into the Schrödinger equation for $\psi_n(\mathbf{k}, \mathbf{r})$. We write the Hamiltonian as the sum of the kinetic energy operator and a periodic potential:

$$\left(\frac{p^2}{2m_0} + V\right)\psi_n(\mathbf{k}, \mathbf{r}) = e^{i\mathbf{s} \cdot \mathbf{r}} \sum_j A_{nj} \left(\frac{p^2}{2m_0} + \frac{\hbar}{m_0}\mathbf{s} \cdot \mathbf{p} + \frac{\hbar^2}{2m_0}s^2 + V\right)\psi_j(\mathbf{k_0}, \mathbf{r})$$

$$= e^{i\mathbf{s} \cdot \mathbf{r}} \sum_j A_{nj} \left[E_j(\mathbf{k_0}) + \frac{\hbar^2}{2m_0}s^2 + \frac{\hbar}{m_0}\mathbf{s} \cdot \mathbf{p}\right]\psi_j(\mathbf{k_0}, \mathbf{r}) \tag{1.33}$$

$$= E_n(\mathbf{k}) e^{i\mathbf{s} \cdot \mathbf{r}} \sum_j A_{nj} \psi_j(\mathbf{k_0}, \mathbf{r}).$$

Since the $\psi_j(\mathbf{k}_0, \mathbf{r})$ form an orthonormal set, we multiply (1–33) by $\exp(-i\mathbf{s} \cdot \mathbf{r})\psi_{j'}^*(\mathbf{k}_0, \mathbf{r})$ and integrate over the entire crystal. The momentum matrix element $\mathbf{p}_{jj'}$ is defined by

$$\mathbf{p}_{jj'} = \int_{\substack{\text{entire}\\\text{crystal}}} \psi_j^*(\mathbf{k}_0, \mathbf{r})\left(\frac{\hbar}{i}\nabla\right)\psi_{j'}(\mathbf{k}_0, \mathbf{r})\, d^3r. \tag{1.34}$$

Then (1–33) reduces to:

$$\left(E_j(\mathbf{k}_0) + \frac{\hbar^2 s^2}{2m_0}\right)A_{nj} + \frac{\hbar}{m_0}\mathbf{s} \cdot \sum_{j'} \mathbf{p}_{jj'}A_{nj'} = E_n(\mathbf{k})A_{nj} \tag{1.35}$$

An equation of the form (1.35) is obtained for each value of j. Evidently we have an infinite set of simultaneous linear equations for the coefficients A_{nj}. The condition that these equations have a nontrivial solution is that the determinant whose general element is of the form

$$H_{jj'} - E_n(\mathbf{k})\delta_{jj'} \tag{1.36}$$

with

$$H_{jj'} = \left[E_j(\mathbf{k}_0) + \frac{\hbar^2 s^2}{2m_0}\right]\delta_{jj'} + \frac{\hbar}{m_0}\mathbf{s} \cdot \mathbf{p}_{jj'} \tag{1.37}$$

must vanish. This is, of course, equivalent to the problem of diagonalizing the matrix representing the Hamiltonian on the basis of the $\chi_j(\mathbf{k}, \mathbf{r})$ whose elements have the form given in (1.37). The off-diagonal matrix elements contain $\mathbf{s}$, and may be treated as perturbations when $\mathbf{s}$ is small. Under the assumption that the band of interest is not degenerate at $\mathbf{k}_0$, ordinary second order perturbation theory is applicable:

$$E_n(\mathbf{k}) = E_n(\mathbf{k}_0) + \frac{\hbar}{m_0}\mathbf{s} \cdot \mathbf{p}_{nn} + \frac{\hbar^2 s^2}{2m_0} + \frac{\hbar^2}{m_0^2}\sum_{j(j\neq n)} \frac{(\mathbf{s} \cdot \mathbf{p}_{nj})(\mathbf{s} \cdot \mathbf{p}_{jn})}{E_n(\mathbf{k}_0) - E_j(\mathbf{k}_0)} + \cdots \tag{1.38}$$

The perturbation theory has given a Taylor series expansion of the energy as a function of $\mathbf{k}$. The convergence of the series is governed by the energy denominators which appear in (1.38). If these are small, the second order term in the energy will be large, and higher terms in the series will be also important. Under these circumstances, it is desirable to diagonalize the portion of the effective Hamiltonian (1.37) which connects the nearly degenerate states.

Let us consider first a situation in which band n has an extremum at $\mathbf{k}_0$, so that term linear in s in (1.38) vanishes. Let s_α and s_β be rectangular components of s with respect to some fixed axes, and $p_{jj'}^\alpha$, etc., be corresponding components of the matrix element. We may differentiate Eq. (1.38) twice and obtain

$$\frac{m_0}{\hbar^2} \frac{\partial^2 E_n}{\partial s_\alpha \partial s_\beta} = \frac{m_0}{\hbar^2} \frac{\partial^2 E_n}{\partial k_\alpha \partial k_\beta} = \delta_{\alpha\beta} + \frac{1}{m_0} \sum_j \frac{(p_{nj}^\alpha p_{jn}^\beta + p_{nj}^\beta p_{jn}^\alpha)}{E_n(\mathbf{k}_0) - E_j(\mathbf{k}_0)} \tag{1.39}$$

It is convenient to define a reciprocal effective mass tensor through

$$\left(\frac{m_0}{m^*}\right)_{\alpha\beta} = \frac{m_0}{\hbar^2} \frac{\partial^2 E}{\partial k_\alpha \partial k_\beta} \tag{1.40}$$

Then we may write in (1.39)

$$\left(\frac{m_0}{m^*}\right)_{\alpha\beta} = \delta_{\alpha\beta} + \frac{1}{m_0} \sum_j \frac{(p_{nj}^\alpha p_{jn}^\beta + p_{nj}^\beta p_{jn}^\alpha)}{E_n(\mathbf{k}_0) - E_j(\mathbf{k}_0)} \tag{1.41a}$$

The result (1.41a) is the sum rule for the effective mass (sometimes referred to as the "f" sum rule). The diagonal elements of (1.41a) simplify to

$$\left(\frac{m_0}{m^*}\right)_{\alpha\alpha} = 1 + \frac{2}{m_0} \sum_j \frac{|p_{nj}^\alpha|^2}{E_n(\mathbf{k}_0) - E_j(\mathbf{k}_0)} \tag{1.41b}$$

The interaction of a given level with lower lying levels, or core states ($E_j < E_n$) tends to decrease the effective mass, while the interaction with higher states tends to increase it.

The first term in (1.38) gives rise to a linear dependence of energy on wave vector going away from $\mathbf{k}_0$. This term vanishes, however, for certain states at symmetry points of the Brillouin zone. In that case, the energy is quadratic in s near $\mathbf{k}_0$, as has been discussed. Symmetry considerations can usually be employed to determine whether the linear term vanishes. Consider a general matrix element of the form

$$\int \psi_i^*(\mathbf{k}_0, \mathbf{r}) \mathbf{O} \psi_j(\mathbf{k}_0, \mathbf{r}) \, d^3r$$

where **O** is any operator. In order that this integral not be zero, it is necessary that the integrand contain a scalar component. The functions $\psi_i(\mathbf{k_0}, \mathbf{r})$, $\psi_j(\mathbf{k_0}, \mathbf{r})$ and the operator **O** must transform according to definite irreducible representations of the point group of $\mathbf{k_0}$, say $\Gamma(i)$, $\Gamma(j)$, $\Gamma(\mathbf{O})$. The integrand then transforms according to the direct product $\Gamma(i) \times \Gamma(\mathbf{O}) \times \Gamma(j)$.

The concept of direct product is best introduced with reference to some matrix representation. The direct product of an $n \times n$ matrix whose elements are a_{ij} and an $m \times m$ matrix whose elements are b_{kl} is an $mn \times mn$ matrix with elements $a_{ij} b_{kl}$ where now i and k designate the row, and j and l the column. The representation of the group formed by constructing the direct products of the matrices of two irreducible representations will generally be reducible. In the present case, if the representation of $\Gamma(i) \times \Gamma(\mathbf{O}) \times \Gamma(j)$ contains $\Gamma(l)$, where $\Gamma(l)$ is the symmetric representation of the point group of $\mathbf{k_0}$, the matrix element need not vanish. In Eq. (1.34), the presence of the operator V indicates that we require $\Gamma(\mathbf{O}) = \Gamma(\mathbf{v})$, where $\Gamma(\mathbf{v})$ is the representation of a vector.

At a general point in the zone, all the representations are the same, and the linear dependence will usually exist. At symmetry points where there are a number of representations the linear dependence will often vanish. In particular, if the group of $\mathbf{k_0}$ contains the inversion, the wave functions will be eigenfunctions of definite parity, and the product $\psi^* V \psi$ will be an odd function whether ψ is odd or even, and the integral must vanish. For instance, $V_\mathbf{k} E = 0$ for all representations at $\varGamma$, H, and N in the body-centered cubic lattice. One expects to find maxima and minima of bands only at points of symmetry or along a symmetry axis where the components of the gradient perpendicular to that axis must vanish.

If the original state at $\mathbf{k_0}$ is degenerate, the perturbation will remove the degeneracy, at least in some directions. This means that in going from a point of high symmetry to a point of lower symmetry, the energy bands are split. It is then necessary to use degenerate perturbation theory. If the momentum operator has matrix elements connecting the members of the degenerate set, the degeneracy will be removed in first order, and the split bands will go away from the symmetry point with a nonzero slope. From the previous argument, it follows that this can only occur in the vicinity of symmetry points for which the point group

1.7 THE EFFECTIVE MASS

does not contain the inversion. In the most interesting cases (Shockley, 1950), however, the momentum operator has nonvanishing matrix elements only between the degenerate subset and states of different energy. The degeneracy is then removed in second order, and the perturbation theory appropriate for this case must be employed.

In this case, the energy is obtained by diagonalizing a rather different type of matrix. Suppose for simplicity that at $\mathbf{k}_0$ we have a doubly degenerate pair of states designated, with reference to Eqs. (1.35), by indices $n = 0, 1$ [energy $E_0(\mathbf{k}_0)$]. The coefficients A_0, A_1 will be the only ones which are of order zero in s. (Only one index need be retained on the quantities A.) We may then solve Eq. (1.35) for $A_j (j \neq 0, 1)$ in terms of A_0 and A_1, retaining terms of first order in s, only,

$$A_{j \atop (j \neq 0, 1)} = \frac{\hbar}{m_0} \mathbf{s} \cdot \left[\frac{\mathbf{p}_{j0} A_0}{E_0(\mathbf{k}_0) - E_j(\mathbf{k}_0)} + \frac{\mathbf{p}_{j1} A_1}{E_0(\mathbf{k}_0) - E_j(\mathbf{k}_0)} \right] \quad (1.42)$$

We use the result (1.42) to eliminate the $A_j (j \neq 0, 1)$ from those of Eqs. (1.35) which determine A_0 and A_1. Two equations for A_0 and A_1 are obtained, the first of which is

$$\left[E_0(\mathbf{k}_0) - E + \frac{\hbar^2 \mathbf{s}^2}{2m_0} + \frac{\hbar^2}{m_0^2} \sum_{j \neq 0, 1} \frac{(\mathbf{s} \cdot \mathbf{p}_{0j})(\mathbf{s} \cdot \mathbf{p}_{j0})}{E_0(\mathbf{k}_0) - E_j(\mathbf{k}_0)} \right] A_0 \quad (1.43)$$

$$+ \frac{\hbar^2}{m_0^2} \left[\sum_{j \neq 0, 1} \frac{(\mathbf{s} \cdot \mathbf{p}_{0j})(\mathbf{s} \cdot \mathbf{p}_{j1})}{E_0(\mathbf{k}_0) - E_j(\mathbf{k}_0)} \right] A_1 = 0$$

A similar equation is, of course, obtained in which the indices 0 and 1 are interchanged. The energy E is found as the solution of a (2×2) determinantal equation

$$\det |H_{lm} - E \delta_{ln}| = 0 \quad (1.44)$$

The problem is equivalent to that of finding the eigenvalues of an effective Hamiltonian, H, whose general element is

$$H_{ln} = \left(E_0(\mathbf{k}_0) + \frac{\hbar^2 \mathbf{s}^2}{2m_0} \right) \delta_{ln} + \frac{\hbar^2}{m_0^2} \sum_{j \neq l, n} \frac{(\mathbf{s} \cdot \mathbf{p}_{lj})(\mathbf{s} \cdot \mathbf{p}_{jn})}{E_0(\mathbf{k}_0) - E_j(\mathbf{k}_0)} \quad (1.45)$$

In the present case $l, n = 0, 1$; but in the case of a larger degenerate set at $\mathbf{k}_0$, Eqs. (1.44) and (1.45) still apply except for the obvious modifications that the order of the determinantal equation is equal to the degree of degeneracy of the set, and the sum in (1.45) excludes all members of the set. Finally, note that if the diagonal momentum matrix elements are not zero, a term $(\hbar/m_0)\mathbf{s} \cdot \mathbf{p}_{ll}\, \delta_{ln}$ is to be added to (1.45).

The matrix elements H_{ln} of the effective Hamiltonian are quadratic in $\mathbf{s}$. When the determinantal equation is solved, only the quadratic dependence of E on $\mathbf{s}$ is to be retained, since the perturbation theory has carried only to second order. The dependence of E on $\mathbf{s}$ will, however, be more complicated than that resulting from (1.38). The energy surfaces in the neighborhood of a degeneracy may be severely warped.

The functional dependence of the sum in (1.45) on the components of $\mathbf{s}$ can usually be determined from symmetry considerations. Consequently we can often set up the equation analogous to (1.44) except for some unknown constants when the matrix elements are unknown, and so determine the general form of the energy surfaces. Kane (1956a, b) has applied this procedure, suitably modified to include spin orbit coupling, in a semiempirical analysis of the degenerate valence bands near $\mathbf{k} = \mathbf{0}$ in germanium, silicon, and indium antimonide. A determination of the form of the energy surfaces near this point of degeneracy is important in the analysis of several experiments.

For some purposes, particularly in the theory of tunneling, it is important to determine the properties of the energy as a function of a complex wave vector $\mathbf{k}$. The mathematical analysis is based on a study of perturbation problem we have already discussed. The problem has been treated by Blount (1962a), some of whose results will be quoted here. First, consider a band which is not degenerate at a (real) point $\mathbf{k}_0$. The energy is an analytic function of complex $\mathbf{k}$ in a region about $\mathbf{k}_0$. Then the energy may be determined as a function of $\mathbf{k}$ from the Taylor expansion (1.38), which is convergent in this region. The region of analyticity is terminated by surfaces of branch points which must exist when the imaginary components of $\mathbf{k}$ are large. In the case of a group of bands which are degenerate at $\mathbf{k}_0$, the behavior of the energy in complex $\mathbf{k}$ space is much more complicated. It suffices here to state that $E(\mathbf{k})$ will not usually be analytic, but that $\mathbf{k}_0$ will be a branch point. This follows from consideration of the secular equation (1.44). A detailed discussion has been given by Blount.

Let us now summarize the general consequences of these considerations with reference to our discussion of group theory. In the first place, the energy is a continuous and differentiable function of (real) **k** throughout the Brillouin zone. If energy levels and wave functions have been determined at a point $\mathbf{k}_0$, the energies of states at $\mathbf{k}_0 + \mathbf{s}$, where **s** is small, may be determined by considering the effect of the perturbation $(\hbar/m_0)\mathbf{s} \cdot \mathbf{p}$. Symmetry considerations are not particularly helpful if $\mathbf{k}_0$ is a general point of the zone. If $\mathbf{k}_0$ is a symmetry point, and $\mathbf{k}_0 + \mathbf{s}$ is a general point, all the degeneracy which may be present at $\mathbf{k}_0$ is removed. If the point group of $\mathbf{k}_0 + \mathbf{s}$ is a subgroup of the group of $\mathbf{k}_0$, but still contains more than the identity, as occurs on going away from $\mathbf{k} = 0$ along a symmetry axis, the wave functions at $\mathbf{k}_0 + \mathbf{s}$ transform according to the subgroup. If the appropriate representation of the group of $\mathbf{k}_0$ is reducible as a representation of the subgroup at $\mathbf{k}_0 + \mathbf{s}$, the degeneracy at $\mathbf{k}_0$ will be removed at least in part. Of course, if the groups at $\mathbf{k}_0$ and $\mathbf{k}_0 + \mathbf{s}$ are the same, the degeneracy will also be the same.

Information concerning the connection of bands and the splitting of degeneracies can be obtained by determining how the representations of the point group of $\mathbf{k}_0$ are expressed in terms of direct sums of the representations at $\mathbf{k}_0 + \mathbf{s}$. These results are summarized in compatibility tables. The nature of this procedure may be appreciated from the following simple argument, which relates the representations at Γ to those along the 100 axis, Δ, in a cubic lattice. We see from inspection of the basis functions previously given for the representations that a function which has Γ_1 symmetry will go into one with Δ_1 symmetry. Functions with p, d, etc., character will be mixed with the original s-like function. Next consider the triply degenerate state Γ_{15}. We see from the basis functions given in Table IV that those functions which transform as y or z will go into the representation Δ_5; the one which transforms as x will go to Δ_1. Thus the triply degenerate Γ_{15} level will be split along this axis into the doubly degenerate Δ_5 and the nondegenerate Δ_1. Similarly, the triply degenerate Γ_{25}' splits into Δ_2' and Δ_5, while the doubly degenerate Γ_{12} splits into Δ_1 and Δ_2, as may again be seen by comparing basis functions. We can also see that the subscript notation for the states gives the appropriate compatibility relations. Unfortunately the compatibility results cannot always be stated so concisely when other axes are considered. The most important compatibility relations are summarized in Tables

XIII and XIV below. Other sets of relations, i.e., between P and Λ, and N and Σ may be immediately deduced from the character tables.

A fundamental question arises in considering the connection of bands concerning the circumstances under which bands may cross. This problem has been analyzed carefully by Herring (1937b). Two important cases must be considered.

TABLE XIII

COMPATIBILITY RELATIONS BETWEEN Γ AND Δ, Λ, Σ

Γ_1	Γ_2	Γ_{12}	Γ_{15}	$\Gamma_{25}{}'$	$\Gamma_1{}'$	$\Gamma_2{}'$	$\Gamma_{12}{}'$	$\Gamma_{15}{}'$	Γ_{25}
Δ_1	Δ_2	$\Delta_1 \Delta_2$	$\Delta_1 \Delta_5$	$\Delta_2{}' \Delta_5$	$\Delta_1{}'$	$\Delta_2{}'$	$\Delta_1{}' \Delta_2{}'$	$\Delta_1{}' \Delta_5$	$\Delta_2 \Delta_5$
Λ_1	Λ_2	Λ_3	$\Lambda_1 \Lambda_3$	$\Lambda_1 \Lambda_3$	Λ_2	Λ_1	Λ_3	$\Lambda_2 \Lambda_3$	$\Lambda_2 \Lambda_3$
Σ_1	Σ_4	$\Sigma_1 \Sigma_4$	$\Sigma_1 \Sigma_3 \Sigma_4$	$\Sigma_1 \Sigma_2 \Sigma_3$	Σ_2	Σ_3	$\Sigma_2 \Sigma_3$	$\Sigma_2 \Sigma_3 \Sigma_4$	$\Sigma_1 \Sigma_2 \Sigma_4$

TABLE XIV

COMPATIBILITY RELATIONS BETWEEN X AND Δ, Z, S

X_1	X_2	X_3	X_4	X_5	$X_1{}'$	$X_2{}'$	$X_3{}'$	$X_4{}'$	$X_5{}'$
Δ_1	Δ_2	$\Delta_2{}'$	$\Delta_1{}'$	Δ_5	$\Delta_1{}'$	$\Delta_2{}'$	Δ_2	Δ_1	Δ_5
Z_1	Z_1	Z_4	Z_4	$Z_3 Z_2$	Z_2	Z_2	Z_3	Z_3	$Z_1 Z_4$
S_1	S_4	S_1	S_4	$S_2 S_3$	S_2	S_3	S_2	S_3	$S_1 S_4$

I. It is possible for energy bands belonging to different representations to cross. Such a crossing, for instance on a symmetry axis, produces an *accidental degeneracy* at the point of contact. This term is applied in order to distinguish these contacts of energy levels from those required by symmetry considerations.

It is necessary to apply the degenerate perturbation procedure previously discussed in order to determine the dependence of the energy on $\mathbf{s} = \mathbf{k} - \mathbf{k_0}$ ($\mathbf{k_0}$ is the point of contact) in the neighborhood of the contact. In the usual cases, the degeneracy is removed in first order (except possibly in special directions) in $\mathbf{s}$ on going away from the point of contact.

II. It is unlikely there will be accidental degeneracies between bands of the same symmetry. By this we mean that (except for certain rather specialized possibilities discussed in detail by Herring) if such contact occurs for some specific crystal potential, it will be removed by almost any small change in the potential. Since all states at general points on the Brillouin zone have the same symmetry, it follows that isolated accidental degeneracies of bands at general points of the zone for crystals with a center of inversion are vanishingly improbable.

1.8 The Density of States

A quantity of fundamental interest in band theory is the number of electron states in an interval of energy. This is specified by the density of states function $G(E)$ which we define as follows: $G(E)\,dE$ is the number of states per volume Ω of the crystal (for each direction of the electron spin separately) with energies between E and $E + dE$. From elementary considerations, it follows that

$$G(E) = \frac{\Omega}{(2\pi)^3} \frac{d}{dE} \int d^3k \tag{1.46}$$

The integral is to be taken over the volume of k-space bounded by a surface of constant energy E. Let dS be an element of area on the energy surface. The perpendicular distance between two surfaces characterized by energies E and $E + dE$ is $dE/|\nabla_\mathbf{k} E|$. The volume of the region of k-space between energy surfaces is $\int (dS/|\nabla_\mathbf{k} E|)\,dE$. Thus

$$G(E) = \frac{\Omega}{(2\pi)^3} \int \frac{dS}{|\nabla_\mathbf{k} E|} \tag{1.47}$$

The integral in (1.47) goes over the surface of constant energy, E.

Equation (1.47) may be evaluated immediately for free electrons for which $E = \hbar^2 k^2/2m$ so that the energy surfaces are spherical. We have

$$|\nabla_\mathbf{k} E| = \frac{\hbar^2 k}{m} = \hbar \left(\frac{2E}{m}\right)^{1/2}$$

Then (for unit volume)

$$G(E) = \frac{m}{2\pi^2 \hbar^3} (2mE)^{1/2} \qquad (1.48)$$

According to the principles of Fermi statistics, the Fermi energy, E_F, at the absolute zero of temperature is determined by the condition that

$$2\int_0^{E_F} G(E)\, dE = n \qquad (1.49)$$

in which n is the number of electrons per unit volume and the factor of 2 on the left takes account of the two possible orientations of electron spin. We find that

$$E_F = \frac{\hbar^2}{2m}(3\pi^2 n)^{2/3} \qquad (1.50)$$

The average energy per electron, $\mathscr{E}$, is given by

$$\mathscr{E} = \frac{2}{n}\int_0^{E_F} G(E) E\, dE \qquad (1.51)$$

In the case of free electrons we find, from (1.48) and (1.50), after some manipulation, that

$$\mathscr{E} = \frac{3}{5} E_F \qquad (1.52)$$

It would be very difficult to evaluate (1.47) directly to determine the density of states in most actual problems in band theory. Some general conclusions as to the nature of the density of states function may be drawn, however, from very general arguments based on this equation.

It is evident from (1.47) that $G(E)$ may have some sort of singularity if the integration includes a point at which $|\nabla_\mathbf{k} E| = 0$. Such points are called *critical points*. We have seen in the previous sections that the symmetry of the crystal may require that $|\nabla_\mathbf{k} E|$ vanish at certain points. Hence, some critical points will be required to exist by symmetry reasons. In fact, there are several kinds of critical points.

1.8 THE DENSITY OF STATES

Let us consider the expansion of the energy in powers of $s = k - k_0$ as given by (1.36) for a case in which the linear term in the equation vanishes. The energy is evidently a quadratic form in the components of s; we may without loss of generality imagine we have chosen our coordinate system to coincide with the principal axes of this form, so that we have

$$E(k) = E(k_0) + \sum_{i=1}^{3} \alpha_i s_i^2 \qquad (1.53)$$

A point k_0, where such an expansion is possible, is called an "analytic critical point." There are four possible types of analytic critical points, (designated P_i according to the values of the coefficients α). These are enumerated as follows:

P_0: α_1, α_2, α_3 all negative (maximum)
P_1: α_1, α_2 negative, α_3 positive (saddle point)
P_2: α_1 negative, α_2, α_3 positive (saddle point)
P_3: α_1, α_2, α_3 all positive (minimum)

There are fundamental considerations of a topological nature which relate the number of critical points of the several kinds. These relations, first obtained by Morse (1938), were applied to the frequency distribution function for lattice vibrations by Van Hove (1953) and by J. C. Phillips (1956). The energy may be considered to be a multiply periodic function of wave vector in the reciprocal lattice. For the purposes of illustrating the argument in a simple manner, let us consider (Montroll, 1954) a two-dimensional square reciprocal lattice, which is shown in Fig. 4. Suppose that a simple energy band exists in this system, which has a maximum and a minimum of energy for the points in the cell shown in the diagram. Let us imagine a set of curves connecting the minima in adjacent cells as shown. On each curve there is a point at which the energy is a relative maximum. The locus of such relative maxima may be obtained: we suppose it is the solid curve passing through the maxima. On this curve, there is a lowest relative maximum: this point is a saddle point. Similarily, on drawing curves connecting the absolute maxima in two cells, one obtains a locus of relative minima, and thereby finds another saddle point which is the highest relative minimum. Evidently two saddle points must exist. (Actually, if the absolute maximum occurred at a point of low symmetry as shown in the diagram, it would have to be repeated at seven other points inside the cell.)

The fundamental result of the previous discussion is that the numbers of the critical points of the various type are not independent if the $E(\mathbf{k})$

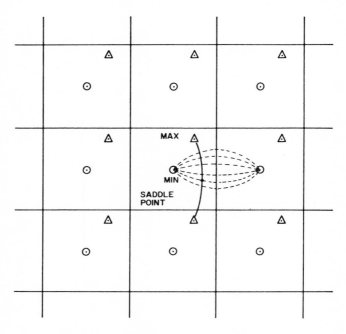

FIG. 4. Critical points in a square lattice.

function is multiply periodic. Let N_j be the number of critical points of type P_j (we are still considering only analytic critical points): the relations which must hold between the N_j are (for a three-dimensional situation) according to Morse:

$$
\begin{aligned}
N_0 &\geqslant 1 \\
N_1 - N_0 &\geqslant 2 \\
N_2 - N_1 + N_0 &\geqslant 1 \\
N_3 - N_2 + N_1 - N_0 &= 0
\end{aligned}
\tag{1.54}
$$

The minimum set of critical points in a given band is thus one maximum, one minimum, and three saddle points of each kind (P_1 and P_2).[5]

We will now study the behavior of the density of states near a critical point following the treatment of Wannier (1959). It is convenient to define new variables $r_i = |\alpha_i|^{1/2} s_i$ in (1.53) so that

$$E(\mathbf{k}) - E(\mathbf{k}_0) = \sum_{i=1}^{3} \varepsilon_i r_i^2, \quad \text{with } \varepsilon_i = \pm 1 \qquad (1.55)$$

The behavior in the vicinity of a minimum or maximum of energy can then be found as was done for the case of free electrons in (1.48). The density of states is the proportional to $(E - E_0)^{1/2}$ near the bottom of the band or to $(E_0 - E)^{1/2}$ near the top (where E_0 is the energy of the minimum or maximum, respectively). The saddle points have to be handled with greater care. These introduce discontinuities in the derivative of the density of states with respect to energy. In order to have a concrete example, let us assume

$$E(\mathbf{k}) - E(\mathbf{k}_0) = r_1^2 + r_2^2 - r_3^2 \qquad (1.56)$$

When $E(\mathbf{k}) > E(\mathbf{k}_0)$, the surfaces of constant energy may be represented as hyperboloids of one sheet; when $E(\mathbf{k}) = E(\mathbf{k}_0)$ the surface becomes a cone passing through the saddle point; for $E(\mathbf{k}) < E(\mathbf{k}_0)$ the energy surfaces are hyperboloids of two sheets. The area of the surfaces of constant energy changes very rapidly when $E(\mathbf{k})$ passes through $E(\mathbf{k}_0)$; this increase in area is responsible for the discontinuity of the derivative of the density of states. To determine $G(E)$, put

$$|E(\mathbf{k}) - E(\mathbf{k}_0)| = a^2$$

---

[5] The simple expression

$$E = E_0 + E_1(\cos k_x a + \cos k_y a + \cos k_z a); \quad E_1 < 0$$

which applies to a single s band in a simple cubic lattice in an extreme tight binding approximation exhibits the minimum number of critical points. The minimum occurs at the zone center Γ, three saddle points of type P_2 occur at the face center X, three saddle points of type P_1 occur at the middle of an edge M, and the maximum occurs at the corner R.

$$\left.\begin{array}{l} r_1 = a \cosh \zeta \cos \phi \\ r_2 = a \cosh \zeta \sin \phi \\ r_3 = a \sinh \zeta \end{array}\right\} \quad \text{for } E(\mathbf{k}) > E(\mathbf{k}_0) \tag{1.57a}$$

$$\left.\begin{array}{l} r_1 = a \sinh \zeta \cos \phi \\ r_2 = a \sinh \zeta \sin \phi \\ r_3 = a \cosh \zeta \end{array}\right\} \quad \text{for } E(\mathbf{k}) < E(\mathbf{k}_0) \tag{1.57b}$$

The surfaces of constant energy are then surfaces of constant a. The density of states is determined by expressing the volume element d^3k in the coordinates specified in (1.57) with the use of the Jacobian determinant. The integrals must be limited to a finite region in the vicinity of the saddle point. This may be done by requiring that

$$r_1^2 + r_2^2 + r_3^2 = a^2 (\sinh^2 \zeta + \cosh^2 \zeta) = R^2 \tag{1.58}$$

where R is some fixed number. Thus, for $E(\mathbf{k}) > E(\mathbf{k}_0)$, $0 \leqslant \sinh \zeta \leqslant [\frac{1}{2}(R^2 a^{-2} - 1)]^{1/2}$, and for $E(\mathbf{k}) < E(\mathbf{k}_0)$, $1 \leqslant \cosh \zeta \leqslant [\frac{1}{2}(R^2 a^{-2} + 1)]^{1/2}$. Thus, for $E(\mathbf{k}) > E(\mathbf{k}_0)$,

$$G(E) \propto (R^2 - a^2)^{1/2} = \sqrt{R^2 + E(\mathbf{k}_0) - E(\mathbf{k})} \tag{1.59}$$

while, for $E(\mathbf{k}) < E(\mathbf{k}_0)$,

$$G(E) \propto (R^2 + a^2)^{1/2} - a = \sqrt{(R^2 + E(\mathbf{k}_0) - E(\mathbf{k}))} - \sqrt{E(\mathbf{k}_0) - E(\mathbf{k})}$$

The function $\sqrt{R^2 + E(\mathbf{k}_0) - E(\mathbf{k})}$ is continuous and has continuous derivatives at the saddle point, so it is evident that $G(E)$ has a discontinuity in its first derivative. Similar considerations may easily be applied near a P_1 point.

The energy surfaces may be much more complicated than allowed by (1.53) in the vicinity of a point of degeneracy. We will not analyze these cases in detail. Phillips has shown that an index i and a topologial weight, q, may be assigned to each such nonanalytic critical point so that the relations (1.54) remain valid [q is the number of times the critical point is to be counted in applying (1.54)]. These relations are used in the following manner: one determines the critical points of each type required by the symmetry considerations of the previous section, and tests to see if the relations (1.54) are satisfied. If they are, no further critical points are required; if not, additional critical points must exist, and have to be located. Once the critical points are located, the behavior of the density of states as a function of energy can be determined in the neighborhood of the point. The construction of the density of states is greatly simplified with this information (see, for instance Callaway and Hughes, 1962).

1.9 The Fermi Surface in the Free Electron Approximation

At the absolute zero of temperature, the occupied states fill some bounded region of **k**-space; all states whose energy is less than the Fermi energy are full. The surface bounding the occupied region is called the Fermi surface. It can be shown that the existence of a Fermi surface across which the occupation number of the one-electron states is discontinuous is an (almost) exact result of the quantum theory of a many-fermion system (Luttinger and Ward, 1960). This discussion applies to metals, which are characterized by the existence of vacant states immediately adjacent to occupied states. In insulators to which the one-electron approximation may be applied, the energy bands are full. In such a case there is no Fermi surface for there are no vacant states adjacent in energy to the occupied states.

There are several experiments which measure characteristics of Fermi surfaces in metals: for instance, its area. Consequently a fundamental problem of band theory is to determine the Fermi surface for a given metal. This problem is a difficult one, being beset with the combined difficulties of an energy level calculation and the construction of the density of states. Consequently, it is of especial interest that it is possible to give at least a qualitative account of Fermi surfaces in many metals in terms of the free electron approximation. It would seem on first consideration that the free electron model would require the Fermi surface to be spherical. This, however, is true only for monovalent materials. When there is more than one electron per atom, the Brillouin zone structure must be taken into account. If we imagine gradually adding electrons to the "empty lattice," the Fermi surface is at first a sphere which expands until it is in contact with the zone boundary. This occurs at the face center N for the body-centered cubic lattice; at the point L for the face-centered cubic. The Fermi surface is in a sense reflected inward from the boundary: the first band continues to fill, while states in the higher bands are occupied. It is convenient to consider the Fermi surface to have portions in the higher bands. If there are enough electrons to fill states at a point where a band degeneracy occurs, states in several

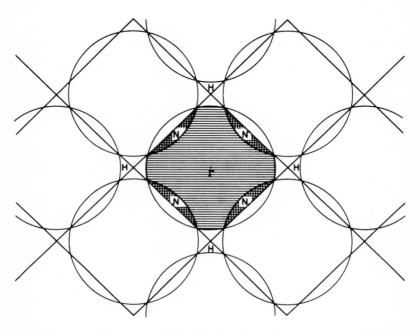

FIG. 5. Construction of the Fermi surface in the free electron approximation. The $k_x k_y$ plane in the reciprocal lattice of a body-centered cubic crystal is considered. The circles are drawn for a radius corresponding to three electrons per atom. The singly shaded area in the central zone represents a region in which states in the first band only are occupied; the doubly shaded area represents occupation of both first and second bands, and the unshaded area represents an unoccupied region.

bands may begin to fill at the same time. Since the free electron approximation yields simple algebraic expressions for the energy (Section 1.4) it is possible, using the principles discussed here, to make explicit constructions of the Fermi surface. This has been done by Harrison (Harrison,

1960b) for the body-centered cubic, face-centered cubic, and hexagonal close-packed structures.

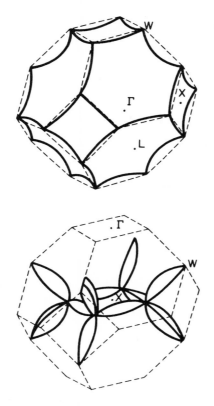

FIG. 6. Fermi surfaces in the free electron approximation. The Fermi surface in the free electron approximation is shown for a face-centered cubic crystal with three electrons per atom. The first band is full. The upper drawing shows the portion of the Fermi surface in the second band; the lower shows the portion in the third band, following Harrison (1960b). There are also small pockets of occupied states around the corners, W, in the fourth band. Note that the lower drawing is centered on X.

The actual construction of the surfaces is facilitated in an extended zone scheme. We can draw about each reciprocal lattice point a sphere whose volume corresponds to the appropriate number of electrons considered. These spheres will intersect when the number of electrons is sufficiently large. These intersections can occur only on the surfaces of a Brillouin zone. The constant energy surface belonging to a particular band changes spheres at such an intersection.

Consider a point $\mathbf{k}$, lying within the central Brillouin zone. If it is contained within a single sphere, a free electron state at that point in the first band is occupied. If it lies within two spheres, states belonging to the first and second bands are occupied, and so on. If it does not lie within a sphere, the state is unoccupied. As an illustration, Fig. 5 shows the construction for the $k_x k_y$ plane in the reciprocal lattice for the body-centered cubic structure for the case of three electrons per atom. Small regions around the corner H are unoccupied, and in the second band, regions around the face centers are occupied. In Fig. 6, some of the Fermi surfaces constructed by Harrison are shown.

1.10 Spin Orbit Coupling

In this section, the role of spin orbit coupling in band theory is considered. Spin orbit coupling is a feature appearing in an approximate form of the Dirac equation which results from elimination of the small components of the wave function (see L. I. Schiff, 1955, p. 332). We begin by considering the Dirac equation for one particle in a periodic potential $V(\mathbf{r})$:

$$[-c\boldsymbol{\alpha}\cdot\mathbf{p} - \beta mc^2 + V]\psi_\mathbf{k} = E_\mathbf{k}\psi_\mathbf{k} \qquad (1.60)$$

The argument of Section 1.1 establishing Bloch's theorem is still valid since it depends only on the translation invariance of the Hamiltonian. Hence, we can write

$$\psi_\mathbf{k} = e^{i\mathbf{k}\cdot\mathbf{r}} u_\mathbf{k}(\mathbf{r}) \qquad (1.61)$$

where $u_\mathbf{k}$ is a four-component spinor. The general procedure of Section 1.7 concerning the calculation of effective masses is still valid. One must replace Eq. (1.35) by

$$E_j(\mathbf{k}_0)A_{nj} - \hbar c \mathbf{s} \cdot \sum_{j'} \boldsymbol{\alpha}_{jj'} A_{nj'} = E_n(\mathbf{k})A_{nj} \qquad (1.62)$$

1.10 SPIN ORBIT COUPLING

where

$$\alpha_{jj'} = \int \psi_j^*(\mathbf{k}_0, \mathbf{r}) \, \alpha \psi_{j'}(\mathbf{k}_0, \mathbf{r}) \, d^3r \tag{1.63}$$

and the integration includes summation over spinor indices.

The energy bands which are derived from the Dirac equation (1.60) differ most strikingly from those calculated in the nonrelativistic theory in that some degeneracies present in the latter case are removed. This results because the transformation properties of a spinor (either four-component or two-component) are quite different from those of a scalar wave function. The splitting of degeneracies is given correctly in a qualitative sense when one considers the approximate "Pauli" equation which contains the spin orbit coupling term. Most analysis has been based on this equation, rather than the Dirac equation.

The standard approximation procedure which was mentioned previously reduced the Dirac equation to the following:

$$\left[\left(1 - \frac{E-V}{2mc^2}\right) \frac{\mathbf{p}^2}{2m} + V - \frac{\hbar^2}{4m^2 c^2} \nabla V \cdot \nabla + \frac{\hbar}{4m^2 c^2} \boldsymbol{\sigma} \cdot (\nabla V \times \mathbf{p}) \right] \psi_\mathbf{k} = E_\mathbf{k} \psi_\mathbf{k} \tag{1.64}$$

in which $\psi_\mathbf{k}$ is now a two-component Pauli spinor. It is seldom possible (or desirable) to solve (1.64) exactly: instead one regards the relativistic corrections

$$-\frac{(E-V)}{2mc^2} \frac{\mathbf{p}^2}{2m} - \frac{\hbar^2}{4m^2 c^2} \nabla V \cdot \nabla + \frac{\hbar}{4m^2 c^2} \boldsymbol{\sigma} \cdot (\nabla V \times \mathbf{p})$$

as perturbations. Effective mass theory is often discussed in terms of (1.64), although all the relativistic corrections except the spin orbit coupling term are frequently omitted. In place of Eqs. (1.35) or (1.62), one has

$$\left[E_j(\mathbf{k}_0) + \left(1 - \frac{E}{2mc^2}\right) \frac{\hbar^2 \mathbf{s}^2}{2m} \right] A_{nj} \tag{1.65}$$

$$+ \sum_{j'} \left(\frac{\hbar^2 \mathbf{s}^2}{4m^2 c^2} V_{jj'} + \frac{\hbar}{m} \mathbf{s} \cdot \boldsymbol{\Pi}_{jj'} \right) A_{nj'} = E_n(\mathbf{k}) A_{nj}$$

where

$$\Pi_{jj'} = \int \psi_{j'}(\mathbf{k}_0, \mathbf{r})^* \times \tag{1.66}$$

$$\left[\left(1 - \frac{E-V}{2mc^2}\right)\mathbf{p} - \frac{i\hbar}{4mc^2}\nabla V + \frac{\hbar}{4mc^2}(\boldsymbol{\sigma} \times \nabla V)\right]\psi_j(\mathbf{k}_0, \mathbf{r})\,d^3r$$

and the integration includes summation over spinor indices. $V_{jj'}$ is an interband matrix element of the potential.

We now wish to investigate the symmetry properties of the one-electron wave functions when these functions are not scalars but Pauli spinors. The translational symmetry properties are identical, but fundamental differences arise when rotations are included. There are two different quantum mechanical operators which correspond to the same transformation of points in space. It is these operators, rather than the physical transformations themselves, which form the group whose irreducible representations are required. This "double valuedness" has far reaching consequences. We will first examine its origin.

Suppose u is a constant spinor in some coordinate system (a subscript μ now indicates the spinor index: $\mu = 1, 2$). Let $\mathbf{R}$ be a pure rotation characterized by Euler angles α, β, γ. In the new coordinate system the corresponding spinor is

$$u_\mu' = \sum_{\lambda=1}^{2} D^{1/2}_{\mu\lambda}(\mathbf{R}) u_\lambda \tag{1.67}$$

in which the matrix $D^{1/2}(\mathbf{R})$ is

$$D^{1/2}(\mathbf{R}) = \begin{pmatrix} \cos\beta/2\, e^{-i(\alpha+\gamma)/2} & -\sin\beta/2\, e^{i(\gamma-\alpha)/2} \\ \sin\beta/2\, e^{i(\alpha-\gamma)/2} & \cos\beta/2\, e^{i(\alpha+\gamma)/2} \end{pmatrix} \tag{1.68}$$

The matrix $D^{1/2}(\mathbf{R})$ belongs to the $D^{1/2}$ representation of the rotation group. The double valuedness of the representation is manifested in the appearance of half-angles in the matrix (1.68). If one of the Euler angles is increased by 2π, the transformation is unchanged, but the representation matrix changes sign. In particular, the matrices

$$\begin{pmatrix} 1 & 0 \\ 0 & 1 \end{pmatrix} \quad \text{and} \quad \begin{pmatrix} -1 & 0 \\ 0 & -1 \end{pmatrix}$$

both correspond to the identity.

1.10 SPIN ORBIT COUPLING

The wave functions of interest to us are not constant spinors, but rather may be represented by

$$\psi_{i\mu} = \phi_i^{(\mu)}(\mathbf{r}) v_\mu \tag{1.69}$$

in which v is a "Pauli" spinor in some coordinate system and $\phi_i^{(\mu)}(\mathbf{r})$ is an ordinary function of position which is a basis function for the lth irreducible representation of a point group, $\Gamma^{(l)}$ (the extension to space groups will be considered subsequently). Under the rotation $\mathbf{R}$, $\phi_i^{(\mu)}$ transforms as follows:

$$P_\mathbf{R}\, \phi_i^{(\mu)} = \sum_{j=1}^{n} [\Gamma^{(l)}(\mathbf{R})]_{ji}\, \phi_j^{(\mu)}(\mathbf{r}) \tag{1.70}$$

In this equation $P_\mathbf{R}$ is the operator which induces the operation $\mathbf{R}$, $[\Gamma^{(l)}(\mathbf{R})]_{ji}$ is the jith element of the matrix representing the operation $\mathbf{R}$ in representation $\Gamma^{(l)}$, and the sum over j includes the n functions which form a basis for $\Gamma^{(l)}$. To determine the transformation properties of $\psi_{i\mu}$, we must consider transformation of both space and spin variables. Let the operator $Q_\mathbf{R}$ induce the transformation $\mathbf{R}$ on the function $\psi_{i\mu}$. $Q_\mathbf{R}$ may be regarded as a product of two operators: the $P_\mathbf{R}$, given previously, and $S_\mathbf{R}$, which acts on the spinor v in accord with (1.67). Then

$$Q_\mathbf{R}\, \psi_{i\mu} = \sum_{j=1}^{n} \sum_{\lambda=1}^{2} [\Gamma^{(l)}(\mathbf{R})]_{ji}\, D^{1/2}(\mathbf{R})_{\lambda\mu}\, \phi_j\, v_\lambda \tag{1.71}$$

$$= \sum_{(j\lambda)} [\Gamma^{(l)}(\mathbf{R}) \times D^{1/2}(\mathbf{R})]_{j\lambda, \mu i}\, \psi_{j\lambda}$$

We infer from (1.71) that $\psi_{i\mu}$ transforms according to the direct product representation: the representation whose matrices are direct products of those belonging to $\Gamma^{(l)}$ and those belonging to $D^{1/2}$. The double valuedness of $D^{1/2}$ implies that the direct product representation is also double valued.

Let us consider the group G formed by the operators $Q_\mathbf{R}$ which correspond to the transformations $\mathbf{R}$ which belong to a point group g. The group G contains twice as many elements as g does: G is homomorphic, rather than isomorphic to g. The general principles of quantum theory require that the one-electron wave functions form bases for the irreducible representations of G. The representations of g may be trivially extended to form representations of G, merely by assigning the same matrix to represent both operators of G which correspond to the same operation, **R**, of g. There are, however, additional representations of G. These have the property that the two matrices E and $\bar{E}$ which correspond in G to the identity, ε, of g have characters which differ in sign. One-electron wave functions which include spin transform according to these additional representations of G. This follows from (1.71), since the elements of the direct product representation $\Gamma^{(l)}(\mathbf{R}) \times D^{1/2}(\mathbf{R})$ contain the half-angle functions of $D^{1/2}(\mathbf{R})$, so that the elements of $\bar{E}$ differ in sign from those of E.

The direct product representation may, however, be reducible as a representation of G. In this case, one can write symbolically

$$\Gamma^{(l)} \times D^{1/2} = \sum_i c_i^{(l)} \gamma^{(i)} \qquad (1.72)$$

in which the γ^i are irreducible representations of G. The physical interpretation of this equation is that it expresses the splitting of a degenerate state $\Gamma^{(l)}$ by spin orbit coupling into states of symmetry $\gamma^{(i)}$. Only the additional representations of G occur on the right side of (1.75). This is true since if the character table itself is considered as a matrix,[6] the character table appropriate to a direct product group is the direct product of the matrix character tables of the groups which are factors of the product group (Murnaghan, 1938).

Up to this point, only proper rotations have been considered. The extension of the results to point groups which contain improper rotations is easily accomplished when one observes that any such group is either isomorphic to a group containing proper rotations only or else is formed by supplementing a group containing proper rotations only by the product of its operations with the inversion J. Further, a Pauli spinor is invariant under inversion: the representation matrix in $D^{1/2}$ for the inversion is

---

[6] A character table may be considered to be a matrix whose rows designate classes and whose columns designate irreducible representations.

just the unit matrix. Hence for any operation $\mathbf{R}$, we may set $D^{1/2}(\mathbf{R}) = D^{1/2}(J\mathbf{R})$. It is then possible to apply (1.71) to obtain the direct product representation of a point group containing improper rotations.

The determination of the representations of simple space groups is essentially similar to the discussion given in Section 1.6. To each transfornation $\{\boldsymbol{\alpha}|t\}$ previously considered, there correspond two quantum operators. The description of the Brillouin zone is unaltered, and one must determine, as before, the group of the wave vector. Bloch spinors of the form $e^{i\mathbf{k}\cdot\mathbf{r}}u_{\mathbf{k}}(\mathbf{r})$, where $u_{\mathbf{k}}$ is a spinor, are basis functions for the irreducible representations. The point group of the wave vector is now a double group, and a wave function must transform according to one of the "additional" representations of this group.

We will not explore the additional representations of double point groups in detail. Character tables for these representations for the simple, body-centered, and face-centered cubic structures, the diamond, and the hexagonal closed-packed lattices are given in the work of Elliott (1954). The number of additional representations is usually smaller than the number of representations in the corresponding single point group. For instance, the double group of the center of the zone, Γ, in the simple, body-centered, and face-centered cubic lattices contains sixteen classes; hence only six additional representations. Thus, the classification of electron states is in a sense less detailed when spin is included. Fortunately, in many interesting applications, the spin orbit coupling can be regarded as a perturbation on a level structure calculated ignoring the spin. The sixfold degenerate states at Γ, H, P (Γ_{15}, $\Gamma_{25}{}'$, etc.) are each split into a fourfold and a twofold degenerate state; but the originally fourfold degenerate states at these points (Γ_{12}, $\Gamma_{12}{}'$, P_3) are not split. The fourfold degenerate states at X, W, and L are split into two doubly degenerate states each. The 100 axis, Δ, is the only symmetry axis for which there are two distinct irreducible representations. As a consequence, accidental degeneracies which might be predicted along symmetry axes in a calculation without spin orbit coupling will be removed when spin is included, except possibly along the Δ axis. Both the fourfold degenerate states Δ_5 and Δ_3 are split. In the case of weak spin orbit coupling, the removal of the accidental degeneracies may in some circumstances produce regions of rapid variation of $E(\mathbf{k})$, and hence small effective masses, in a narrow band.

1.11 Time Reversal Symmetry

In addition to the spatial translations, rotations, and reflections, it is necessary also to consider time inversion. This operation is of a different nature from those previously considered in that it is represented by an antiunitary operator, which we denote by $\mathbf{Q}$. (An operator $\mathbf{Q}$ is antilinear if it satisfies

$$\mathbf{Q}(a\psi + b\phi) = a^* \mathbf{Q}\psi + b^* \mathbf{Q}\phi \tag{1.73}$$

for arbitrary functions ψ and ϕ. The operator is antiunitary if in addition to (1.73), we have

$$(\psi, \phi) = (\mathbf{Q}\psi, \mathbf{Q}\phi) \tag{1.74}$$

Application of the time reversal operator twice to a wave function must restore the original state; but it does not follow that $\mathbf{Q}^2$ is the unit operator. Instead, it can be shown (Wigner, 1959) that

$$\mathbf{Q}^2 = \pm I \tag{1.75}$$

The plus sign applies in a theory in which spin is neglected, or to a system containing an even number of spin $\frac{1}{2}$ particles; the minus sign applies to a system with an odd number of spin $\frac{1}{2}$ particles (in particular to one-electron theory including spin). Further, it can be shown that, for a Hamiltonian operator H which may be considered to be a function of position, momentum, and spin operators:

$$\mathbf{Q}H(\mathbf{r}, \mathbf{p}, \mathbf{s})\mathbf{Q}^{-1} = H(\mathbf{r}, -\mathbf{p}, -\mathbf{s}) \tag{1.76}$$

Except when an external magnetic field is present, the Hamiltonian of a physical system will be unchanged by the simultaneous transformations $\mathbf{p} \to -\mathbf{p}$; $\mathbf{s} \to -\mathbf{s}$, so that the time reversal operator commutes with the Hamiltonian.

In the simple Schrödinger theory in which spin is neglected, the operator $\mathbf{Q}$ is just the operator $\mathbf{K}$ of complex conjugation

$$\mathbf{Q} = \mathbf{K}, \qquad \mathbf{Q}\phi = \phi^* \tag{1.77}$$

If spin is included, the operator must anticommute with each of the Pauli spin operators. This is accomplished through the choice

$$\mathbf{Q} = \sigma_y \mathbf{K} \tag{1.78}$$

1.11 TIME REVERSAL SYMMETRY

in which

$$\sigma_y = \begin{pmatrix} 0 & -i \\ i & 0 \end{pmatrix}$$

(It will be observed that σ_y anticommutes with the real operators σ_x, σ_z. The inclusion of **K** ensures that **Q** anticommutes with σ_y also.)

We will now show that time reversal requires, for a crystal, that $E(\mathbf{k}) = E(-\mathbf{k})$ regardless of the spatial symmetry of the system (Kramers' theorem), and that a double degeneracy of the band system throughout the Brillouin zone must exist if the potential has a center of inversion. We consider a Hamiltonian including spin orbit coupling

$$H = \frac{\mathbf{p}^2}{2m} + V(\mathbf{r}) + \frac{\hbar}{4m^2 c^2} \boldsymbol{\sigma} \cdot (\nabla V \times \mathbf{p}) \tag{1.79}$$

[The additional terms present in (1.64) do not affect the argument and are therefore neglected.] The eigenfunctions of (1.79) are of the Bloch form

$$H\psi_\mathbf{k} = E(\mathbf{k})\psi_\mathbf{k}; \qquad \psi_\mathbf{k} = e^{i\mathbf{k}\cdot\mathbf{r}} u_\mathbf{k}$$

where $u_\mathbf{k}$ is a Pauli spinor. Application of the time reversal operator leads to (since **Q** commutes with H)

$$QH\psi_\mathbf{k} = HQ\psi_\mathbf{k} = H\sigma_y \psi_\mathbf{k}^* = E(\mathbf{k})\sigma_y \psi_\mathbf{k}^* \tag{1.80}$$

Hence $\sigma_y \psi_\mathbf{k}^*$ is an eigenfunction of H with eigenvalue $E(\mathbf{k})$. But

$$\sigma_y \psi_\mathbf{k}^* = e^{-i\mathbf{k}\cdot\mathbf{r}} w_\mathbf{k}(\mathbf{r})$$

in which $w_\mathbf{k}$ is a Pauli spinor with the full periodicity of the potential. Evidently $\sigma_y \psi_\mathbf{k}^*$ is a Bloch function for a state of wave vector $-\mathbf{k}$, and we have

$$E(\mathbf{k}) = E(-\mathbf{k}) \tag{1.81}$$

Next, consider the equation satisfied by $u_\mathbf{k}$ for the Hamiltonian of (1.79). It is

$$\left[\frac{\mathbf{p}^2}{2m} + V(\mathbf{r}) + \frac{\hbar}{m}\mathbf{k}\cdot\mathbf{p} + \frac{\hbar}{4m^2 c^2}\boldsymbol{\sigma}\cdot(\nabla V \times \mathbf{p}) + \frac{\hbar^2}{4m^2 c^2}\mathbf{k}\cdot(\boldsymbol{\sigma} \times \nabla V)\right] u_\mathbf{k}(\mathbf{r})$$
$$= E(\mathbf{k}) u_\mathbf{k}(\mathbf{r}) \tag{1.82}$$

Let us apply the inversion operator $\mathbf{J}$ to this equation. Then $\mathbf{p}$ changes into $-\mathbf{p}$ and $\mathbf{r}$ into $-\mathbf{r}$. We get

$$\left[\frac{\mathbf{p}^2}{2m} + V(-\mathbf{r}) - \frac{\hbar}{m}\mathbf{k}\cdot\mathbf{p} + \frac{\hbar}{4m^2c^2}\boldsymbol{\sigma}\cdot(\nabla V(-\mathbf{r})\times\mathbf{p}) \right. \quad (1.83)$$

$$\left. - \frac{\hbar^2}{4m^2c^2}\mathbf{k}\cdot(\boldsymbol{\sigma}\times\nabla V(-\mathbf{r}))\right]u_{\mathbf{k}}(-\mathbf{r}) = E(\mathbf{k})\,u_{\mathbf{k}}(-\mathbf{r})$$

If $V(\mathbf{r}) = V(-\mathbf{r})$, we see that this equation is the same as (1.82) with $\mathbf{k}$ changed into $-\mathbf{k}$. Hence, if the potential has inversion symmetry,

$$u_{-\mathbf{k}}(\mathbf{r}) = u_{\mathbf{k}}(-\mathbf{r}) \quad (1.84)$$

The solution of (1.82) for wave vector $-\mathbf{k}$ is required by (1.81) to have the same energy as the solution for $\mathbf{k}$. But, at a general point in the zone, the wave function is not an eigenfunction of the inversion, so $u_{\mathbf{k}}(-\mathbf{r})$ is a different function from $u_{\mathbf{k}}(\mathbf{r})$. Hence these are two different wave functions corresponding to the same wave vector which have the same energy. Thus we conclude that, if the crystal potential has inversion symmetry, there is a double degeneracy of the band structure throughout the zone.

Representation theory is greatly complicated by the presence of an antiunitary operator in the group. If $\mathbf{U}$ is a spatial transformation, the group is augmented by including transformations of the form $\mathbf{QU}$. Then one finds that the representation matrices which transform the eigenfunctions under the operations of the group do not form representations of the group, but rather are corepresentations. For details of the theory of corepresentations, the book by Wigner (1959) should be consulted. Time inversion does not lead to any further classification of eigenvalues (no additional quantum members). Extra degeneracies may, however, be produced. The circumstances in which extra degeneracies may arise have been discussed by Herring (1937a) and Elliott (1954).

Chapter 2

Methods of Calculation

2.1 General Discussion

In this chapter we shall survey the principal methods of calculating an energy band system for a particular material. We shall first assume that the effective crystal potential for which the one-electron wave equation is to be solved is known. Our interest will be in the procedure according to which this equation is solved, subject to the appropriate boundary conditions. Subsequently we will explore the problem of the determination of the crystal potential in the Hartree-Fock approximation. Reviews of the general principles and methods of band theory have been given by Jones (1960), Pincherle (1960), and Reitz (1955).

It is convenient to use the so-called atomic units in an energy band problem. We set $\hbar = 1$. As unit of length we take the Bohr radius of hydrogen $a_0 = \hbar^2/me^2$. If this is to have the numerical value 1, we must choose $me^2 = 1$. A convenient unit of energy is the Rydberg, which is the ionization energy of hydrogen, $e^2/2a_0$. In order for this to have the numerical value 1, we must have $e^2 = 2$, from which it follows that $m = \tfrac{1}{2}$. Finally, the velocity of light can be determined from the dimensionless relation $\hbar c/e^2 = 137.037$ to have the numerical value $c = 274.074$. In cgs units, the present unit of length is $a_0 = 5.2917 \times 10^{-9}$ cm; also note that 1 Rydberg = 13.6049 ev.

The one-electron Schrödinger equation now has the form

$$(-\nabla^2 + V_{\mathbf{k},i}^{j,n}(\mathbf{r}))\psi_{\mathbf{k},i}^{j,n} = E_{\mathbf{k},i}^{j,n}\psi_{\mathbf{k},i}^{j,n} \tag{2.1}$$

in which we refer to the nth state belonging to the jth row of the ith irreducible representation of the group of the wave vector $\mathbf{k}$. Except in cases where confusion would otherwise result, we will suppress some or all of the indices, i, j, n. If the potential V in (2.1) is derived from the Hartree-Fock equations, it will depend on the state considered, and thus bears a full set of indices. For the present we will ignore these complications and consider the potential V to be the same for all states. The boundary conditions for this equation are implied by the translational symmetry of the function, which is expressed by Bloch's theorem [Eq. (1.8)]

$$\psi_\mathbf{k}(\mathbf{r} + \mathbf{R}_i) = e^{i\mathbf{k}\cdot\mathbf{R}_i}\psi_\mathbf{k}(\mathbf{r})$$

In addition $\psi_\mathbf{k}$ has the rotation and reflection symmetry implied by the representation to which it belongs.

The rather diverse methods of finding solutions to (2.1) have this in common: for a three-dimensional problem, they all involve expansion of the unknown function in sets of known functions, e.g., plane waves, products of radial functions and spherical harmonics, solutions of an atomic self-consistent field problem, etc.

It is generally possible to choose the functions appearing in the expansion in such a way that some of the requirements on $\psi_\mathbf{k}$ are satisfied initially (for instance we may expand $\psi_\mathbf{k}$ in terms of a set of functions which satisfy Bloch's theorem). The remaining requirements on the wave function are then to be satisfied by the choice of the coefficients of the functions in the expansion. The various methods of calculation differ among themselves in the choice of the particular condition which is initially satisfied, and in the choice of the set of functions used for the expansion.

2.2 Plane Wave Expansions

A choice of functions for expansion which appears very simple and direct is a set of plane waves whose wave vectors are reciprocal lattice vectors. We have seen in Section 1.2 that a function of the form $\exp[i(\mathbf{k} + \mathbf{K}_n)\cdot\mathbf{r}]$, where $\mathbf{K}_n$ is any reciprocal lattice vector, satisfies Bloch's theorem for wave vector $\mathbf{k}$. Rather than proceed directly to the expansion (1.15), it is desirable generally to form symmetrical linear combinations of plane waves which transform according to one of the irreducible representations of the group of the wave vector.[1] Such a combination is expressed as

$$\Phi_{ki}^{jl} = \sum_{K_n} C_{ki}^{jl}(K_n) \, e^{i(k+K_n)\cdot r} \qquad (2.2)$$

The sum in (2.2) runs over selected groups of reciprocal lattice vectors which are chosen in the following way: Let α be an operator in the point group of k. Then, for a given K_n, only those functions are included which have the wave vectors $\alpha(k + K_n)$ where α runs over the operations of the point group. The superscript l indicates that Φ_{ki}^{jl} is the lth such combination of this type. We will now show how the coefficients are determined, following Wigner (1959).[2]

Let $[\alpha]_{i,mj}$ be the mjth matrix element of the matrix representation of the operator α in the ith irreducible representation of a group. A function Φ_i^j is said to transform according to the jth row of the ith irreducible representation if, for each α in the group,

$$\alpha \Phi_i^j = \sum_m [\alpha]_{i,mj} \Phi_i^m \qquad (2.3)$$

The sum over m in (2.3) runs from 1 to $d(i)$ where $d(i)$ is the dimension of the irreducible representation.[3] There are $d(i)$ functions Φ_i^m (including Φ_i^j), all of which satisfy an equation similar to (2.3). Equation (2.3) is now multiplied by $[\alpha]_{i',m'j'}$ and the result is summed over all the operations of the group. We have

---

[1] L. Eyges (1961, 1962) has attacked the periodic potential problem by expressing the Fourier coefficient of the wave function as a product of a function dependent on $|k + K_n|$ only times a Kubic harmonic (in cubic lattices). His procedure is equivalent to the one described here.

[2] A general discussion of the construction of symmetrized linear combinations of plane waves has been given by Schlosser (1962) for cubic lattices.

[3] The action of α on the function Φ is the following: Φ is a function of the position vector r. The operator α acting on r sends r into r': $r' = \alpha r$. Then $\alpha \Phi(r) = \Phi(\alpha^{-1} r)$. It is required by (2.3) that $\Phi(\alpha^{-1} r)$ be a linear combination of the original $\Phi(r)$ belonging to the i'th representation.

CHAPTER 2. METHODS OF CALCULATION

$$\sum_\alpha [\alpha]_{i', m'j'} \alpha \Phi_i{}^j = \sum_\alpha \sum_m [\alpha]_{i'm'j'} [\alpha]_{i, mj} \Phi_i{}^m$$

In order to evaluate the right-hand side of this equation, use is made of the following theorem (for proof, see Wigner, 1959, p. 79):

$$\sum_\alpha [\alpha]_{i', m'j'} [\alpha]_{i, mj} = \frac{h}{d(i)} \delta_{ii'} \delta_{jj'} \delta_{mm'} \tag{2.4}$$

(where h is the order of the group). Hence we have

$$\sum_\alpha [\alpha]_{i', m'j'} \alpha \Phi_i{}^j = \frac{h}{d(i)} \delta_{ii'} \delta_{jj'} \Phi_i{}^{m'} \tag{2.5a}$$

and in particular,

$$\Phi_i{}^j = \frac{d(i)}{h} \sum_\alpha [\alpha]_{i, jj} \alpha \Phi_i{}^j \tag{2.5b}$$

Now let F be an arbitrary function which can be expressed as a linear combination of functions Φ_i^j belonging to the various rows of the irreducible representations of the group

$$F = \sum_{i,j} a_i{}^j \Phi_i{}^j \tag{2.6}$$

where the a_i^j are coefficients. Let us form

$$\sum_\alpha [\alpha]_{i,jj} \alpha F = \sum_{i'j'} a_{i'}{}^{j'} \sum_\alpha [\alpha]_{i,jj} \alpha \Phi_{i'}{}^{j'} \tag{2.7}$$

$$= \frac{h}{d(i)} \sum_{i'j'} a_{i'}{}^{j'} \delta_{ii'} \delta_{jj'} \Phi_{i'}{}^{j'} = \frac{h}{d(i)} a_i{}^j \Phi_i{}^j$$

Equation (2.7) expresses the result we desire: If we take an arbitrary function F, and form the sum over all the operations of the group of $[\alpha]_{i,jj} \alpha F$, the result, if not zero, is a function transforming according to the jth row of the ith irreducible representation.

2.2 PLANE WAVE EXPANSIONS

This procedure is quite general, and may be applied to form symmetrized linear combinations of plane waves, spherical harmonics, or other functions. For the particular case of plane waves, we note that

$$\alpha \exp(i\mathbf{k} \cdot \mathbf{r}) = \exp(i\mathbf{k} \cdot \alpha^{-1}\mathbf{r}) = \exp(i\alpha^{-1}\mathbf{k} \cdot \mathbf{r})$$

Since the Φ_i^j are, in this case, linear combinations of plane waves, a single plane wave can also be regarded as a linear combination of different $\Phi_{i'}^{j'}$, so that (2.7) is applicable. We then have

$$\Phi_{ki}^{jl} \propto \sum_\alpha [\alpha]_{i,jj} \exp[i\alpha(\mathbf{k} + \mathbf{K}_n^{(l)}) \cdot \mathbf{r}] \qquad (2.8)$$

In the particular case of a one-dimensional representation the matrix element $[\alpha]_{i,jj}$ is just the character of the operation in that representation.

As an illustration of the procedure, we obtain the linear combination of plane waves whose wave vectors are of the type $(2\pi/a)(1, 1, 0)$ transforming according to (the row) xy in the Γ_{25}' representation pertaining to the Brillouin zone of the body-centered cubic lattice. Consider an orthogonal basis of functions transforming like xy, xz, yz. On this basis, matrices may be constructed representing the operators of Table I. For example the operator $(y, -x, z)$ is represented by the matrix

$$\begin{pmatrix} -1 & 0 & 0 \\ 0 & 0 & 1 \\ 0 & -1 & 0 \end{pmatrix}$$

and the operator $(z, x, -y)$ by

$$\begin{pmatrix} 0 & 1 & 0 \\ 0 & 0 & -1 \\ -1 & 0 & 0 \end{pmatrix}$$

Only the operators which have a nonzero (xy, xy) element contribute to the sum. If we let α be the operator $(y, -x, z)$, then for $\mathbf{K}_n = (2\pi/a)(1, 1, 0)$ we get $\alpha \mathbf{K}_n = (2\pi/a)(1, -1, 0)$ and have a contribution to (2.8):

$$- e^{2\pi i/a(x-y)}$$

After going through the table of operators, we find the desired combination:

$$4\left[e^{2\pi i/a(x+y)} + e^{-2\pi i/a(x+y)} - e^{2\pi i/a(x-y)} - e^{-2\pi i/a(x-y)}\right]$$
$$= 8\left[\cos\frac{2\pi}{a}(x+y) - \cos\frac{2\pi}{a}(x-y)\right] \qquad (2.9)$$

There is a simpler procedure for determining the coefficients of a symmetrized combination of plane waves which is useful in many cases. This rests on the fact that the Fourier transform of a spherical harmonic is proportional to the same spherical harmonic in **k** space. Let **K** be a vector in **k** space with spherical components K, K_θ, K_ϕ, with reference to some fixed axes. Then

$$\int e^{i\mathbf{K}\cdot\mathbf{r}} Y_{lm}(\theta,\phi)\, d\omega = 4\pi i^l j_l(Kr) Y_{lm}(K_\theta, K_\phi) \qquad (2.10)$$

The basis functions for the various representations of the wave vector groups given in Chapter I are linear combinations of spherical harmonics of the same l but differing m. Hence, the Fourier transform of such a combination in ordinary space is proportional to the same combination of spherical harmonics in **k** space, and the constant of proportionality depends only on l and the magnitude of **k**. For example, a basis function proportional to xy leads to a plane wave expansion in which the coefficients are proportional to $K_x K_y$, as will be seen to be the case in (2.9).

For example, suppose we wish to make a linear combination of plane waves with wave vectors of the type $(2\pi/a)(1, 0, 0)$ which transform according to $x^2 - y^2$ in the representation H_{12}. The coefficients in the expansion will be proportional to $k_x^2 - k_y^2$. Hence the combination will be

$$e^{2\pi i x/a} + e^{-2\pi i x/a} - e^{2\pi i y/a} - e^{-2\pi i y/a} = 2\left(\cos\frac{2\pi x}{a} - \cos\frac{2\pi y}{a}\right) \quad (2.11)$$

Application of Eq. (2.8) also will yield a combination proportional to (2.11). Finally, the reader should note that if we make Taylor series expansions about the origin of the combinations (2.9) and (2.11), the first nonvanishing term is proportional to xy in the case of (2.9) and to $(x^2 - y^2)$ in the case of (2.11). Higher order terms will be proportional to other combinations of spherical harmonics which transform in the appropriate way [for instance, $x^4 - y^4$ in the case of (2.11)]. The expansion of symmetrized linear combination of plane waves furnishes a practical method for the construction of Kubic harmonics (Flower et al., 1960).

One must also note that it will sometimes be possible to make more than one symmetrized linear combination of plane waves which transforms according to a particular row of an irreducible representation from plane waves of the same type. This may easily be seen if we consider for example the construction of a plane wave combination transforming like x in the representation Γ_{15} of the full cubic group using plane waves of the type $(2\pi/a)$ (3, 2, 1). Evidently there will be three such combinations according as $k_x = \pm 3, \pm 2, \pm 1$. One must be careful, in any application, to make sure that all pertinent combinations have been obtained.

A table of the symmetrized linear combinations of plane waves which transform according to the irreducible representations which contain s, p, and d functions in their expansion is given for a point of full cubic symmetry in Appendix I.

We may now proceed with the calculation of energy levels. The wave function is expressed as a linear combination of the functions $\Phi^{jl}_{\mathbf{k}i}$

$$\psi^{j}_{\mathbf{k}i} = \sum_{l} a^{jl}_{\mathbf{k}i} \Phi^{jl}_{\mathbf{k}i} \qquad (2.12)$$

Equation (2.12) is substituted into the one-electron Schrödinger equation (2.1). We then multiply both sides of the result by $\Phi^{j'l'}_{\mathbf{k}'i'}$ and integrate over the entire crystal (which consists of N unit cells of volume Ω_0). Functions which belong to different representations of the space group, or to different rows of the same representation are orthogonal. In addition, we shall normalize our functions $\Phi^{jl}_{\mathbf{k}i}$ (for instance, by dividing by the square root of the sum of the squares of the coefficients C^l of the waves in the combination) so that

$$\int \Phi^{j'l'*}_{\mathbf{k}'i'} \Phi^{jl}_{\mathbf{k}i} d^3r = N\Omega_0 \, \delta_{i'i} \, \delta_{j'j} \, \delta_{\mathbf{k}'\mathbf{k}} \, \delta_{ll'} \qquad (2.13)$$

Since all of the plane waves in any symmetrized combination have wave vectors of the same magnitude, it follows that

CHAPTER 2. METHODS OF CALCULATION

$$-\nabla^2 \Phi^l = \mathbf{h}_n^{(l)2} \Phi^l \tag{2.14}$$

where

$$\mathbf{h}_n^{(l)} = \mathbf{k} + \mathbf{K}_n^{(l)}$$

The combinations Φ are ordinarily listed in order of increasing $|\mathbf{h}^{(l)}|$. We obtain from (2.1)

$$N\Omega_0 a_{\mathbf{k}'i'}^{jl'}(\mathbf{h}^2 - E) + \sum_l a_{\mathbf{k}j}^{il} \int \Phi_{\mathbf{k}'i'}^{jl'*} V \Phi_{\mathbf{k}i}^{jl} d^3r = 0$$

The crystal potential does not have any (nonvanishing) matrix elements between functions which belong to different irreducible representations or to different rows of the same representation of the space group. Hence we have

$$a_{\mathbf{k}i}^{jl'}(\mathbf{h}^2 - E) + \sum a_{\mathbf{k}i}^{jl} V_{l'l} = 0 \tag{2.15}$$

in which

$$V_{l'l} = \frac{1}{N\Omega_0} \int \Phi_{\mathbf{k}i}^{jl'*} V \Phi_{\mathbf{k}i}^{jl} d^3r \tag{2.16}$$

There is one such equation for each value of l. Now let us examine the structure of the quantities $V_{l'l}$:

$$V_{l'l} = \sum_{n,n'} \frac{C^{l'}(\mathbf{K}_{n'})^* C^l(\mathbf{K}_n)}{N\Omega_0} \int \exp[i(\mathbf{h}_n^{(l)} - \mathbf{h}_{n'}^{(l')}) \cdot \mathbf{r}] V(\mathbf{r}) d^3r \tag{2.17}$$

Although the wave vectors $\mathbf{h}_n^{(l)}$, $\mathbf{h}_{n'}^{(l')}$ are generally not reciprocal lattice vectors, their difference must be a lattice vector, say $\mathbf{K}_s$. Since the potential $V(\mathbf{r})$ is periodic in space, it can be expressed as a sum of potentials defined within one unit cell of the lattice only.

$$V(\mathbf{r}) = \sum_{\mathbf{R}_n} V'(\mathbf{r} - \mathbf{R}_n) \tag{2.18}$$

$V'(\mathbf{r} - \mathbf{R}_n)$ is different from zero only in the unit cell centered at $\mathbf{R}_n$. It follows from the periodicity of the potential that all the $V'(\mathbf{r} - \mathbf{R}_N)$ are the same except for location. The integral in (2.17) has the form

$$\sum_{\mathbf{R}_n} \int e^{i\mathbf{K}_s \cdot \mathbf{r}} V'(\mathbf{r} - \mathbf{R}_n) \, d^3r = \sum_{\mathbf{R}_n} e^{i\mathbf{K}_s \cdot \mathbf{R}_n} \int e^{i(\mathbf{K}_s \cdot (\mathbf{r} - \mathbf{R}_n))} V'(\mathbf{r} - \mathbf{R}_n) \, d^3r$$

The integral on the right-hand side has the same value in each cell of the lattice. Further, $\exp(i\mathbf{K}_s \cdot \mathbf{R}_n) = 1$ for all $\mathbf{R}_n$. Let us define a "Fourier Coefficient" of potential $V(\mathbf{K}_s)$ by

$$V(\mathbf{K}_s) = \frac{1}{\Omega_0} \int e^{i\mathbf{K}_s \cdot \mathbf{r}} V'(\mathbf{r}) \, d^3r \qquad (2.19)$$

(the integral is evaluated in the cell at the origin). Then (2.19) becomes $N\Omega_0 V(\mathbf{K}_s)$. The matrix element $V_{l'l}$ is a linear combination of Fourier coefficients of potential.

$$V_{l'l} = \sum_s b_{l'l}(\mathbf{K}_s) V(\mathbf{K}_s) \qquad (2.20)$$

The coefficients $b(\mathbf{K}_s)$ are determined from the coefficients C_n^l of the plane waves appearing in the expression for Φ^l.

$$b_{l'l} = \sum_{n,n'} C^{l'*}(\mathbf{K}_{n'}) C^l(\mathbf{K}_n) \qquad (2.21)$$

where the summation is restricted to values n and n' such that $\mathbf{h}_{n'} - \mathbf{h}_n = \mathbf{K}_s$. After substitution of (2.20), Eq. (2.15) becomes

$$a_{\mathbf{k}i}^{jl'}(\mathbf{h}^2 - E) + \sum_{l'} a_{\mathbf{k}i}^{jl} \left[\sum_s b_{l'l}(\mathbf{K}_s) V(\mathbf{K}_s) \right] = 0 \qquad (2.22)$$

A nontrivial solution of the infinite set of linear equations (2.22) exists if and only if the determinant of the coefficients of the unknown quantities $a_{\mathbf{k}i}^{jl}$ is zero:

$$\det \left| (\mathbf{h}^2 - E) \delta_{ll'} + \sum_s b_{l'l}(\mathbf{K}_s) V(\mathbf{K}_s) \right| = 0 \qquad (2.23)$$

Alternatively, we may say that it is necessary to diagonalize the matrix representing the Hamiltonian on the basis of functions which are symmetrized linear combinations of plane waves belonging to a particular row of a given irreducible representation. The matrix elements are, effectively,

$$H_{l'l} = \mathbf{h}^{(l)2}\,\delta_{ll'} + \sum_s b_{l'l}(\mathbf{K}_s)V(\mathbf{K}_s) \qquad (2.24)$$

As an illustration of the determination of the coefficients $b_{l'l}$, consider the expectation value of the crystal potential calculated with the previously determined functions transforming according to Γ_{25}'. If we normalize these functions, our expectation value is

$$\tfrac{1}{4}\int |(e^{2\pi i/a(x+y)} + e^{-2\pi i/a(x+y)} - e^{2\pi i/a(x-y)} + e^{-2\pi i/a(x-y)})|^2\, V(\mathbf{r})\,d^3r$$

This gives a contribution to (2.24) which is

$$V(0) - 2V\left[\frac{2\pi}{a}(2,0,0)\right] + V\left[\frac{2\pi}{a}(2,2,0)\right] \qquad (2.25)$$

In practice, the appropriate linear combination of Fourier coefficients of potential may be determined easily and systematically, once the symmetrized functions Φ have been determined, from a table of the differences of the wave vectors of the plane waves which contribute to the expansion.

2.3 Plane Wave Expansions: An Example

As an example of the calculation of energy bands with the use of plane wave expansions, we will determine the energy of the state Γ_{25}' in the body-centered cubic lattice (Callaway, 1959). The crystal potential considered will be that of a lattice of point charges of atomic number Z and lattice parameter a. This model system is particularly well adapted to a calculational technique based on plane wave expansions since the Fourier coefficients of the crystal potential, $V(\mathbf{K}_s)$ are very simple. To determine these coefficients, consider Poisson's equation for the potential energy function. In atomic units, this equation is

2.3 PLANE WAVE EXPANSIONS: AN EXAMPLE

$$\nabla^2 V = 8\pi\rho \tag{2.26}$$

Fourier expansions for the potential and the number density ρ are introduced.

$$V = \sum_n V(\mathbf{K}_n)\, e^{-i\mathbf{K}_n \cdot \mathbf{r}}, \rho = \sum_n \rho(\mathbf{K}_n)\, e^{-i\mathbf{K}_n \cdot \mathbf{r}} \tag{2.27}$$

$V(\mathbf{K}_n)$ is given by (2.19) and similarly,

$$\rho(\mathbf{K}_n) = \frac{1}{\Omega_0} \int \rho(\mathbf{r})\, e^{i\mathbf{K}_n \cdot \mathbf{r}}\, d^3r \tag{2.28}$$

The integral in (2.28) includes the unit cell at the origin only. Ω_0 is the volume of this cell. With these substitutions, we get from (2.26)

$$V(\mathbf{K}_n) = -\frac{8\pi}{\mathbf{K}_n^2}\, \rho(\mathbf{K}_n) \tag{2.29}$$

The case $\mathbf{K}_n = 0$ must be considered separately. In spite of appearances, Eq. (2.29) possesses a finite limit in this case.

Equation (2.29) is quite general. For the model considered, within the cell centered at $\mathbf{R}_n = 0$, we have

$$\rho(\mathbf{r}) = Z[\delta(\mathbf{r}) - 1/\Omega_0] \tag{2.30}$$

Hence $\rho(\mathbf{K}_n) = Z/\Omega_0$, and

$$V(\mathbf{K}_n) = -\frac{8\pi Z}{\mathbf{K}_n^2 \Omega_0} \tag{2.31}$$

For the body-centered lattice, $\Omega_0 = a^3/2$, and $\mathbf{K}_n^2 = (4\pi^2/a^2)\mathbf{n}^2$ where $\mathbf{n}^2 = n_1^2 + n_2^2 + n_3^2$. Then

$$V(\mathbf{K}_n) = -\frac{4Z}{\pi a \mathbf{n}^2} \tag{2.32}$$

We define $V(0)$ as the limit of $V(\mathbf{K}_n)$ as $\mathbf{K}_n \to 0$. Thus

$$V(0) = -8\pi \lim_{\mathbf{K}_n \to 0} \frac{\rho(\mathbf{K}_n)}{\mathbf{K}_n^2} \tag{2.33}$$

To obtain this, we expand $e^{i\mathbf{K} \cdot \mathbf{r}}$ in powers of K, retaining terms through second order

$$\rho(\mathbf{K}) = \frac{1}{\Omega_0} \left[\int \rho(\mathbf{r}) \, d^3r + i \int \rho(\mathbf{r}) \, (\mathbf{K} \cdot \mathbf{r}) \, d^3r - \frac{K^2}{2} \int \rho(\mathbf{r}) r^2 \cos^2 \theta \, d^3r \right]$$

(Here θ is the angle between $\mathbf{K}$ and $\mathbf{r}$.) Since the charge distribution is neutral, the first integral in this expansion vanishes — since it has inversion symmetry, the second term also vanishes. In the third term, we put $\cos^2 \theta = \frac{1}{3} + \frac{2}{3} P_2 (\cos \theta)$ (P_2 is the second Legendre polynomial). P_2 is a basis function for the Γ_{12} representation in a cubic crystal; hence if the charge density has cubic symmetry (Γ_1) the integral of this function must vanish. Thus

$$\lim_{\mathbf{K} \to 0} \frac{\rho(\mathbf{K})}{K^2} = - \frac{1}{6\Omega_0} \int \rho(\mathbf{r}) r^2 \, d^3r$$

and

$$V(0) = + \frac{4\pi}{3\Omega_0} \int \rho(\mathbf{r}) r^2 \, d^3r \tag{2.34}$$

For the model considered, this integral becomes

$$V(0) = - \frac{4\pi Z}{3\Omega_0^2} \int r^2 \, d^3r \tag{2.35}$$

This integral can be worked out exactly for cubic lattices. In the case of the body centered cubic lattice, the result is:

$$\Omega_0^{-1} \int r^2 \, d^3r = (\sqrt{221}/100) a^2$$

Then

$$V(0) = - 1.2454 \, Z/a \tag{2.36}$$

It is now necessary to determine the matrix elements of the potential between the symmetrized combinations of plane which belong to Γ_{25}'. The lowest (in kinetic energy) such combination was determined in the preceding section. The reciprocal lattice vectors which have to be considered include those of 200, 211, 220, 310, 222, and 321 types, etc. The combinations are given in Appendix I. Note that in the case of the 211 and 321 waves there are two and three linearly independent, orthogonal combinations, respectively, of plane waves which transform according to each of the rows of the Γ_{25}' representation. The 200 waves do not contribute. The matrix elements are evaluated in the fashion discussed in the previous section. For example, the expectation value of the potential using the combination formed from plane waves is found [see Eq. (2.25)] to be $-0.7679\, Z/a$.

For arbitrary values of the parameters Z and a, it is necessary to determine the eigenvalue by numerical techniques. Let us, however, consider the problem in the light of perturbation theory, in which the crystal potential is considered as a perturbation. We use the notation of the previous section. The perturbed energy of the state formed from the lth linear combination of plane waves is, to second order

$$E^{(l)} = \mathbf{h}^{(l)2} + V_{ll} + \sum_{l' \neq l} \frac{|V_{ll'}|^2}{\mathbf{h}^{(l)2} - \mathbf{h}^{(l')2}} \tag{2.37}$$

The orders of magnitude of successive terms in the perturbation expansion may easily be determined. The kinetic energy $\mathbf{h}^{(l)2}$ is of order $4\pi^2/a^2$. The matrix elements of the potential are proportional to Z/a; hence the second order term is of order Z^2. Each successive term in the expansion introduces a matrix element and another energy denominator in each order, and thus gives an additional factor Za. The nth order term (counting the kinetic energy as of order zero) is proportional to $(1/a^2)(Za)^n$. Evidently the quantity aE/Z is a function of the single variable Za:

$$aE/Z = f(Za) \tag{2.38}$$

Perturbation theory yields a power series expansion of the function $f(Za)$ (lowest term is of order $1/Za$). These results are general. For the case of the lowest state belonging to Γ_{25}', evaluation of the coefficients in the perturbation series gives

$$E = \frac{8\pi^2}{a^2} - 0.7679\frac{Z}{a} - 0.00509\, Z^2 + \ldots \tag{2.39}$$

The rapid decrease of the coefficients with increasing order should be noted. A numerical calculation of the energy of this state by matrix diagonalization shows that the first three terms of the series as given in (2.39) are a good approximation to the energy up to $Za \approx 25$. There are two reasons for the apparently good convergence of the series: In the first place, the minimum energy denominator in (2.37) is $8\pi^2/a^2$; secondly, a considerable amount of cancellation occurs in the computation of matrix elements such as (2.25). This cancellation would not occur if we had considered an s-like state. However, the plane wave expansion procedure does cease to be useful for large Za. Under these circumstances, the energy of a state must approach a limit independent of a ($-Z^2/9$ in the present case), corresponding to a hydrogenic wave function. Also, the bandwidth must fall off exponentially as $Za \to \infty$.

2.4 Orthogonalized Plane Waves

A very serious difficulty usually arises in the application of the plane wave expansion procedure to less artificial problems: The convergence of the expansions may be poor; plane waves of large kinetic energy being important in the wave function. Further, the expansion for a state of wave vector **k** will converge to the state of lowest energy for that wave vector, and this will usually be an uninteresting state. For example, consider sodium. The states which are important in the study of the electronic properties are those related to 3s and 3p free atom wave functions. A plane wave expansion will generally, however, converge to a 1s state. In principle, it would be possible to circumvent this problem by finding higher eigenvalues of the Hamiltonian matrix, but to obtain any degree of accuracy, it would be necessary to employ such large matrices that the calculation would be impractical.

A way out of this difficulty was proposed by Herring (1940). Consider a crystal with one kind of atoms only. Suppose that wave functions for the core states (1s, 2s, 2p in the case of sodium) are known (perhaps from a self-consistent field calculation for the free atom, the assumption being made that these are unaltered in the solid) or that they can be calculated by other methods. Let a core function pertaining to state j be designated $u_j(\mathbf{r})$. In the designation appropriate to the free atom j stands for the three quantum numbers n, l, m, with l the angular momentum quantum number and m the z-component of angular momentum. These functions are assumed to be orthonormal. It is possible to combine such "free atom" functions to form functions possessing the periodicity required by Bloch's theorem: Let $\mathbf{R}_n$ be a direct lattice vector; then the required functions are

$$\phi_{\mathbf{k},j}(\mathbf{r}) = \frac{1}{\sqrt{N}} \sum_{\mathbf{R}_n}' e^{i\mathbf{k}\cdot\mathbf{R}_n} u_j(\mathbf{r} - \mathbf{R}_n) \tag{2.40}$$

Observe that

$$\phi_{\mathbf{k},j}(\mathbf{r} + \mathbf{R}_l) = \frac{1}{\sqrt{N}} \sum_{\mathbf{R}_n}' e^{i\mathbf{k}\cdot\mathbf{R}_n} u_j(\mathbf{r} + \mathbf{R}_l - \mathbf{R}_n)$$

$$= \frac{e^{i\mathbf{k}\cdot\mathbf{R}_l}}{\sqrt{N}} \sum_{\mathbf{R}_n}' e^{i\mathbf{k}\cdot(\mathbf{R}_n - \mathbf{R}_l)} u_j[\mathbf{r} - (\mathbf{R}_n - \mathbf{R}_l)]$$

$$= e^{i\mathbf{k}\cdot\mathbf{R}_l} \phi_{\mathbf{k},j}(\mathbf{r}) \tag{2.41}$$

The last step follows since the sum over $\mathbf{R}_n$ may be replaced by a sum over $\mathbf{R}_n - \mathbf{R}_l$ which is identical with that in (2.40). N is the number of atoms in the crystal and the factor $1/\sqrt{N}$ ensures normalization of the $\phi_{\mathbf{k},j}$. Such functions are often called Bloch functions. They are solutions of the Schrödinger equation only if the overlap of functions on different atoms can be ignored. (This will usually be a good approximation for core functions.)

Next, observe that a plane wave of wave vector $\mathbf{k}$ can be made orthogonal by the Schmidt process to the core functions of an equivalent $\mathbf{k}$ (orthogonality to functions of inequivalent $\mathbf{k}$ is guaranteed by the general theory). Let us denote a function constructed in this way by $X_{\mathbf{k}}$, an orthogonalized plane wave.

$$X_{\mathbf{k}} = \frac{e^{i\mathbf{k}\cdot\mathbf{r}}}{(N\Omega_0)^{1/2}} - \sum_j \mu_{\mathbf{k}j} \phi_{\mathbf{k},j}(\mathbf{r}) \tag{2.42}$$

The condition that

$$\int \phi_{\mathbf{k}l}^*(\mathbf{r}) X_{\mathbf{k}}(\mathbf{r}) \, d^3r = 0 \tag{2.43}$$

determines the "orthogonality coefficient" μ_{kl} as follows:

$$\int \phi_{kl}^*(\mathbf{r}) X_\mathbf{k}(\mathbf{r}) \, d^3r = \frac{1}{\Omega_0^{1/2} N} \sum_n \int e^{i\mathbf{k} \cdot (\mathbf{r} - \mathbf{R}_n)} u_l^*(\mathbf{r} - \mathbf{R}_n) \, d^3r - \qquad (2.44)$$

$$\sum_j \mu_{kj} \sum_{n,n'} \frac{e^{i\mathbf{k} \cdot (\mathbf{R}_n - \mathbf{R}_{n'})}}{N} \int u_l^*(\mathbf{r} - \mathbf{R}_{n'}) u_j(\mathbf{r} - \mathbf{R}_n) \, d^3r$$

This equation may be greatly simplified if one makes use of the orthonormality of the "atomic" functions u. It is also assumed that these functions do not overlap:

$$\int u_l^*(\mathbf{r} - \mathbf{R}_{n'}) u_j(\mathbf{r} - \mathbf{R}_n) \, d^3r = \delta_{nn'} \delta_{jl} \qquad (2.45)$$

Also, the integral in the first term on the right-hand side of (2.44) is the same in each cell. Hence, (2.43) yields

$$\mu_{kl} = \frac{1}{\Omega_0^{1/2}} \int_\infty e^{i\mathbf{k} \cdot \mathbf{r}} u_l^*(\mathbf{r}) \, d^3r \qquad (2.46)$$

An orthogonalized plane wave is a function which behaves as a plane wave at large distances from an atom, but possesses the rapidly varying character of an atomic wave function near any nucleus. Since both of these characteristics must be possessed by electron wave functions in solids, the orthogonalized plane waves are very useful functions for use in wave function expansions.

As an example of the construction of an orthogonalized plane wave, let us orthogonalize the plane wave with $\mathbf{k} = 0$ to a $1s$ hydrogenic function. The orthogonality coefficient is $\mu_{0,0}$:

$$u_{1s} = \left(\frac{\alpha^3}{\pi}\right)^{1/2} e^{-\alpha r} \qquad (2.47)$$

$$\mu_{0,0} = \frac{1}{\Omega_0^{1/2}} \left(\frac{\alpha^3}{\pi}\right)^{1/2} \int e^{-\alpha r} \, d^3r = 8 \left(\frac{\pi}{\alpha^3 \Omega_0}\right)^{1/2}$$

Then, in the cell centered at $\mathbf{R}_n = 0$,

$$X_0 = \frac{1}{(N\Omega_0)^{1/2}} (1 - 8 e^{-\alpha r}) \qquad (2.48)$$

2.4 ORTHOGONALIZED PLANE WAVES

This function possesses the single radial node characteristic of $2s$ functions (the node here occurs at $\alpha r = 2.07$, while in the $2s$ hydrogenic function similar to (2.47) it occurs at $\alpha r = 2$). For $r \gg 1/\alpha$, the function X_0 becomes flat. We shall see in subsequent sections that this behavior is characteristic of s-like states at the center of the zone.

Symmetrized linear combinations of orthogonalized plane waves which transform according to a particular irreducible representation can be formed with exactly the same coefficients as in the case of ordinary plane waves which was previously discussed. The wave function which is being determined is conveniently expanded in a series of symmetrized linear combination of orthogonalized plane waves. Some degree of complication arises because the orthogonalized plane waves are not mutually orthogonal. Evidently it is necessary to consider the simultaneous diagonalization of the matrices of the Hamiltonian and of unity on the basis of these functions.

We now consider the determination of the fundamental matrix elements: We define

$$(X_{\mathbf{k}'}|H|X_{\mathbf{k}}) = \int X_{\mathbf{k}'}^* H X_{\mathbf{k}} \, d^3r = \frac{1}{N\Omega_0} \int e^{-i\mathbf{k}' \cdot \mathbf{r}} (-\nabla^2 + V) e^{i\mathbf{k} \cdot \mathbf{r}} d^3r -$$

$$\frac{1}{N\Omega_0^{1/2}} \sum_j \left[\mu_{\mathbf{k}'j'}^* \int \phi_{\mathbf{k}'j'}^* H e^{i\mathbf{k} \cdot \mathbf{r}} d^3r + \mu_{\mathbf{k}j} \int e^{-i\mathbf{k}' \cdot \mathbf{r}} H \phi_{\mathbf{k}j}(\mathbf{r}) \, d^3r \, + \right.$$

$$\sum_{jj'} \mu_{\mathbf{k}j} \mu_{\mathbf{k}'j'}^* \int \phi_{\mathbf{k}'j'}^* H \phi_{\mathbf{k}j} \, d^3r \quad (2.49)$$

The symbol $(f|g)$ will occasionally be used to represent the scalar product:

$$(f|g) = \int f^* g \, d^3r$$

Successful application of expansions in orthogonalized plane waves require that the core functions $\phi_{\mathbf{k}j}$ be eigenfunctions of the same Hamiltonian as is used for the valence states of interest. This is essential since otherwise the valence electron states which one is attempting to construct will contain a portion of the real core states, and hence yield an incorrect energy value. Such errors will produce an energy value which is too low. The question of error will be considered in more detail subsequently. The ϕ_{kj} are assumed to be orthonormal eigenfunctions of H

$$H\phi_{kj} = E_j \phi_{kj}; \qquad (\phi_{kj}|\phi_{k'j'}) = \delta_{k,k'+K_n} \delta_{jj'} \qquad (2.50)$$

Also observe that

$$\frac{1}{N\Omega_0}\int e^{-i\mathbf{k}'\cdot\mathbf{r}}(-\nabla^2 + V) e^{i\mathbf{k}\cdot\mathbf{r}} d^3r = \mathbf{k}^2 \delta_{\mathbf{k}\mathbf{k}'} + V(\mathbf{k}-\mathbf{k}') \qquad (2.51)$$

[$\mathbf{k} - \mathbf{k}'$ must be a reciprocal lattice vector; otherwise $(X_{\mathbf{k}'}|H|X_{\mathbf{k}})$ is automatically zero.] Now (2.49) simplifies to

$$(X_{\mathbf{k}'}|H|X_{\mathbf{k}}) = \mathbf{k}^2 \delta_{\mathbf{k}\mathbf{k}'} + V(\mathbf{k}-\mathbf{k}') - \sum_j \mu^*_{\mathbf{k}'j} \mu_{\mathbf{k}j} E_j \qquad (2.52)$$

We can then deduce immediately that

$$(X_{\mathbf{k}'}|X_{\mathbf{k}}) = \delta_{\mathbf{k}\mathbf{k}'} - \sum_j \mu^*_{\mathbf{k}'j} \mu_{\mathbf{k}j} \qquad (2.53)$$

The products of the orthogonality coefficients $\mu_{\mathbf{k}j}$ in (2.52) are usually positive while the energies E_j and the Fourier coefficients $V(\mathbf{k} - \mathbf{k}')$ are negative. Hence the orthogonality terms tend to reduce the contribution from the potential energy to the matrix elements. In this respect, they have an effect similar to that which would be produced by a repulsive potential.

The orthogonality coefficients $\mu_{\mathbf{k}j}$ are actually rather complicated expressions. The sum over j includes core states characterized by quantum numbers n, l, m. If wave functions for states of differing azimuthal quantum number m but fixed n and l have the same radial wave function, the sum over m can be carried out. It is then possible to express the second term in (2.53) in terms of radial integrals: Let us put

$$u_j(\mathbf{r}) = R_{nl}(\mathbf{r}) Y_{lm}(\theta, \phi) \qquad (2.54)$$

in which Y_{lm} is a normalized spherical harmonic, and R_{nl} is a radial function, normalized so that

2.4 ORTHOGONALIZED PLANE WAVES

$$\int_0^\infty |R_{nl}(\mathbf{r})|^2 r^2 \, dr = 1$$

Then, by expanding $e^{i\mathbf{k}\cdot\mathbf{r}}$ in spherical harmonics and using the addition theorem, it is found that

$$\sum_j \mu_{\mathbf{k}'j}^* \mu_{\mathbf{k}j} = \frac{4\pi}{\Omega_0} \sum_{n,l} (2l+1) P_l(\cos\theta_{\mathbf{k}\mathbf{k}'}) I_{knl} I_{k'nl} \qquad (2.55)$$

In this equation $\theta_{\mathbf{k}\mathbf{k}'}$ is the angle between the wave vectors $\mathbf{k}$ and $\mathbf{k}'$, P_l is a Legendre polynomial, and

$$I_{knl} = \int_0^\infty j_l(kr) R_{nl}(r) r^2 \, dr \qquad (2.56)$$

[j_l is a spherical Bessel function: $j_l(x) = \sqrt{\pi/2x}\, J_{l+1/2}(x)$.]

In the practical application of the OPW method, the core functions u_j are hardly ever known exactly. It is important, then, to show what sort of error is induced in the energy value of a valence electron state by orthogonalization to incorrect core functions. We follow the analysis of Herring. Suppose, then, that the exact eigenfunctions of the Hamiltonian are denoted as $\psi_{\mathbf{k}s}$ (energy values $E_{\mathbf{k}s}$). The subscript j is used to denote core state. Let a particular approximate valence electron wave function be $\psi'_{\mathbf{k}0}$ (an approximation to $\psi_{\mathbf{k}0}$), this being the function for which the expectation value of the Hamiltonian, $E'_{\mathbf{k}0}$, is a minimum, subject to the condition of orthogonality to the approximate core functions $\phi_{\mathbf{k}j}$. The functions $\psi_{\mathbf{k}s}$ form a complete set, and all functions are normalized.

Then we can expand $\psi'_{\mathbf{k}0}$ in the $\psi_{\mathbf{k}s}$

$$\psi'_{\mathbf{k}0} = \sum_s (\psi_{\mathbf{k}s}|\psi'_{\mathbf{k}0})\psi_{\mathbf{k}s} \qquad (2.57\text{a})$$

Similarly, we have for the energy

$$E'_{\mathbf{k}0} = (\psi'_{\mathbf{k}0}|H|\psi'_{\mathbf{k}0}) = \sum_s |(\psi'_{\mathbf{k}s}|\psi'_{\mathbf{k}0})|^2 E_{\mathbf{k}s} \qquad (2.57\text{b})$$

The difference between the "exact" energy, $E_{\mathbf{k}0}$, and this is given by

$$E_{\mathbf{k}0} - E'_{\mathbf{k}0} = \sum_s |(\psi_{\mathbf{k}s}|\psi'_{\mathbf{k}0})|^2 (E_{\mathbf{k}0} - E_{\mathbf{k}s}) \qquad (2.58)$$

Only the core states have energies lower than $E_{\mathbf{k}0}$. Then we overestimate the error if we include only core states, j, in the sum in (2.58)

$$E_{\mathbf{k}0} - E'_{\mathbf{k}0} \leqslant \sum_j |(\psi_{\mathbf{k}j}|\psi'_{\mathbf{k}0})|^2 (E_{\mathbf{k}0} - E_{\mathbf{k}j})$$

Since $\psi'_{\mathbf{k}0}$ is orthogonal to the approximate core function $\phi'_{\mathbf{k}j}$, we may apply Bessel's inequality in the form

$$|(\psi_{\mathbf{k}j}|\psi_{\mathbf{k}j})|^2 = 1 \geqslant |(\psi_{\mathbf{k}j}|\psi'_{\mathbf{k}0})|^2 + |(\psi_{\mathbf{k}j}|\phi_{\mathbf{k}j})|^2 \qquad (2.59)$$

Hence we have

$$E_{\mathbf{k}0} - E'_{\mathbf{k}0} \leqslant \sum_j [1 - |(\psi_{\mathbf{k}j}|\phi_{\mathbf{k}j})|^2](E_{\mathbf{k}0} - E_{\mathbf{k}j}) \qquad (2.60)$$

Since the scalar product $(\psi_{\mathbf{k}j}|\phi_{\mathbf{k}j})$ is analogous to the cosine of the angle between vectors, we can see that, in a sense, the error produced by orthogonalization to inaccurate core functions is a second order correction. But, in practical cases, the energy differences in (2.60) which are important may be quite large (of the order of 10 to 50 rydbergs); hence it does not take large departures of $\phi'_{\mathbf{k}j}$ from $\psi_{\mathbf{k}j}$ to produce significant effects. This question is a most serious one in the application of the OPW procedure.

2.5 The Pseudopotential

J. C. Phillips and L. Kleinman (1959) have given a discussion of the Orthogonalized Plane Wave method which shows clearly the significance of the orthogonalization terms. Let $\psi_{\mathbf{k}i}$ be the exact (valence) electron wave for the ith irreducible representation of wave vector $\mathbf{k}$ (we suppress the other indices) and let $\phi_{\mathbf{k}j}$ be a wave function for a core state with the

2.5 THE PSEUDOPOTENTIAL

same wave vector (say a Bloch function). Since the valence function $\psi_{\mathbf{k}i}$ must be orthogonal to the $\phi_{\mathbf{k}j}$, it is convenient to write

$$\psi_{\mathbf{k}i} = v_{\mathbf{k}i} + \sum_j a^i_{\mathbf{k}j} \phi_{\mathbf{k}j} \qquad (2.61)$$

where

$$a^i_{\mathbf{k}j} = -(\phi_{\mathbf{k}i}|v_{\mathbf{k}j}) = -\int \phi^*_{\mathbf{k}i}(\mathbf{r}) v_{\mathbf{k}j}(\mathbf{r}) \, d^3r \qquad (2.62)$$

The orthogonality of $\psi_{\mathbf{k}i}$ to the core states is insured by (2.62). We may expect $v_{\mathbf{k}i}$ to be a "smooth" function even near a nucleus since the rapid variation of $\psi_{\mathbf{k}i}$ in that region really is due to the requirements of orthogonality. We are going to determine the equation satisfied by $v_{\mathbf{k}i}$.

The core function $\phi_{\mathbf{k}j}$ in (2.61) must be an eigenfunction of the crystal Hamiltonian: $H\phi_{\mathbf{k}j} = E_{\mathbf{k}j} \phi_{\mathbf{k}j}$. Then if $H\psi_{\mathbf{k}i} = E_{\mathbf{k}i} \psi_{\mathbf{k}i}$ we get:

$$Hv_{\mathbf{k}i} + \sum_j a^i_{\mathbf{k}j} (E_{\mathbf{k}j} - E_{\mathbf{k}i}) \phi_{\mathbf{k}j} = E_{\mathbf{k}i} v_{\mathbf{k}i} \qquad (2.63)$$

We now define a "pseudopotential" V_p by

$$V_p = \sum_j a^i_{\mathbf{k}j} (E_{\mathbf{k}j} - E_{\mathbf{k}i}) \frac{\phi_{\mathbf{k}j}}{v_{\mathbf{k}i}} \qquad (2.64)$$

Then (2.63) takes a familiar form:[4]

$$(H + V_p) v_{\mathbf{k}i} = E_{\mathbf{k}i} v_{\mathbf{k}i} \qquad (2.65)$$

The repulsive pseudopotential so distorts the actual crystal potential contained in H that the wave function of lowest energy for the modified

---

[4] The repulsive potential may also be represented in terms of an integral operator. For an arbitrary function $g(r)$, we have

$$V_p g(\mathbf{r}) = \int V_p(\mathbf{r}, \mathbf{r}') g(\mathbf{r}') \, d^3r'$$

where

$$V_p(\mathbf{r}, \mathbf{r}') = \sum_j (E - E_{\mathbf{k}j}) \phi^*_{\mathbf{k}j}(\mathbf{r}') \phi_{\mathbf{k}j}(\mathbf{r})$$

problem (2.65) is the smooth function $v_{\mathbf{k}i}$. A qualitative idea of the form of the pseudopotential can be obtained if we make the following crude assumptions. Suppose only one core function $\phi_{\mathbf{k}j}$ contributes with any importance to (2.64). Let this have the simple form, in any cell

$$\phi_{\mathbf{k}j} = B_j r^n e^{-sr} \tag{2.66a}$$

In this region, it will be a reasonable approximation to put

$$v_{\mathbf{k}i} = C_i r^p e^{-tr} \tag{2.66b}$$

Then

$$V_p = A r^{n-p} e^{-(s-t)r} \tag{2.67}$$

where

$$A = (E_{\mathbf{k}i} - E_{\mathbf{k}j}) (v_{\mathbf{k}i}|\phi_{\mathbf{k}j}) B_j / C_i \tag{2.68}$$

Further, one will often have, approximately, $p - n = 1$. Then the pseudopotential will be a repulsive Yukawa potential, which is a form that has been used in studies of the alkali metals (Callaway, 1958b). Phillips and Kleinman have employed a form of the OPW method with the orthogonalization terms replaced by a repulsive potential for extensive calculations of energy bands in semiconductors. It must be noted that the pseudopotential depends in detail upon the wave function and energy of the state being investigated. Hence, if a pseudopotential is not regarded as merely an empirical representation of some complex physics, it has to be determined in a self-consistent way for each state.

Bassani and Celli (1961) have observed that the total potential which is the sum of the ordinary potential and the repulsive pseudopotential may be sufficiently weak so that it may be treated successfully by second order perturbation theory. A perturbation calculation of energy levels similar to that of Section 2.3 is then possible without the previously troublesome restriction that only states orthogonal by reasons of symmetry to core states could be considered. For further discussion of the pseudo-potential, see Cohen and Heine[5] (1961), and Brown (1962).

---

[5] These authors observe that the repulsive potential defined by (2.64) is not unique. Any combination of the $\phi_{\mathbf{k}j}$ may be added to $v_{\mathbf{k}i}$ in (2.61) without altering $\psi_{\mathbf{k}i}$. This indeterminacy may be utilized to make $v_{\mathbf{k}i}$ as smooth as possible.

2.6 The Cellular Method

An alternative approach to the problem of calculation of electronic energy levels in solids is the cellular method. Historically, it was the first procedure to be developed for this purpose (Wigner and Seitz, 1933, 1934; Slater, 1934; Shockley, 1937). In outline the method is extremely simple: the Schrödinger equation is solved in one atomic cell subject to the boundary conditions on the wave function and its derivatives implied by Bloch's theorem. For some states in simple crystals the boundary conditions are sufficiently simple so that a solution can be obtained with very little labor. (This is particularly true of states at the bottom of the lowest valence electron band in cubic metals.) For many states of interest, however, the complexity of the boundary condition causes serious mathematical difficulties.

We discuss the cellular method here in accord with the work of Von der Lage and Bethe (1947). It is first assumed that the crystal potential is spherically symmetric within any given polyhedral cell. This assumption is made so that it will be possible to separate variables in the Schrödinger equation (such an assumption, which cannot be strictly correct, is not required in procedures based on plane wave expansions). The one-electron wave function for a state belonging to the ith irreducible representation of wave vector $\mathbf{k}$ can be expressed as a sum of products of radial wave functions $R_l(E, r)$ and spherical harmonics $Y_{lm}(\theta, \phi)$

$$\psi_{\mathbf{k}i} = \sum_l {}' A^i_{\mathbf{k},l} \left[\sum_m {}' C^{lm}_{\mathbf{k}i} Y_{lm}(\theta, \phi) \right] R_l(E, r) \tag{2.69}$$

The radial functions satisfy the radial wave equation

$$\frac{1}{r^2} \frac{d}{dr}\left(\frac{r^2 dR_l}{dr}\right) + \left(E - V - \frac{l(l+1)}{r^2}\right) R_l = 0 \tag{2.70}$$

in which V is the (spherically symmetric) potential. The combination of spherical harmonics of different m values for a given l is of course chosen to produce a function transforming according to a row of the particular irreducible representation (i) considered. These functions are conveniently called Kubic harmonics; some of them have already been

78 CHAPTER 2. METHODS OF CALCULATION

given in Chapter I as basis functions for the irreducible representations there discussed. Extensive tables of Kubic harmonics have been given by Bell (1953) and by Altmann (1956).

The boundary conditions on the wave function are obtained as follows: Let **A** be a point on some face of the atomic polyhedron, and let **B** be a

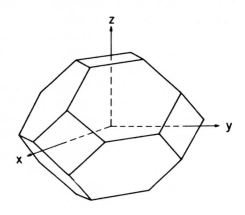

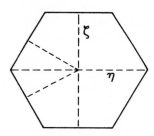

FIG. 7. Wigner-Seitz cell for the body-centered cubic lattice. The lower drawing shows the lines of symmetry in a hexagonal face used in applying the boundary conditions in the cellular method.

point perpendicularly opposite on a parallel face. These points are separated by a lattice translation vector **T**. Since Bloch's theorem asserts that

$$\psi_{\mathbf{k}}(\mathbf{r} + \mathbf{T}) = e^{i\mathbf{k} \cdot \mathbf{T}} \psi_{\mathbf{k}}(\mathbf{r})$$

we have that
$$\psi_{\mathbf{k}}(\mathbf{A}) = e^{i\mathbf{k}\cdot\mathbf{T}}\,\psi_{\mathbf{k}}(\mathbf{B}) \tag{2.71}$$
It is usually possible to restate the boundary conditions to refer to values of the function at a single point.

For example, let us consider states in the body-centered cubic lattice. The atomic cell for this lattice is shown in Fig. 7. The parallel square faces are located on planes x, y, or $z = \pm a/2$ (a is the lattice parameter) while the hexagonal faces have normal vectors $(a/4)$ $(\pm 1, \pm 1, \pm 1)$. Hence, the translation $\mathbf{T}$ is of the type $(a, 0, 0)$ for points on the square faces and $(a/2)(1, 1, 1)$ for the hexagonal faces. First let us consider states pertaining to the center of the Brillouin zone: $\mathbf{k} = \mathbf{0}$. From (2.71) $\psi_0(\mathbf{A}) = \psi_0(\mathbf{B})$: such functions are said to be periodic with the periods of all pairs of parallel face of the cell. We will discuss in detail only the case in which ψ belongs to the Γ_1 representation and is unchanged by any operation of the full cubic group. Then consider the function at a point close to a square face of the cell, say $(\tfrac{1}{2}a + \varepsilon, y, z)$. If this corresponds to $\mathbf{A}$, then $\mathbf{B} = (-\tfrac{1}{2}a + \varepsilon, y, z)$. Hence

$$\psi(\tfrac{1}{2}a + \varepsilon, y, z) = \psi(-\tfrac{1}{2}a + \varepsilon, y, z)$$

However, the operation $(-x, y, z)$ leaves ψ unaltered. Hence

$$\psi(\tfrac{1}{2}a + \varepsilon, y, z) = \psi(\tfrac{1}{2}a - \varepsilon, y, z)$$

We conclude that the normal derivative of this function vanishes on the square faces. A similar argument shows that the normal derivative also vanishes at the center of a hexagonal face, and along certain lines in this face (but not everywhere): Let (x, y, z) be the coordinates of a point on a hexagonal face. Then, on applying the lattice translation $(-\tfrac{1}{2}a, -\tfrac{1}{2}a, -\tfrac{1}{2}a)$ followed by the operation $(-y, -x, -z)$ we find

$$\psi(x, y, z) = \psi(x - \tfrac{1}{2}a, y - \tfrac{1}{2}a, z - \tfrac{1}{2}a) = \psi(\tfrac{1}{2}a - y, \tfrac{1}{2}a - x, \tfrac{1}{2}a - z)$$

But $(\tfrac{1}{2}a - y, \tfrac{1}{2}a - x, \tfrac{1}{2}a - z)$ are the coordinates of a point in the same hexagonal face which is symmetrically situated across the line $\zeta = 0$ illustrated in Fig. 7. Again, considering a point close to (x, y, z) displaced along the normal to the face by an infinitesimal amount, we have

$$\psi(x + \varepsilon, y + \varepsilon, z + \varepsilon) = \psi(\tfrac{1}{2}a - y - \varepsilon, \tfrac{1}{2}a - x - \varepsilon, \tfrac{1}{2}a - z - \varepsilon)$$

We conclude that the function ψ is symmetric about the line $\zeta = 0$ and the normal derivative is antisymmetric about this line.

Boundary conditions may be deduced in a similar manner for functions belonging to other representations. At the point $H = (2\pi/a)\,(1, 0, 0)$ the wave function is periodic with the periods of pairs of square faces, but is antiperiodic with periods of pairs of hexagonal faces. For the representation H_1, this implies that, while the normal derivative still vanishes on the square faces, the wave function itself must vanish at the center of a hexagonal face and on the lines $\zeta = 0$. The normal derivative on the hexagonal face is symmetric about the line $\zeta = 0$.

The complexity of the boundary conditions should be evident from this discussion. A standard procedure has been to satisfy the boundary conditions at a small number of selected points — one additional point for each radial function in the expansion (2.69). This procedure does not, however, converge readily (Howarth and Jones 1952). Further, for a general potential, it is not clear whether the expansion (2.69) of the wave functions converges in the region outside of the largest sphere which can be inscribed in the atomic cell, since the potential will not be spherically symmetric in this region (Ham, 1954, 1960).

A simple approximation which may be used for a few interesting states is the following: For the purpose of satisfying the boundary conditions, the polyhedral cell is replaced by the sphere of equal volume (spherical approximation). Under these circumstances only a single term is required in (2.69) for states at $\mathbf{k} = \mathbf{0}$. Functions of even l are then periodic in the crystal and even under inversion; hence these must have vanishing radial derivative on the sphere. Functions of odd l must vanish on the sphere. These simplified boundary conditions are reasonably accurate only for states near the bottom of a simple s band. In this approximation the actual crystal structure is irrelevant; only the volume of the cell is important.

2.7 The Effective Mass

An interesting and important application of the cellular method is involved in the determination of effective masses and wave functions for states near the bottom of the lowest valence electron band in a cubic structure (Bardeen, 1938a). The procedure used here is that of Silverman (1952), and differs from that of Section 1.7 in that, here, the differential equation of first order perturbation theory is solved, subject to the

2.7 THE EFFECTIVE MASS

boundary conditions imposed by the spherical approximation to the cellular method. A variational principle for the determination of the components of the effective mass tensor at any point in the Brillouin zone has been given by Cohen and Ham (1960).

The wave function of a state near the bottom of the band (which is assumed to be at $\mathbf{k} = 0$) is written in the Bloch form. All indices other than the wave vector are suppressed, and spin-orbit coupling is neglected.

$$\psi_{\mathbf{k}} = e^{i\mathbf{k}\cdot\mathbf{r}} u_{\mathbf{k}}$$

It is found by substitution in the Schrödinger equation that the function $u_{\mathbf{k}}$ satisfies the differential equation (atomic units)

$$(-\nabla^2 + V)u_{\mathbf{k}} - 2i\mathbf{k}\cdot\nabla u_{\mathbf{k}} = (E(\mathbf{k}) - \mathbf{k}^2)u_{\mathbf{k}} \tag{2.72}$$

We express $u_{\mathbf{k}}$ as a power series in $\mathbf{k}$; the coefficients are functions of position which have to be determined:

$$u_{\mathbf{k}}(\mathbf{r}) = \sum_{n=0}^{\infty} u_n(\mathbf{r}) k^n \tag{2.73a}$$

Since $u_{\mathbf{k}}$ is periodic in the atomic cell, the boundary conditions for it are the same as for a wave function of $\mathbf{k} = 0$.

The energy is also expanded in powers of k:

$$E(\mathbf{k}) = \sum_{n=0}^{\infty} E_{2n} k^{2n} \tag{2.73b}$$

Odd powers of k are absent in (2.73b) because $E(\mathbf{k})$ has inversion symmetry; on account of the spherical boundary conditions terms possessing cubic symmetry, rather than spherical symmetry [such as a possible $E_4^{(2)}(k_x^2 k_y^2 + k_y^2 k_z^2 + k_x^2 k_z^2 - \frac{1}{5}k^4)$], will not appear. When Eqs. (2.73a) and (2.73b) are substituted into (2.72), and coefficients of like powers of k are equated to zero, a set of equations is obtained for the functions u_n. They are:

$$(-\nabla^2 + V - E_0)u_0 = 0 \tag{2.74a}$$

$$(-\nabla^2 + V - E_0)u_1 = 2i\hat{\mathbf{k}}\cdot\nabla u_0 \tag{2.74b}$$

$$(-\nabla^2 + V - E_0)u_2 = 2i\hat{\mathbf{k}} \cdot \nabla u_1 + (E_2 - 1)u_0 \tag{2.74c}$$

$$(-\nabla^2 + V - E_0)u_n = 2i\hat{\mathbf{k}} \cdot \nabla u_{n-1} + (E_2 - 1)u_{n-2} + E_4 u_{n-4} + \cdots \tag{2.74d}$$

In Eqs. (2.74), $\hat{\mathbf{k}}$ is a unit vector in the direction of $\mathbf{k}$. It is assumed that the solution of (2.74a) is known (subject to the boundary condition $(du_0/dr)_{r=r_s} = 0$, where r_s is the radius of the sphere of equal volume).

It is possible to determine solutions of the remaining equations by inspection. The particular integral of (2.74b) can be seen to be $-i\hat{\mathbf{k}} \cdot \mathbf{r}u_0$. Evidently u_1 is an odd function; and thus must satisfy the boundary condition $u_1(r_s) = 0$. This may be accomplished by adding a multiple of a $p(l=1)$ solution, whose radial part is denoted by f_p, of the homogeneous equation [obtained by dropping the right side of (2.74b)]. Then

$$u_1 = i\hat{\mathbf{k}} \cdot \mathbf{r}[(f_p/r) - u_0] \tag{2.75}$$

Similarly, it may be verified that a particular solution of (2.74c) is

$$-i\hat{\mathbf{k}} \cdot \mathbf{r}u_1 + \tfrac{1}{2}(\hat{\mathbf{k}} \cdot \mathbf{r})^2 u_0 + E_2 g$$

where g satisfies

$$(-\nabla^2 + V - E_0)g = u_0 \tag{2.76a}$$

A function g which satisfies (2.76a) may be formally determined by differentiating the equation $(-\nabla^2 + V - E)u_0(E, r) = 0$ with respect to E and setting $E = E_0$:

$$(-\nabla^2 + V - E_0)\frac{\partial u_0}{\partial E} = u_0 \tag{2.76b}$$

The function u_2 is even under inversion. It must, consequently, satisfy the condition $(\partial u_2/\partial r)_{r_s} = 0$. In order to satisfy this condition it is first necessary to separate it into s and d functions. This may be done with the use of the identity $\cos^2\theta = \tfrac{2}{3}P_2 + \tfrac{1}{3}$, where P_2 is the second Legendre polynomial. Then we may write

$$u_2 = (\tfrac{2}{3}rf_p - \tfrac{1}{3}r^2 u_0 + c_d f_d)P_2 + \left(\tfrac{1}{3}rf_p - \tfrac{1}{6}r^2 u_0 + E_2\frac{\partial u_0}{\partial E} + c_s u_0\right) \tag{2.77}$$

The function f_d is the radial part of the d solution of the homogeneous equation. If we differentiate the multiplier of P_2 with respect to r, the boundary condition determines c_d

$$c_d = \tfrac{2}{3} r_s \left(\frac{df_p}{dr}\right)_{r_s} \left(\frac{df_d}{dr}\right)_{r_s}^{-1} \tag{2.78}$$

However, the boundary condition on the s part of u_2 cannot be satisfied by a choice of c_s, since $(\partial u_0/\partial r)_{r_s}$ is already zero. Also $\partial u_0/\partial E$ is uniquely determined by the condition that $u_0(E, r)$ be normalized for all energies: If

$$\int_\Omega u_0^2(E, r) \, d^3r = 1$$

then

$$\int_\Omega u_0 \frac{\partial u_0}{\partial E} d^3r = 0 \tag{2.79}$$

Hence we may set $c_s = 0$, since its only role is to "renormalize" u_0. Thus we see that the boundary condition on u_2 in fact determines E_2: we have

$$E_2 = -r_s/3 \left(\frac{df_p}{dr}\right)_{r_s} \left(\frac{\partial^2 u_0}{\partial r \, \partial E}\right)_{r_s, E_0}^{-1} \tag{2.80}$$

The quantity $(\partial^2 u_0/\partial r \, \partial E)_{r_s, E_0}$ can be determined from (2–76b) with the aid of Green's theorem. We multiply both sides of (2–76b) by u_0, integrate over the cell volume, and apply Green's theorem:

$$\int_\Omega u_0 [-\nabla^2 + V - E_0] \frac{\partial u_0}{\partial E} d^3r = \int_\Omega \frac{\partial u_0}{\partial E} [-\nabla^2 + V - E_0] u_0 \, d^3r + \tag{2.81}$$

$$\int_s \left[\left(\frac{\partial u_0}{\partial r}\right)_{r_s} \left(\frac{\partial u_0}{\partial E}\right)_{r_s} - u_0 \left(\frac{\partial^2 u_0}{\partial r \, \partial E}\right)_{r_s, E_0}\right] ds = \int u_0^2 \, d^3r = 1.$$

In (2.81), ds is the element of area, and the second integral on the right side is taken over the surface of the sphere of equal volume. It is convenient to define the function u_0 and its derivatives so that the integral over all solid angles is unity (this may be done by including a normalized spherical harmonic $Y_0^0 = (4\pi)^{-1/2}$ in the definition). With the use of (2.74a) and the boundary conditions, we obtain

$$\left(\frac{\partial^2 u_0}{\partial r \, \partial E}\right)_{r_s, E_0} = -\frac{1}{r_s^2 \, u_0(r_s)} \tag{2.82}$$

Then (2.80) may be re-expressed as

$$E_2 = \frac{r_s^3 \, u_0^2(r_s)}{3} \frac{r_s}{f_p(r_s)} \left(\frac{df_p}{dr}\right)_{r_s} \tag{2.83}$$

(Here we have used $r_s \, u_0(r_s) = f_p(r_s)$.) If we introduce the functions $R_0 = r u_0$, $P = r f_p$, which are the ones usually calculated, we have

$$E_2 = \frac{r_s \, R_0^2}{3} \left[\frac{r}{P}\frac{dP}{dr} - 1\right]_{r_s} \tag{2.84}$$

The expression for c_d may also be converted into a more convenient form

$$c_d = -\frac{2 E_2}{r_s \, R_0(r_s)} \left(\frac{df_d}{dr}\right)_{r_s}^{-1} \tag{2.85}$$

Silverman has continued this procedure through the calculation of E_4. His result is:

$$E_4 = \frac{2}{5} r_s^2 \, E_2 - \frac{4}{15} \frac{E_2^2}{\gamma} r_s^2 \left(\frac{r}{f_d}\frac{df_d}{dr}\right)^{-1} + \tag{2.86}$$

$$\frac{\gamma E_2}{u_0(r_s)} \left[\frac{E_2}{\gamma}\left(\frac{\partial u_0}{\partial E}\right)_{r_s, E_0} - r_s \, u_0(r_s) \frac{\int_0^{r_s} P^2(E_0, r)\, dr}{P^2(E_0, r_s)}\right]$$

where

$$\gamma = \frac{1}{3} r_s^3 \, u_0^2(r_s) = -\frac{r_s \, u_0(r_s)}{3}\left(\frac{\partial^2 u_0}{\partial r \, \partial E}\right)_{r_s, E_0}^{-1} \tag{2.87}$$

and

$$\frac{\int_0^{r_s} P^2(E_0, r)\, dr}{P^2(E_0, r_s)} = -\frac{d}{dE}\left[\frac{1}{P}\frac{dP}{dr}\right]_{E_0, r_s} \tag{2.88}$$

2.7 THE EFFECTIVE MASS

An important property of these formulas is that, in spite of first appearances, they are independent of the normalization of the wave function, and depend in fact only on logarithmic derivatives of the functions on the surface of the atomic sphere. In particular, using the relation $f_p(r_s) = r_s u_0(r_s)$ Eq. (2.80) can be put into a form explicitly independent of the normalization:

$$E_2 = -\frac{1}{3}\frac{r_s^2}{f_p(r_s)}\left(\frac{df_p}{dr}\right)_{r_s} u_0(r_s) \left(\frac{\partial^2 u_0}{\partial r\, \partial E}\right)_{r_s, E_0}^{-1} \qquad (2.89\text{a})$$

It is convenient to introduce the notation

$$\phi_l = \frac{r_s}{f_l(r_s)}\left(\frac{df_l}{dr}\right)_{r_s}, \qquad \phi_l' = -\left(\frac{\partial \phi_l}{\partial E}\right)_{E_0}, \qquad \phi_l'' = \left(\frac{\partial^2 \phi_l}{\partial E^2}\right)_{E_0} \qquad (2.90)$$

Then (2.89a) becomes

$$E_2 = \frac{r_s^2}{3}\frac{\phi_1}{\phi_0'} \qquad (2.89\text{b})$$

and Brooks has shown (Brooks, 1958) that E_4 can be expressed in the normalization independent form:

$$E_4 = \frac{2r_s^4}{15}\frac{\phi_1}{\phi_0'}\left[1 - \frac{2}{3}\frac{\phi_1}{\phi_2} - \frac{5}{6}\frac{\phi_1}{\phi_0'} + \frac{5}{12}\frac{\phi_0'' \phi_1}{(\phi_0')^2}\right] \qquad (2.91)$$

Finally, it is interesting to note that this procedure can be carried through for the Dirac equation which describes the (relativistic) motion of one electron in the periodic potential, V (Callaway et al., 1957). Relativistic effects can be important in solid state physics because the valence electron penetrates the core into a region where the electrostatic potential is very strong. The effective mass in this case is determined from an expression similar to (2.84) which contains in place of the single function P the large component radial wave functions for the p states of $j = \frac{3}{2}(P_{3/2})$ and $j = \frac{1}{2}(P_{1/2})$. The function R_0 goes over to the similar large component of the $j = \frac{1}{2}$, s state function $(R_{1/2})$

$$E_2 = \frac{r_s R_{1/2}^2}{3}\left[\frac{2}{3}\frac{r_s}{P_{3/2}}\frac{dP_{3/2}}{dr} + \frac{r_s}{3P_{1/2}}\frac{dP_{1/2}}{dr} - 1\right]_{r_s} \qquad (2.92)$$

2.8 Variational Methods

There have been a number of attempts to avoid the difficulties involved with point matching of boundary conditions in the cellular method (Kohn, 1952; Howarth and Jones, 1952; Ham, 1954; Altmann, 1958a, b). One of the most interesting of these approaches is the variational method of Kohn and Rostoker (see Kohn and Rostoker, 1954; Ham and Segall, 1961; see also Korringa, 1947). We will examine this procedure in some detail:

Consider the Schrödinger equation in the form

$$(H - E)\psi = 0, \quad \text{with} \quad H = -\nabla^2 + V.$$

The customary variational principle for this problem is

$$\delta I = 0 \quad \text{where} \quad I = \int \psi^*(H - E)\psi \, d^3r = 0 \qquad (2.93)$$

The integral in (2.93) is to be evaluated in an atomic cell. Suppose we make an arbitrary variation of ψ by replacing ψ in (2.93) by ψ', where $\psi' = \psi + \delta\psi$ and ψ satisfies both the Schrödinger equation and the boundary condition. We find with the use of Green's theorem.

$$\delta I = \int (\delta\psi \, \nabla\psi^* - \psi^* \nabla \, \delta\psi) \cdot d\mathbf{s} \qquad (2.94)$$

The integral here goes over the surface of the atomic cell. The quantity, δI, will vanish in general only if $\delta\psi$ obeys the correct boundary conditions prescribed by Bloch's theorem (for then the contributions from conjugate points on the cell boundary will cancel[6]). It is possible to find a different functional which does vanish for arbitrary variations of ψ (Kohn, 1952).

The variational principle used by Kohn and Rostoker is based on an integral equation which is equivalent to the Schrödinger equation: To determine the energy of states at a particular point $\mathbf{k}$ inside the Brillouin zone, we find the Green's function which satisfies the equation

$$(\nabla^2 + E)G_\mathbf{k}(\mathbf{r} - \mathbf{r}') = \delta(\mathbf{r} - \mathbf{r}') \qquad (2.95)$$

---

[6] Conjugate points are pairs of points separated by a lattice translation vector.

2.8 VARIATIONAL METHODS

subject to the boundary conditions required by Bloch's theorem

$$G_\mathbf{k}(\mathbf{r} + \mathbf{R}_n) = e^{i\mathbf{k}\cdot\mathbf{R}_n} G_\mathbf{k}(\mathbf{r}) \qquad (2.96)$$

(As usual, $\mathbf{R}_n$ is any lattice translation vector.)

The standard technique for calculating Green's functions involves expansion in eigenfunctions of the homogeneous equation which satisfy the appropriate boundary conditions. In general, if the eigenfunctions of V^2 are distinguished by an index j (eigenvalues are E_j), we have (Goertzel and Tralli, 1960)

$$G(\mathbf{r}, \mathbf{r}') = -\sum_j \psi_j^*(\mathbf{r}') \frac{1}{E_j - E} \psi_j(\mathbf{r}) \qquad (2.97)$$

In the present case,

$$\psi_j(\mathbf{r}) = \frac{1}{\Omega_0^{1/2}} e^{i(\mathbf{k} + \mathbf{K}_n)\cdot\mathbf{r}} \qquad (2.98)$$

where $\mathbf{K}_n$ is any translation vector in the reciprocal lattice, and Ω_0 is the volume of the atomic cell. Then

$$G_\mathbf{k}(\mathbf{r} - \mathbf{r}') = -\frac{1}{\Omega_0} \sum_{\mathbf{K}_n} \frac{e^{i(\mathbf{k} + \mathbf{K}_n)\cdot(\mathbf{r} - \mathbf{r}')}}{(\mathbf{k} + \mathbf{K}_n)^2 - E} \qquad (2.99)$$

If we operate on $G(\mathbf{r}, \mathbf{r}')$ with $(V^2 + E)$, it will be verified that (2.95) is satisfied since, according to the closure property of the eigenfunctions,

$$\sum_j \psi_j^*(\mathbf{r}')\psi_j(\mathbf{r}) = \sum_{\mathbf{K}_n} \frac{e^{i(\mathbf{k} + \mathbf{K}_n)\cdot(\mathbf{r} - \mathbf{r}')}}{\Omega_0} = \delta(\mathbf{r} - \mathbf{r}')$$

In addition $G(\mathbf{r} - \mathbf{r}')$ has the properties

$$G(\mathbf{r}' - \mathbf{r}) = G^*(\mathbf{r} - \mathbf{r}') \qquad (2.100a)$$

and

$$G(\mathbf{r} + \mathbf{R}_N - \mathbf{r}') = e^{i\mathbf{k}\cdot\mathbf{R}_N} G(\mathbf{r} - \mathbf{r}') \qquad (2.100b)$$

The Schrödinger equation may be put in the form

$$(V^2 + E)\psi_\mathbf{k}(\mathbf{r}) = V\psi_\mathbf{k}(\mathbf{r}) \qquad (2.101)$$

A formal solution is

$$\psi_{\mathbf{k}}(\mathbf{r}) = \int_\Omega G_{\mathbf{k}}(\mathbf{r} - \mathbf{r}') V(\mathbf{r}') \psi_{\mathbf{k}}(\mathbf{r}')\, d^3r' \qquad (2.102)$$

This equation may be verified by direct substitution into (2.101). The energy value E is involved in the integral equation through its appearance in the expression for the Green's function. That $\psi_{\mathbf{k}}$ satisfies the proper boundary conditions is guaranteed by Eq. (2.100b).

The integral equation (2.102) may be derived from the variational principle

$$\delta \Lambda = 0 \qquad (2.103)$$

where

$$\Lambda = \int_\Omega \psi^*(\mathbf{r}) V(\mathbf{r}) \psi(\mathbf{r})\, d^3r - \iint_{\Omega\Omega'} \psi^*(\mathbf{r}) V(\mathbf{r}) G_{\mathbf{k}}(\mathbf{r} - \mathbf{r}') V(\mathbf{r}') \psi(\mathbf{r}')\, d^3r'\, d^3r \qquad (2.104)$$

This variational principle has the important property that $\delta\Lambda$ vanishes in first order for all variations from the solution of (2.102) regardless of whether or not the variations satisfy the boundary conditions. Further, for the exact solution of the problem

$$\Lambda(\psi, \mathbf{k}, E) = 0 \qquad (2.105)$$

If an approximate Λ is computed from some trial function ψ_t, and Eq. (2.105) is solved for the energy, the error in the energy is of second order compared to that of the trial function.

A convenient method of employing the variational principle is to choose a trial function which is a linear combination of a finite number of basis functions with undetermined coefficients

$$\psi_{\mathbf{k}}(\mathbf{r}) = \sum_n^N C_{n\mathbf{k}}\, \phi_{n\mathbf{k}} \qquad (2.106)$$

Let us define a set of quantities Λ_{nl}

$$\Lambda_{nl} = \int \phi_{n\mathbf{k}}^*(\mathbf{r}) V(\mathbf{r}) \phi_{l\mathbf{k}}(\mathbf{r}) - \iint \phi_{n\mathbf{k}}^*(\mathbf{r}) V(\mathbf{r}) G_{\mathbf{k}}(\mathbf{r} - \mathbf{r}') V(\mathbf{r}') \phi_{l\mathbf{k}}(\mathbf{r}')\, d^3r\, d^3r' \qquad (2.107)$$

2.8 VARIATIONAL METHODS

An approximate Λ is

$$\Lambda = \sum_{n,l}^{N} C_{n\mathbf{k}}^* \Lambda_{nl} C_{l\mathbf{k}} \tag{2.108}$$

The stationary condition on Λ requires that the partial derivatives of Λ with respect to the coefficients C vanish, or that, for each n

$$\sum_{l}^{N} \Lambda_{nl} C_{l\mathbf{k}} = 0 \tag{2.109}$$

A set of N linear homogeneous equations is obtained. The condition for a nontrivial solution is that

$$\det \Lambda_{nl} = 0 \tag{2.110}$$

This equation may be used to determine the energy for a state of given $\mathbf{k}$ (or the wave vector $\mathbf{k}$ for which the wave function will have a prescribed energy). If the functions $\phi_{n\mathbf{k}}$ belong to a complete set, the energy obtained will approach the correct energy as the number of functions increase.

In the simple case in which the $\phi_{n\mathbf{k}}$ are plane waves

$$\phi_{n\mathbf{k}} = \Omega_0^{-1/2} e^{i(\mathbf{k}+\mathbf{K}_n) \cdot \mathbf{r}}$$

the Λ_{nl} have the form:

$$\Lambda_{nl} = V(\mathbf{K}_l - \mathbf{K}_n) + \sum_{\mathbf{K}_j}' \frac{V(\mathbf{K}_j - \mathbf{K}_n) V(\mathbf{K}_l - \mathbf{K}_j)}{(\mathbf{k} + \mathbf{K}_j)^2 - E} \tag{2.111}$$

One might wish to use (2.110) with (2.111) in place of the plane wave method previously discussed, but in general the sum in the second term of (2.111) would be quite difficult to evaluate.

The possibilities of practical application of the Kohn-Rostoker method are limited by the difficulty in evaluating the second term in Eq. (2.107). It is possible to bring this principle to the aid of the cellular method, however, if one may assume that the potential in an atomic cell is constant outside of the inscribed sphere (radius r_i). By proper choice of the zero of energy, the potential may be made to vanish outside this sphere

$$V(r) = 0 \quad \text{for} \quad r \geqslant r_i \qquad (2.112)$$

Physical potentials will not have this property; it is hoped that not too much violence is done to whatever real problem is being studied by this approximation.

It is then convenient to take a trial function of a form similar to (2.69):

$$\psi_{\mathbf{k},t} = \sum_{l,m}' C_{\mathbf{k},lm} Y_{lm}(\theta, \phi) R_l(E, r) \qquad (2.113)$$

in which the radial function R_l satisfied the radial equation (2.70). Then $\psi_{\mathbf{k},t}$ satisfies the Schrödinger equation (energy E) for $r < r_i$

$$[\nabla^2 + E]\psi_{\mathbf{k},t} = V\psi_{\mathbf{k},t} \quad \text{for} \quad r < r_i \qquad (2.114)$$

However, the coefficients $C_{\mathbf{k},lm}$ have not yet been determined so that the boundary conditions may be approximately satisfied.

In accord with (2.110), it is now necessary to determine the "matrix elements" of Λ, which are to be constructed using functions $R_l(r)Y_{lm}(\theta, \phi)$ as a basis. This derivation is quite lengthy, and will not be presented in complete detail here. For a more complete treatment, the interested reader should consult the paper of Kohn and Rostoker.

The restriction to potentials which vanish outside of an inscribed sphere makes possible the conversion of the expression (2.104) for Λ to a form which does not depend explicitly on the potential but involves surface integrals over the inscribed sphere. Some care has to be employed in the procedure because the Green's function is singular when its argument is zero. It is desirable to consider a sphere whose radius is slightly smaller than that of the inscribed sphere; $r_i - \varepsilon$. Later, we shall let $\varepsilon \to 0$. First we observe that, on account of (2.114)

$$\int_{\Omega'} G(\mathbf{r} - \mathbf{r}')V(\mathbf{r}')\psi(\mathbf{r}')\,d^3r' = \int G(\mathbf{r} - \mathbf{r}')(\nabla'^2 + E)\psi(\mathbf{r}')\,d^3r'$$

The operator ∇' differentiates functions with respect to coordinates $\mathbf{r}'$. Green's theorem enables us to write

2.8 VARIATIONAL METHODS

$$\int G(\mathbf{r}-\mathbf{r}')(\nabla'^2+E)\psi(\mathbf{r}')\,d^3r' = \int (\nabla'^2+E)G(\mathbf{r}-\mathbf{r}')\psi(\mathbf{r}')\,d^3r' \qquad (2.115)$$

$$+ \int [G(\mathbf{r}-\mathbf{r}')\nabla'\psi(\mathbf{r}') - \psi(\mathbf{r}')\nabla' G(\mathbf{r}-\mathbf{r}')] \cdot d\mathbf{s}'$$

$$= \psi(\mathbf{r}) + \int\limits_{r_i - \varepsilon} \left[G(\mathbf{r}-\mathbf{r}')\frac{\partial \psi(\mathbf{r}')}{\partial r'} - \psi(\mathbf{r}')\frac{\partial G(\mathbf{r}-\mathbf{r}')}{\partial r'} \right] ds'$$

Thus

$$\psi(\mathbf{r}) - \int G(\mathbf{r}-\mathbf{r}')V(\mathbf{r}')\psi(\mathbf{r}')\,d^3r' \qquad (2.116)$$

$$= - \int\limits_{r_i - \varepsilon} \left[G(\mathbf{r}-\mathbf{r}')\frac{\partial \psi(\mathbf{r}')}{\partial r'} - \psi(\mathbf{r}')\frac{\partial G(\mathbf{r}-\mathbf{r}')}{\partial r'} \right] ds'$$

The volume integrals in this equation include the interior of the sphere of radius $r_i - \varepsilon$; the surface integrals pertain to the surface of this sphere. If we let $\varepsilon \to 0$, we find from (2.116) that for potentials which vanish beyond the inscribed sphere the basic integral equation (2.102) reduces to

$$\int\limits_{r_i} \left[G(\mathbf{r}-\mathbf{r}')\frac{\partial \psi(\mathbf{r}')}{\partial r'} - \psi(\mathbf{r}')\frac{\partial G(\mathbf{r}-\mathbf{r}')}{\partial r'} \right] ds' = 0 \qquad (2.117)$$

We return to the determination of Λ from Eq. (2.104). Since there are two integrations, a second sphere of slightly smaller radius, which is taken to be $r_i - 2\varepsilon$, is also required. Then, by an argument similar to that leading to (2.116), we have for $r' < r$:

$$\int \psi^*(\mathbf{r})V(\mathbf{r})G(\mathbf{r}-\mathbf{r}')\,d^3r = \int\limits_u (\nabla^2 + E)\psi^*\, G(\mathbf{r}-\mathbf{r}')\,d^3r \qquad (2.118)$$

$$= \int\limits_{r_i - 2\varepsilon} \left[\frac{\partial \psi^*}{\partial r} G(\mathbf{r}-\mathbf{r}') - \psi^* \frac{\partial G(\mathbf{r}-\mathbf{r}')}{\partial r} \right] ds$$

The expressions (2.116) and (2.118) are substituted into (2.104)

$$\Lambda = \int_\Omega \psi^*(\mathbf{r}) V(\mathbf{r}) \left[\psi(\mathbf{r}) - \int G(\mathbf{r} - \mathbf{r}') V(\mathbf{r}') \psi(\mathbf{r}') \, d^3 r' \right] d^3 r \qquad (2.119)$$

$$= -\int \psi^*(\mathbf{r}) V(\mathbf{r}) \left[\int\limits_{r_i - \varepsilon} \left\{ G(\mathbf{r} - \mathbf{r}') \frac{\partial \psi(\mathbf{r}')}{\partial r'} - \psi(\mathbf{r}') \frac{\partial G(\mathbf{r} - \mathbf{r}')}{\partial r'} \right\} dS' \right] d^3 r$$

$$= \int\limits_{r_i - 2\varepsilon} dS \int\limits_{r_i - \varepsilon} dS' \left[\frac{\partial \psi^*(\mathbf{r})}{\partial r} - \psi^* \frac{\partial}{\partial r} \right] \left[\psi(\mathbf{r}') \frac{\partial G}{\partial \mathbf{r}'} (\mathbf{r} - \mathbf{r}') - G(\mathbf{r} - \mathbf{r}') \frac{\partial \psi(\mathbf{r}')}{\partial r'} \right]$$

In order to compute the "matrix elements" of Λ, we replace ψ^* in (2.119) by $R_{l'} Y^*_{l'm'}(\theta, \phi)$ and ψ by $R_l Y_{lm}(\theta, \phi)$. It is then necessary to express the Green's function (2.99) in terms of spherical waves. This may be done with the use of the expansion for plane waves

$$e^{i\mathbf{K} \cdot \mathbf{r}} = 4\pi \sum_{l,m} i^l j_l(Kr) Y_{lm}(\theta, \phi) Y_{lm}^*(\theta_\mathbf{K}, \phi_\mathbf{K}) \qquad (2.120)$$

In (2.120) θ, ϕ and $\theta_\mathbf{K}, \phi_\mathbf{K}$ are the polar angles of $\mathbf{r}$ and $\mathbf{K}$, respectively, with respect to a fixed set of axes. Then we find

$$G(\mathbf{r} - \mathbf{r}') = -\frac{(4\pi)^2}{\Omega_0} \sum_{\substack{l,m \\ l',m'}} \sum_{K_n} i^{l-l'} \frac{j_l(|\mathbf{K}_n + \mathbf{k}|r) j_{l'}(|\mathbf{K}_n + \mathbf{k}|r')}{(\mathbf{K}_n + \mathbf{k})^2 - E} \qquad (2.121)$$

$$\times Y_{lm}(\theta, \phi) Y^*_{l'm'}(\theta', \phi') Y^*_{lm}(\theta_\mathbf{K}, \phi_\mathbf{K}) Y_{l'm'}(\theta_\mathbf{K}, \phi_\mathbf{K})$$

The angles θ', ϕ' and $\theta_\mathbf{K}, \phi_\mathbf{K}$ now refer to the vectors $\mathbf{r}'$ and $(\mathbf{K}_n + \mathbf{k})$, respectively.

Equation (2.121) is not, however, a convenient form of the Green's function. Within one atomic cell, it must be possible to express the Green's function as the sum of two parts: one which is a particular integral of (2.95) and hence singular for $r = r'$; the other part a solution of the homogeneous equation, regular for $r = r'$.

It can easily be verified that a singular solution of (2.95), valid within the inscribed sphere, is given by

$$G_0(\mathbf{r} - \mathbf{r}') = -\frac{1}{4\pi} \frac{\cos(\kappa |\mathbf{r} - \mathbf{r}'|)}{|\mathbf{r} - \mathbf{r}'|} \qquad (2.122)$$

2.8 VARIATIONAL METHODS

with $\kappa^2 = E$. The expansion of this function in spherical waves is,

$$G_0(\mathbf{r} - \mathbf{r}') = \begin{cases} \kappa \sum_{l,m} j_l(\kappa r) n_l(\kappa r') Y_{lm}(\theta, \phi) Y_{lm}^*(\theta', \phi') & \text{for} \quad r < r' \\ \kappa \sum_{l,m} j_l(\kappa r') n_l(\kappa r) Y_{lm}^*(\theta, \phi) Y_{lm}(\theta', \phi') & \text{for} \quad r > r' \end{cases}$$

(2.123)

In this equation n_l is the spherical Bessel function of the second kind, singular for zero values of its argument[7]

$$n_l(x) = (\pi/2x)^{1/2} J_{-l-1/2}(x)$$

The regular part of $G(r - r')$ must have the form:

$$\sum_{lm} \sum_{l'm'} A_{lm,l'm'} j_l(\kappa r) j_{l'}(\kappa r') Y_{lm}(\theta, \phi) Y_{l'm'}^*(\theta', \phi') \qquad (2.124)$$

in which the $A_{lm,l'm'}$ are constants to be determined so that the boundary conditions are satisfied. (This function is seen to be regular; it is easily shown to be a solution of the homogeneous equation.) Hence we can write for $r > r' < r_i$:

$$G(\mathbf{r} - \mathbf{r}') = \sum_{l,m} \sum_{l'm'} [A_{lm,l'm'} j_l(\kappa r) j_l(\kappa r') + \qquad (2.125)$$

$$\kappa \, \delta_{ll'} \, \delta_{mm'} \, j_l(\kappa r) n_l(\kappa r')] Y_{lm}(\theta, \phi) Y_{l'm'}^*(\theta', \phi')$$

The coefficients $A_{lm,l'm'}$ may be deduced by comparison with (2.121) and (2.125)

---

[7] The expansion (2.123) for G_0 may be deduced from the more familiar expansion

$$-\frac{1}{4\pi} \frac{e^{i\kappa|\mathbf{r} - \mathbf{r}'|}}{|\mathbf{r} - \mathbf{r}'|} = \kappa \sum_{l,m} j_l(\kappa r)[n_l(\kappa r') - ij_l(\kappa r')] Y_{lm}(\theta, \phi) Y_{lm}^*(\theta', \phi') \quad \text{for} \quad r < r'$$

by identifying real and imaginary parts. A similar expression with r and r' interchanged is valid for $r > r'$.

$$A_{lm,l'm'} = -\frac{(4\pi)^2}{\Omega_0} i^{(l-l')} [j_l(\kappa r) j_l(\kappa r')]^{-1} \quad (2.126)$$

$$\times \sum_n \frac{j_l(|\mathbf{K}_n + \mathbf{k}|r) j_{l'}(|\mathbf{K}_n + \mathbf{k}|r') Y_{lm}^*(\theta_\mathbf{K}, \phi_\mathbf{K}) Y_{l'm'}(\theta_\mathbf{K}, \phi_\mathbf{K})}{(\mathbf{K}_n + \mathbf{k})^2 - E}$$

$$- \frac{\kappa \delta_{ll'} \delta_{mm'} n_l(\kappa r')}{j_l(\kappa r)} \quad (r < r' < r_i)$$

The right side of (2.126) must be independent of the values of r and r'. Kohn has given, fortunately, expressions for the $A_{lm,l'm'}$ which are simpler to evaluate. These will not be discussed here.

We must now return to Eq. (2.119) to evaluate the quantities $\Lambda_{lm,l'm'}$, which are "matrix elements of Λ" computed with basis functions $R_l Y_{lm}(\theta, \phi)$. Equation (2.125) is substituted and the integrals over the spheres are performed using the orthonormality of the spherical harmonics. The result is, in the limit $\varepsilon \to 0$:

$$\Lambda_{lm,l'm'} = R_l R_{l'} [L_l j_l - j_l'] [(A_{lm,l'm'} j_{l'}' + \kappa \delta_{ll'} n_{l'}') \quad (2.127)$$

$$- (A_{lm,l'm'} j_{l'} + \kappa \delta_{ll'} \delta_{mm'} n_{l'}) L_{l'}]$$

$$= -R_l R_{l'} [L_l j_l - j_l'][L_{l'} j_{l'} - j_{l'}'] \left[A_{lm,l'm'} + \kappa \delta_{ll'} \delta_{mm'} \frac{n_l L_l - n_l'}{j_l L_l - j_l'} \right]$$

In (2.127),

$$L_l = \frac{1}{R_l} \frac{dR_l}{dr}\bigg|_{r=r_i}, \quad j_l' = \frac{dj_l(Kr)}{dr}\bigg|_{r=r_i}, \quad \text{etc.} \quad (2.128)$$

and all functions are evaluated at r_i.

The determinantal equation (2.110) may now be constructed, and simplified by division of the common factors $-R_l [L_l j_l - j_l']$ from each row and $R_{l'} [L_{l'} j_{l'} - j_{l'}'']$ from each column. The resulting equation,

$$\det \left| A_{lm,l'm'} + \kappa \delta_{ll'} \delta_{mm'} \frac{n_l' - n_l L_l}{j_l' - j_l L_l} \right| = 0 \quad (2.129)$$

gives the required connection between E and k.

The utility of this procedure for energy level calculations is seen to depend on the following two principal factors:

1. The availability of the "structure constants" $A_{lm,l'm'}$ as functions of energy and **k** for the different lattices. Some computations of these quantities have been made by Ham and Segall (1961).

2. The adequacy of the approximation in which the potential in an atomic cell is constant outside the inscribed sphere.

It is, of course, necessary to determine the logarithmic derivative of the radial functions as functions of energy on the inscribed sphere. This is done with reasonable ease by numerical integration of the radial equation. However, these quantities may also be determined, at least for the alkali metals, by extrapolation from spectroscopic data pertaining to the free atom. This procedure will be discussed in Section 2.14.

The Kohn-Rostoker method has been applied extensively to energy level calculations in the alkali metals by Ham (1960). Some of the results of these calculations will be discussed in Section 3.3.

2.9 The Augmented Plane Wave Method

The fundamental reason for the difficulty of energy band calculations is that the only functions which satisfy the boundary conditions imposed by Bloch's theorem in a simple manner are plane waves, but plane wave expansions do not converge readily in the interior of an atomic cell. To get around this difficulty Slater proposed (1937) to expand the wave function in a set of functions composed of plane waves in the outer regions of the atomic cell, and a sum of spherical waves in the interior.[8] Let us consider such a function

$$\Phi_{\mathbf{k}} = a_0\, \varepsilon(r - r_i)\, e^{i\mathbf{k}\cdot\mathbf{r}} + \sum_{l,m} a_{lm}\, \varepsilon(r_i - r)\, Y_{l,m}(\theta, \phi)\, R_l(E, r) \qquad (2.130)$$

The function ε is a unit step function

$$\varepsilon(x) = 1 \quad \text{for} \quad x \geqslant 0 \qquad (2.131)$$
$$\varepsilon(x) = 0 \quad \text{for} \quad x < 0$$

---

[8] The method has been developed further by several authors: see Slater (1953a), Saffern and Slater (1953), Howarth (1955), Leigh (1956), Schlosser (1960).

96   CHAPTER 2. METHODS OF CALCULATION

This plane wave is joined to the spherical waves on a sphere whose radius is r_i (usually, this sphere is the inscribed sphere). It is convenient to choose the coefficients a_{lm} so that the function $\Phi_\mathbf{k}$ is continuous across the sphere. To do this, we expand the plane wave in spherical harmonics according to (2.120). Then we find that

$$a_{lm} = 4\pi i^l\, a_0\, Y_{lm}^*(\theta_\mathbf{k},\, \phi_\mathbf{k}) \frac{j_l(k, r_i)}{R_l(E, r_i)} \qquad (2.132)$$

Although $\phi_\mathbf{k}$ is continuous across the sphere, its normal derivative is discontinuous. A plane wave in general cannot be joined smoothly onto spherical waves in the interior of some region: there must be scattered waves as well.

The function $R_l(E, r)$ is a solution of the radial equation (2.70) for some energy. There are two ways in which this energy E may be chosen. In the first paper by Slater (1937) on this topic, the energy E is left undetermined at this stage; it is later set equal to the energy resulting as the solution of the problem of diagonalizing the matrix of the Hamiltonian on the basis of augmented plane waves. In the subsequent work of Saffern and Slater (1953) whose point of view will be adopted here, the energy E is set equal to the expectation value of the energy of the single augmented plane wave (2.130). In both cases, the potential is required to vanish beyond the inscribed sphere. This restriction is not, however, essential to the APW method.

Some care is required in the computation of the average energy on account of the discontinuity in the derivative of the function on the inscribed sphere. Let $u(\mathbf{r})$ be a continuous function which has a discontinuous gradient on a sphere ($|\mathbf{r}| = r_i$), and let u_i and u_o represent the function inside and outside the surface of discontinuity, respectively. We can write

$$\nabla u = \nabla u_i + (\nabla u_o - \nabla u_i)\varepsilon(r - r_i)$$

The Laplacian of u has a delta function singularity for $r = r_i$, whose strength is given by the discontinuity in the normal derivative. The integral of the Laplacian through an arbitrarily small volume containing the surface of discontinuity is: ($d\mathbf{s}$ lies along the outgoing normal)

$$\int u^* \nabla^2 u\, d^3r = \int u^* [\nabla u_o - \nabla u_i] \cdot d\mathbf{s}$$

2.9 THE AUGMENTED PLANE WAVE METHOD

Hence, the volume integral of the Laplacian through the atomic cell is

$$\int u_i^* \nabla^2 u_i \, d^3r + \int u_o^* \nabla^2 u_o \, d^3r + \int u_o^* \nabla u_o \cdot ds - \int u_i^* \nabla u_i \cdot ds$$

The surface integrals cover the sphere on which the discontinuity occurs.

Since u_0 is a plane wave in this case, the third integral vanishes, and the second integral may be transformed by Green's theorem so that we have

$$\int_\Omega u^* H u \, d^3r = \int u_i^* H u_i \, d^3r + \int u_i^* \nabla u_i \cdot ds + \int_\omega (\nabla u_o)^* \cdot \nabla u_o \, d^3r \quad (2.133)$$

where ω is the volume of the atomic cell outside the inscribed sphere. The radial function R_l is normalized to unity inside the inscribed sphere. Then from (2.133)

$$\int \Phi_k^* (-\nabla^2 + V) \Phi_k \, d^3r = k^2 |a_0|^2 \omega + E \sum_{lm} |a_{lm}|^2 + \text{surface term} \quad (2.134)$$

The normalization condition on Φ_k requires that

$$\int \Phi_k^* \Phi_k \, d^3r = 1 = |a_0|^2 \omega + \sum_{lm} |a_{lm}|^2 \quad (2.135)$$

The surface term has the value [from (2.133)]

$$r_i^2 \sum_{lm} |a_{lm}|^2 \left(\frac{dR_l}{dr}\right)_{r_i} R_l(r_i) \quad (2.136)$$

We substitute a_{lm} from Eq. (2.132) into (2.136). The sum on m may be performed with the use of the addition theorem for spherical harmonics

$$\sum_m Y_{lm}^*(\theta_{k'}, \phi_{k'}) Y_{lm}(\theta_k, \phi_k) = \frac{(2l+1)}{4\pi} P_l(\cos \theta_{kk'}) \quad (2.137)$$

where $\theta_{kk'}$ is the angle between $\mathbf{k}$ and $\mathbf{k}'$. Then (2.136) becomes (since $\mathbf{k} = \mathbf{k}'$ in this case)

$$4\pi r_i^2 |a_0|^2 \sum_l (2l+1) j_l^2(kr_i) \left(\frac{1}{R_l} \frac{dR_l}{dr}\right)_{r_i} \quad (2.138)$$

We identify the integral on the left-hand side of (2.134) with E according to the previous discussion. We then combine (2.134), (2.135), and (2.138) to obtain

$$(E - \mathbf{k}^2)\omega = 4\pi r_i^2 \sum_l (2l + 1) j_l^2(kr_i) \left(\frac{d}{dr} \ln R_l(E, r)\right)_{r_i} \quad (2.139)$$

This equation gives the connection between E and $\mathbf{k}$ for a single augmented plane wave.[9] The energy is actually a multivalued function of $\mathbf{k}^2$: This occurs because the logarithmic derivative of each radial function R varies from $+\infty$ to $-\infty$ (the slope of the logarithmic derivative as a function of energy is always negative, much like a cotangent curve) as the energy increases, being singular when a node of the radial function passes through the inscribed sphere. It may be hoped that the APW of lowest energy will be a good approximation for states of small $\mathbf{k}$, at least in simple metals.

It is now necessary to compute the matrix elements of the Hamiltonian, and of unity between augmented plane waves. The formulas may be simplified considerably through the use of the following result: Consider the integral throughout the inscribed sphere of products of radial functions $R_l(E, r)$, belonging to different energies: We define the quantity $I_l(E_1, E_2)$ through the equation

$$\int_0^{r_i} R_l(E_1, r) R_l(E_2, r) r^2 \, dr = I_l(E_1, E_2) R_l(E_1, r_i) R_l(E_2, r_i) \quad (2.140)$$

The integral may be computed in the following manner: The functions $P_l = rR_l$ satisfy the equations

$$\frac{d^2}{dr} P_l(E_1, r) + \left[E_1 - V - \frac{l(l+1)}{r^2}\right] P_l(E_1, r) = 0 \quad (2.141\text{a})$$

$$\frac{d^2}{dr} P_l(E_2, r) + \left[E_2 - V - \frac{l(l+1)}{r^2}\right] P_l(E_2, r) = 0 \quad (2.141\text{b})$$

---

[9] If the potential is constant but does not vanish beyond the inscribed sphere, we add tn $\mathbf{k}^2$ in (2.139) the quantity $V(0)$, which is the average of the potential in this region.

2.9 THE AUGMENTED PLANE WAVE METHOD

Equation (2.141a) is multiplied by $P_l(E_2, r)$; Eq. (2.141b) is multiplied by $P_l(E_1, r)$, and the equations are subtracted. We have, upon integration

$$I_l(E_1, E_2) = \frac{r_i^2}{E_1 - E_2} \left[\frac{d}{dr} \ln R_l(E_2, r_i) - \frac{d}{dr} \ln R_l(E_1, r) \right]_{r=r_i} \quad (2.142)$$

In the limit $E_2 \to E_1$, we evidently obtain the useful result that:

$$\int_0^{r_i} R_l^2(E, r) r^2 \, dr = - r_i^2 \, R_l^2(E, r_i) \frac{\partial}{\partial E} \left[\frac{d}{dr} \ln R_l(E, r) \right]_{r_i} \quad (2.143)$$

Since the left-hand side of (2.143) is always positive, the statement which was made previously that the slope of the logarithmic derivative as a function of energy is always negative has been proved.

The matrix element of unity between two augmented plane waves (or in other terms, the overlap integral between them) may now be computed. We will designate this quantity by $\langle \mathbf{k}_1 \, E_1 | \mathbf{k}_2 \, E_2 \rangle$. It is only necessary to consider waves such that $\mathbf{k}_1 - \mathbf{k}_2$ is a reciprocal lattice vector in the computation of matrix elements. The integral involves two parts; one coming from the plane wave outside the sphere; the other from the spherical waves in the interior. The former may be written

$$\int_{\text{cell}} - \int_{\text{sphere}} e^{i(\mathbf{k}_1 - \mathbf{k}_2) \cdot \mathbf{r}} = \Omega_0 \, \delta_{\mathbf{k}_1, \mathbf{k}_2} - 4\pi \int_0^{r_i} \frac{\sin |\mathbf{k}_1 - \mathbf{k}_2| r}{|\mathbf{k}_1 - \mathbf{k}_2| r} r^2 \, dr \quad (2.144)$$

$$= \Omega_0 \, \delta_{\mathbf{k}_1, \mathbf{k}_2} - 4\pi r_i^2 \frac{j_1(|\mathbf{k}_1 - \mathbf{k}_2| r_i)}{|\mathbf{k}_1 - \mathbf{k}_2|}$$

$$= \omega \delta_{\mathbf{k}_1, \mathbf{k}_2} - 4\pi r_i^2 \frac{j_1(|\mathbf{k}_1 - \mathbf{k}_2| r_i)}{|\mathbf{k}_1 - \mathbf{k}_2|} (1 - \delta_{\mathbf{k}_1, \mathbf{k}_2})$$

The second contribution is

$$\sum_{l,m}' a_{lm}^*(\mathbf{k}_1, E_1) a_{lm}(\mathbf{k}_2, E_2) I_l(E_1, E_2) R_l(E_1, r_i) R_l(E_2, r_i) \quad (2.145)$$

The sum over m may be performed with the aid of the addition theorem (2.137). When the contributions (2.144) and (2.145) are combined, the result is

$$\langle \mathbf{k}_1 \, E_1 | \mathbf{k}_2 \, E_2 \rangle = \omega \, \delta_{\mathbf{k}_1 \mathbf{k}_2} - 4\pi r_i^2 \frac{j_1(|\mathbf{k}_1 - \mathbf{k}_2| r_i)}{|\mathbf{k}_1 - \mathbf{k}_2|} (1 - \delta_{\mathbf{k}_1, \mathbf{k}_2}) \tag{2.146}$$

$$+ 4\pi \sum_l (2l+1) j_l(k_1 r_i) j_l(k_2 r_i) P_l(\cos \theta_{\mathbf{k}_1, \mathbf{k}_2}) I_l(E_1, E_2)$$

The matrix element of the Hamiltonian between two augmented plane waves is denoted by $\langle \mathbf{k}_1 \, E_1 | H | \mathbf{k}_2 \, E_2 \rangle$. It may be computed according to a procedure similar to that used in the calculation of the average of the energy. From (2.133) we have

$$\langle \mathbf{k}_1 \, E_1 | H | \mathbf{k}_2 \, E_2 \rangle = \int_\Omega \Phi_{\mathbf{k}_1}^* H \Phi_{\mathbf{k}_2} \, d^3 r \tag{2.147}$$

$$= \int u_{1i}^* (-\nabla^2 + V) u_{2i} \, d^3 r + \int u_{1i}^* \nabla u_{2i} \cdot d\mathbf{s} + \int (\nabla u_{1o})^* \cdot \nabla u_{2o} \, d^3 r$$

The first term on the right of (2.147) may be expressed as

$$E_2 \sum_{lm} a_{lm}^*(\mathbf{k}_1, E_1) a_{lm}(\mathbf{k}_2, E_2) I_l(E_1, E_2) R_l(E_1, r_i) R_l(E_2, r_i)$$

The surface term becomes

$$r_i^2 \sum_{l,m} a_{lm}^*(\mathbf{k}_1, E_1) a_{lm}(\mathbf{k}_2, E_2) \left[R_l(E_1, r) \frac{dR_l}{dr}(E_2, r) \right]_{r_i}$$

The plane wave part is just

$$\mathbf{k}_1 \cdot \mathbf{k}_2 \left[\int_{\text{cell}} - \int_{\text{sphere}} \right] e^{i(\mathbf{k}_2 - \mathbf{k}_1) \cdot \mathbf{r}} \, d^3 r$$

When these contributions are combined, the result is

$$\langle \mathbf{k}_1 \, E_1 | H | \mathbf{k}_2 \, E_2 \rangle = \mathbf{k}_1 \cdot \mathbf{k}_2 \left[\omega \, \delta_{\mathbf{k}_1, \mathbf{k}_2} - 4\pi r_i^2 \frac{j_1(|\mathbf{k}_2 - \mathbf{k}_1| r_i)}{|\mathbf{k}_2 - \mathbf{k}_1| r_i} (1 - \delta_{\mathbf{k}_1, \mathbf{k}_2}) \right]$$

$$+ 4\pi r_i^2 \sum_l (2l+1) j_l(k_1 r_i) j_l(k_2 r_i) P_l(\cos \theta_{\mathbf{k}_1 \mathbf{k}_2}) \tag{2.148}$$

$$\times \frac{1}{E_1 - E_2} \left[E_1 \frac{d}{dr} \ln R_l(E_2, r) - E_2 \frac{d}{dr} \ln R_l(E_1, r) \right]_{r = r_i}$$

It should be noted that the energy and overlap matrix elements both vanish in the case in which $k_1 = k_2$ but $E_1 \neq E_2$, provided that (2.139) is satisfied

$$\langle k_1 E_1 | k_1 E_2 \rangle = 0 \quad (E_1 \neq E_2)$$
$$\langle k_1 E_1 | H | k_1 E_2 \rangle = 0 \quad (E_1 \neq E_2) \qquad (2.149)$$

This shows that the augmented plane waves used to represent electrons in the bands of interest for solid state problems are orthogonal to those of core states. Consequently an expansion in augmented plane waves will converge to the required state in the chosen band.

In the application of the augmented plane wave method, it is desirable, as in the case of orthogonalized plane wave expansions, to form symmetrized linear combinations of augmented plane waves whose wave vectors are connected by operations of the group of the wave vector for the particular point in the zone under investigation. The coefficients in the symmetrized combination of augmented plane waves must be the same as those in the combinations previously discussed in Section 2.2. The spherical harmonics participating in the spherical wave portion must also combine to form the appropriate Kubic harmonics, since only these functions can satisfy the correct boundary conditions on the inscribed sphere.

The augmented plane wave method is evidently a most difficult one, and with one exception (Chodorow, 1939) calculations based on this method have utilized modern electronic computing equipment. The method has been applied to copper by Howarth (1955), to iron by Wood (1960, 1962), and to lithium and sodium by Schlosser (1960). The convergence of the method is asserted to be quite rapid. The APW method is similar to the Kohn-Rostoker approach in that it requires both knowledge of the logarithmic derivative of the wave function on the inscribed sphere as a function of energy and a truncation of the potential at the sphere. (The latter requirement, while present in Slater's formulation of the APW method, is not essential.) Further, since only a finite and reasonably small number of spherical waves can, in practice, be included in an APW, the wave function itself actually is discontinuous on the inscribed sphere. Leigh (1956) and Schlosser (1960) have given variational formulations of the APW method which take account of this feature.

2.10 The Tight Binding Approximation

The methods of calculating energy levels which have been discussed previously have a reasonable claim to exactness in that the expansions employed will, in principle, converge to the state of interest as more functions are added. At the same time the calculations are generally quite difficult to perform except at points of high symmetry in the Brillouin zone. The tight binding approximation, in its usual formulation, does not lead to an exact solution of the one-electron Schrödinger equation since wave functions belonging to bound atomic states do not form a complete set of functions. It does provide a reasonably simple technique of approximating the energy of states at general points of the zone.

We saw in Section 2.4, in the discussion of the wave function of core states in connection with OPW method, that a Bloch function $\phi_{\mathbf{k}}$, based on the free-atom wave function $u(\mathbf{r})$

$$\phi_{\mathbf{k}} = \frac{1}{\sqrt{N}} \sum_{\nu} e^{i\mathbf{k}\cdot\mathbf{R}_{\nu}} u(\mathbf{r} - \mathbf{R}_{\nu})$$

has the periodicity required for a state of wave vector $\mathbf{k}$. In the limit of very large atomic separation, these functions would be exact one-electron wave functions. One may suppose that in the case of finite but large atomic separations these functions still are good approximations.

In the case of a nondegenerate atomic level (s-state), the problem is to find the average energy of the function $\phi_{\mathbf{k}}$ above. The Hamiltonian contains a sum of potentials centered on each atom

$$H = H_0(\mathbf{r}) + \sum_{\nu \neq 0} V(\mathbf{r} - \mathbf{R}_{\nu}), \tag{2.150}$$

in which $H_0(\mathbf{r}) = -\nabla^2 + V(\mathbf{r})$.

We suppose that u satisfies

$$H_0 u(\mathbf{r}) = E_0 u(\mathbf{r})$$

The functions $u(\mathbf{r} - \mathbf{R}_{\nu})$ on a given site are normalized. Then

2.10 THE TIGHT BINDING APPROXIMATION

$$\int \phi_\mathbf{k}^* H \phi_\mathbf{k} \, d^3r = N^{-1} \sum_{\nu,\nu''} e^{-i\mathbf{k}\cdot(\mathbf{R}_\nu - \mathbf{R}_{\nu''})} \int u^*(\mathbf{r} - \mathbf{R}_\nu)$$

$$\times \left[H_0 + \sum\nolimits' V(\mathbf{r} - \mathbf{R}_{\nu'}) \right] u(\mathbf{r} - \mathbf{R}_{\nu''}) \, d^3r$$

$$= N^{-1} \left[E_0 \sum_{\nu,\nu''} e^{-i\mathbf{k}\cdot(\mathbf{R}_\nu - \mathbf{R}_{\nu''})} \int u^*(\mathbf{r} - \mathbf{R}_\nu) u(\mathbf{r} - \mathbf{R}_{\nu''}) \, d^3r \right.$$

$$\left. + \sum_{\substack{\nu,\nu',\nu'' \\ \nu' \neq \nu''}} e^{-i\mathbf{k}\cdot(\mathbf{R}_\nu - \mathbf{R}_{\nu''})} \int u^*(\mathbf{r} - \mathbf{R}_\nu) V(\mathbf{r} - \mathbf{R}_{\nu'}) u(\mathbf{r} - \mathbf{R}_{\nu''}) \, d^3r \right]$$

This expression can be simplified. Since the summations run over the entire crystal, one index (say ν'') can be set equal to zero and the $(1/N)$ multiplier deleted. Then

$$\int \phi_\mathbf{k} H \phi_\mathbf{k} \, d^3r = E_0 \left[1 + \sum_{\nu \neq 0} e^{i\mathbf{k}\cdot\mathbf{R}_\nu} \int u^*(\mathbf{r} - \mathbf{R}_\nu) u(\mathbf{r}) \, d^3r \right] \quad (2.151)$$

$$+ \sum_{\substack{\nu,\nu' \\ \nu' \neq 0}} e^{-i\mathbf{k}\cdot\mathbf{R}_\nu} \int u^*(\mathbf{r} - \mathbf{R}_\nu) V(\mathbf{r} - \mathbf{R}_{\nu'}) u(\mathbf{r}) \, d^3r$$

The functions $\phi_\mathbf{k}$ are not normalized. In order to compute the average energy, $\langle E_\mathbf{k} \rangle$, of $\phi_\mathbf{k}$, it is necessary to divide Eq. (2.151) by the integral

$$\int \phi_\mathbf{k}^* \phi_\mathbf{k} \, d^3r = (N^{-1}) \sum_{\nu,\nu'} e^{-i\mathbf{k}\cdot(\mathbf{R}_\nu - \mathbf{R}_{\nu'})} \int u^*(\mathbf{r} - \mathbf{R}_\nu) u(\mathbf{r} - \mathbf{R}_{\nu'}) \, d^3r \quad (2.152)$$

$$= 1 + \sum_{\nu \neq 0} e^{-i\mathbf{k}\cdot\mathbf{R}_\nu} \int u^*(\mathbf{r} - \mathbf{R}_\nu) u(\mathbf{r}) \, d^3r$$

Hence

$$E(\mathbf{k}) = \frac{\int \phi_\mathbf{k}^* H \phi_\mathbf{k} \, d^3r}{\int \phi_\mathbf{k}^* \phi_\mathbf{k} \, d^3r} = E_0 + \frac{\sum\limits_{\substack{\nu,\nu' \\ \nu' \neq 0}} e^{-i\mathbf{k}\cdot\mathbf{R}_\nu} \int u^*(\mathbf{r} - \mathbf{R}_\nu) V(\mathbf{r} - \mathbf{R}_{\nu'}) u(\mathbf{r}) \, d^3r}{1 + \sum\limits_{\nu \neq 0} e^{-i\mathbf{k}\cdot\mathbf{R}_\nu} \int u^*(\mathbf{r} - \mathbf{R}_\nu) u(\mathbf{r}) \, d^3r}$$

$$(2.153)$$

In general, Eq. (2.153) is quite difficult to evaluate on account of the presence of three center integrals in the numerator. A common, but not always valid, approximation is to neglect these altogether (two center approximation): We then have:

$$E(\mathbf{k}) = E_0 + \qquad (2.154)$$

$$\frac{\sum_{\substack{\nu \\ \nu \neq 0}} [\int |u(\mathbf{r})|^2 V(\mathbf{r} - \mathbf{R}_\nu) d^3r + e^{-i\mathbf{k}\cdot\mathbf{R}_\nu} \int u^*(\mathbf{r} - \mathbf{R}_\nu) V(\mathbf{r} - \mathbf{R}_\nu) u(\mathbf{r}) d^3r]}{1 + \sum_{\substack{\nu \neq 0}} e^{-i\mathbf{k}\cdot\mathbf{R}_\nu} \int u^*(\mathbf{r} - \mathbf{R}_\nu) u(\mathbf{r}) d^3r}$$

For an S state wave function, the integrals in (2.154) cannot depend on the direction of $\mathbf{R}_\nu$ (provided that the potential on any site is spherically symmetric). It is then convenient to define quantities S, J, K:

the "overlap integral"
$$S(|\mathbf{R}_\nu|) = \int u^*(\mathbf{r} - \mathbf{R}_\nu) u(\mathbf{r}) d^3r \qquad (2.155a)$$

the "interaction integral"
$$J(|\mathbf{R}_\nu|) = \int u^*(\mathbf{r} - \mathbf{R}_\nu) V(\mathbf{r} - \mathbf{R}_\nu) u(\mathbf{r}) d^3r \qquad (2.155b)$$

$$= \int u^*(\mathbf{r}) V(\mathbf{r}) u(\mathbf{r} - \mathbf{R}_\nu) d^3r$$

the "crystal field integral"
$$K = \sum_{\substack{\nu \\ \nu \neq 0}} \int |u(\mathbf{r})|^2 V(\mathbf{r} - \mathbf{R}_\nu) d^3r \qquad (2.155c)$$

Then

$$E(\mathbf{k}) = E_0 + \frac{K + \sum_\nu J(|\mathbf{R}_\nu|) e^{-i\mathbf{k}\cdot\mathbf{R}_\nu}}{1 + \sum_\nu e^{-i\mathbf{k}\cdot\mathbf{R}_\nu} S(|\mathbf{R}_\nu|)} \qquad (2.156)$$

Let us consider a hypothetical simple cubic crystal of lattice parameter a and include all terms through third neighbors. We find that

$$\sum_{\substack{100 \\ \text{positions}}} e^{-i\mathbf{k}\cdot\mathbf{R}_\nu} = 2[\cos k_x a + \cos k_y a + \cos k_z a]$$

$$\sum_{\substack{110 \\ \text{positions}}} e^{-i\mathbf{k}\cdot\mathbf{R}_\nu} = 4[\cos k_x a \cos k_y a + \cos k_x a \cos k_z a + \cos k_y a \cos k_z a]$$

$$\sum_{\substack{111 \\ \text{positions}}} e^{-i\mathbf{k}\cdot\mathbf{R}_\nu} = 8\cos k_x a \cos k_y a \cos k_z a$$

Equation (2.156) may now be written explicitly as:

$$E = E_0 + [K + 2J(a)(\cos k_x a + \cos k_y a + \cos k_z a) +$$

$$4J(\sqrt{2}a)(\cos k_x a \cos k_y a + \cos k_y a \cos k_z a + \cos k_x a \cos k_z a) +$$

$$8J(\sqrt{3}a)(\cos k_x a \cos k_y a \cos k_z a)][1 + 2S(a)(\cos k_x a + \cos k_y a +$$

$$\cos k_z a) + 4S(\sqrt{2}a)(\cos k_x a \cos k_y a + \cos k_y a \cos k_z a + \cos k_x a \cos k_z a) +$$

$$8S(\sqrt{3}a)(\cos k_x a \cos k_y a \cos k_z a)]^{-1} \qquad (2.157)$$

Equation (2.157), which gives the energy as a function of wave vector for a single cubic lattice, may be adapted to other cubic structures (e.g., body-centered or face-centered cubic) simply by adjusting the lattice parameter and reordering the terms. For example, for the S band in a body-centered cubic crystal, in which the nearest neighbors are in the $a(\frac{1}{2}, \frac{1}{2}, \frac{1}{2})$ positions, the second nearest in the $a(100)$ position, etc., we find for the numerator in (2.157)

$$E_0 + K + 8J(\sqrt{3}\,a/2)(\cos k_x a/2 \cos k_y a/2 \cos k_z a/2) +$$

$$2J(a)(\cos k_x a + \cos k_y a + \cos k_z a) + 4J(\sqrt{2}\,a)(\cos k_x a \cos k_y a +$$

$$\cos k_y a \cos k_z a + \cos k_x a \cos k_z a) \qquad (2.158)$$

A similar expression is valid for the denominator.

If the overlap integrals in the denominator of (2.156) are neglected, this equation has the form of a Fourier expansion of the energy. Such an expansion must exist, but the coefficients need not agree with those determined from calculated interaction integrals $J(|\mathbf{R}_\nu|)$.

If the band being considered is based on free atom states with nonzero angular momentum, the $(2l + 1)$ degeneracy of the atomic states will be removed in the solid. The crystal Hamiltonian, which is given by (2.150) has matrix elements which connect functions on different atoms which pertain to states of different z components of angular momentum. Since this perturbation is associated with a vanishing energy denominator, it must be included at the beginning, whereas the mixing of states of different l values is usually regarded as small, in the tight binding approximation. Evidently it is necessary to employ first order degenerate perturbation theory. When the lack of orthogonality of the atomic wave functions on different atoms is considered, it is necessary to solve a determinantal equation of the form:

$$\det |H_{mm'}(\mathbf{k}) - E I_{mm'}(\mathbf{k})| = 0 \qquad (2.159)$$

The quantities $H_{mm'}$ and $I_{mm'}$ are the matrix elements of the Hamiltonian and of unity on the basis of Bloch functions formed from orthonormal atomic functions whose z components of angular momentum are m and m' respectively. These matrix elements may be determined, formally, in terms of overlap, interaction, and crystal field integrals as in the previous discussion:

$$H_{mm'}(\mathbf{k}) = (N^{-1}) \sum_{\nu, \nu''} e^{-i\mathbf{k}\cdot(\mathbf{R}_\nu - \mathbf{R}_{\nu''})} \int u_m^*(\mathbf{r} - \mathbf{R}_\nu) \times \qquad (2.160)$$

$$\left[H_0 + \sum_{\nu'} V(\mathbf{r} - \mathbf{R}_{\nu'}) \right] u_{m'}(\mathbf{r} - \mathbf{R}_{\nu''}) \, d^3r$$

$$= E_0 \left[\delta_{mm'} + \sum_{\nu \neq 0} e^{-i\mathbf{k}\cdot\mathbf{R}_\nu} \int u_m^*(\mathbf{r} - \mathbf{R}_\nu) u_{m'}(\mathbf{r}) \, d^3r \right] +$$

$$\sum_{\substack{\nu, \nu' \\ \nu' \neq 0}} e^{-i\mathbf{k}\cdot\mathbf{R}_\nu} \int u_m^*(\mathbf{r} - \mathbf{R}_\nu) V(\mathbf{r} - \mathbf{R}_{\nu'}) u_{m'}(\mathbf{r}) \, d^3r$$

$$I_{mm'}(\mathbf{k}) = \delta_{mm'} + \sum_{\nu \neq 0} e^{-i\mathbf{k}\cdot\mathbf{R}_\nu} \int u_m^*(\mathbf{r} - \mathbf{R}_\nu) u_{m'}(\mathbf{r}) \, d^3r \qquad (2.161)$$

2.10 THE TIGHT BINDING APPROXIMATION

We can generalize the notation of Eq. (2.155) in an obvious manner. Let us also include three center integrals by defining

$$J_{mm'}(\mathbf{R}_\nu, \mathbf{R}_{\nu'}) = \int u_m^*(\mathbf{r} - \mathbf{R}_\nu) V(\mathbf{r} - \mathbf{R}_{\nu'}) u_{m'}(\mathbf{r}) \, d^3r \qquad (2.162)$$

for $\mathbf{R}_\nu \neq \mathbf{R}_{\nu'} \neq 0$. We then have

$$H_{mm'}(\mathbf{k}) = E_0 I_{mm'}(\mathbf{k}) + K_{mm'} + \sum_{\nu=0}' e^{-i\mathbf{k}\cdot\mathbf{R}_\nu} \left[J_{mm'}(\mathbf{R}_\nu) + \sum_{\substack{\nu' \\ \nu' \neq \nu}} J_{mm'}(\mathbf{R}_\nu, \mathbf{R}_{\nu'}) \right]$$

(2.163a)

$$I_{mm'} = \delta_{mm'} + \sum_{\nu \neq 0}' e^{-i\mathbf{k}\cdot\mathbf{R}_\nu} S_{mm'}(\mathbf{R}_\nu) \qquad (2.163b)$$

These matrix elements are obviously quite complicated. General formulas are not available. In many applications, three center and overlap integrals have been neglected. Slater and Koster (1954) have tabulated the interaction integrals in the two center approximation for s, p, and d functions appropriate for cubic lattices. This tabulation has been extended to hexagonal structures by Miasek (1957).

It is useful to consider the orders of magnitudes of the particular types of integrals appearing in the tight binding approximation in order to be able to assess the validity of the usual approximations in which three center and overlap integrals are neglected. A systematic study of this sort has been made by Wohlfarth (1953), who has considered a linear chain of hydrogen atoms: his results will be quoted below. We consider hydrogenic wave functions for simplicity with an exponential dependence at large r (proportional say to $e^{-\alpha r}$); and let R_1 be the nearest neighbor distance in the lattice. (For simplicity we may consider the point charge lattice discussed in Section 2.3 — charges of atomic number Z.) The two center interaction integrals between nearest neighbors will be proportional to $(Z/R)f(\alpha R_1)e^{-\alpha R_1}$ where $f(\alpha R_1)$ is a polynomial which will depend on the particular functions involved. The overlap integrals will be of similar structure although the factor Z/R_1 will not be present. Except for this factor of Z/R_1, there is no reason to expect the overlap integrals to be much smaller than the interaction integrals. To the extent that the potential in one cell depends only on the charge distribution within that cell, the three center integrals will have to have an exponential dependence at least as strong as $e^{-2\alpha R_1}$. These, then, may be expected to be small in the limit of strong binding which may or may not be attained in particular case.

In the example considered by Wohlfarth, the potential is not that of neutral cells; rather, he has considered a potential which is a sum of unscreened Coulomb potentials. In this case, some three center integrals have the same exponential dependence as the two center ones. Let a be the separation between nuclei (nuclei are located at points na where n is an integer). The nuclei have unit charge (in units of e). Wohlfarth finds an expression for the bandwidth, E_B, correct within the tight binding approximation, through the order of e^{-2a}.

$$E_B = 4.84ae^{-a}(1 - 0.30a^{-1} - 1.96a^{-2}) \qquad (2.164)$$

If all overlap and three center integrals are neglected, the result is

$$E_B = 8ae^{-a}(1 + a^{-1}) \qquad (2.165)$$

If the overlap integrals between nearest neighbors are included, then

$$E_B = 2.67ae^{-a}(1 - 3a^{-1} - 6a^{-2}) \qquad (2.166)$$

This example suggests that the neglect of overlap and three center integrals in the interest of calculational simplicity is not likely to be justified. Slater and Koster (1954) have, however, proposed that the simplest approximation to the tight binding method, in which only nearest neighbor interaction integrals are included, may have some value as an interpolation scheme: The integrals appearing in the $E(\mathbf{k})$ expressions in this approximation are regarded as parameters to be determined by fitting the results of more accurate band structure calculations at certain points of the zone, or from experimental data. This procedure may be of some utility for narrow bands.

2.11 Wannier Functions

The wave functions employed in the tight binding method are linear combinations of atomic orbitals and are not exact solutions of the Schrödinger equation of the one-electron problem. Calculations with such functions are rendered cumbersome by the lack of orthogonality

2.11 WANNIER FUNCTIONS

of the atomic orbitals centered on different atoms. Wannier (1937) introduced a set of functions which have come to bear his name that are generalizations of atomic orbitals designed to avoid these difficulties. Let $\psi_j(\mathbf{k}, \mathbf{r})$ be an (exact) solution of the one-electron Schrödinger equation for a state of wave vector $\mathbf{k}$ and band index j. These functions are orthonormal in the following sense:

$$\int \psi_{j'}^*(\mathbf{k}', \mathbf{r}) \psi_j(\mathbf{k}, \mathbf{r}) \, d^3r = \delta_{jj'} \delta(\mathbf{k} - \mathbf{k}') \tag{2.167}$$

The integration includes the entire (infinite) crystal.[10] The Wannier function which is centered on a site located at $\mathbf{R}_n$ is defined by

$$a_j(\mathbf{r} - \mathbf{R}_n) = \frac{\Omega^{1/2}}{(2\pi)^{3/2}} \int e^{-i\mathbf{k} \cdot \mathbf{R}_n} \psi_j(\mathbf{k}, \mathbf{r}) \, d^3k = \frac{\Omega^{1/2}}{(2\pi)^{3/2}} \int e^{i\mathbf{k} \cdot (\mathbf{r} - \mathbf{R}_n)} u_j(\mathbf{k}, \mathbf{r}) \, d^3k \tag{2.168a}$$

where Ω is the volume of the unit cell, and $u_j(\mathbf{k}, \mathbf{r})$ is the part of the Bloch function which has the periodicity of the potential. The integration is restricted to $\mathbf{k}$ vectors in the Brillouin zone. In some cases it is desirable to treat $\mathbf{k}$ as a discrete vector

$$\mathbf{k} = \frac{l}{\mathscr{L}} \hat{\mathbf{a}}' + \frac{m}{\mathscr{M}} \hat{\mathbf{b}}' + \frac{n}{\mathscr{N}} \hat{\mathbf{c}}'$$

where $\hat{\mathbf{a}}', \hat{\mathbf{b}}'$, and $\hat{\mathbf{c}}'$ are primitive vectors of the reciprocal lattice; $\mathscr{L}, \mathscr{M}$, and $\mathscr{N}$ are integers specifying the number of unit cells in the crystal in the sense of periodic boundary conditions; and l, m, and n are positive or negative integers whose range is so chosen so that $\mathbf{k}$ lies inside the Brillouin zone. Then the Wannier function is defined by

$$a_j(\mathbf{r} - \mathbf{R}_n) = \frac{1}{\sqrt{N}} \sum_k e^{-i\mathbf{k} \cdot \mathbf{R}_n} \psi_j(\mathbf{k}, \mathbf{r}) \tag{2.168b}$$

where $N = \mathscr{N}\mathscr{L}\mathscr{M}$, and the summation includes all wave vectors in the zone. The Wannier function is a localized function. The exact nature

---

[10] More precisely, the integral in (2.167) may differ from a delta function by a phase factor (Blount, 1962a). This is not important for our purposes.

of the localization is somewhat difficult to demonstrate rigorously (Blount, 1962a), but may be seen qualitatively in the following way: We put $\mathbf{r} = \mathbf{R}_i$ into (2.168a), where $\mathbf{R}_i$ is a lattice vector. Then we have

$$a_n(\mathbf{R}_i - \mathbf{R}_n) = \frac{\Omega^{1/2}}{(2\pi)^{3/2}} \int e^{i\mathbf{k}\cdot(\mathbf{R}_i - \mathbf{R}_n)} u_j(\mathbf{k}, 0)\, d^3k.$$

If $|\mathbf{R}_i - \mathbf{R}_n|$ is large, the exponential oscillates rapidly, and the integral over $\mathbf{k}$ will become small. Hence the Wannier function will become small at large distances.

The orthonormality properties of these functions may be demonstrated as follows:

$$\int a_{j'}^*(\mathbf{r} - \mathbf{R}_n) a_j(\mathbf{r} - \mathbf{R}_m)\, d^3r \qquad (2.169)$$

$$= \frac{\Omega}{(2\pi)^3} \int e^{-i(\mathbf{k}\cdot\mathbf{R}_m - \mathbf{k}'\cdot\mathbf{R}_n)} \psi_{j'}^*(\mathbf{k}', \mathbf{r}) \psi_j(\mathbf{k}, \mathbf{r})\, d^3k'\, d^3k\, d^3r$$

$$= \frac{\Omega}{(2\pi)^3} \delta_{jj'} \int e^{-i\mathbf{k}\cdot(\mathbf{R}_m - \mathbf{R}_n)}\, d^3k$$

$$= \delta_{jj'}\, \delta_{mn}$$

The summation relation (A2.10) has been employed in the last step of (2.169). Wannier functions which are centered on different sites, or which belong to different bands, are orthogonal.

The equation which defines the Wannier function may be inverted to determine $\psi_j(\mathbf{k}, \mathbf{r})$. We multiply by $e^{i\mathbf{k}'\cdot\mathbf{R}_n}$ and sum over lattice sites n.

$$\sum_n e^{i\mathbf{k}'\cdot\mathbf{R}_n} a_j(\mathbf{r} - \mathbf{R}_n) = \frac{\Omega^{1/2}}{(2\pi)^{3/2}} \int \sum_n e^{i(\mathbf{k}' - \mathbf{k})\cdot\mathbf{R}_n} \psi_j(\mathbf{k}, \mathbf{r})\, d^3k$$

The sum on the right-hand side is given in (A 2.9b). The result is[11]

$$\psi_j(\mathbf{k}, \mathbf{r}) = \frac{\Omega^{1/2}}{(2\pi)^{3/2}} \sum_n e^{i\mathbf{k}\cdot\mathbf{R}_n} a_j(\mathbf{r} - \mathbf{R}_n) \qquad (2.170)$$

---

[11] If the Wannier function is defined through (2.168b) a factor $N^{-1/2}$ replaces $(2\pi)^{3/2}/\Omega^{1/2}$ in equation (2.170).

2.11 WANNIER FUNCTIONS

In the case of free electrons, for which the wave functions are

$$\psi(\mathbf{k}, \mathbf{r}) = \frac{1}{(2\pi)^{3/2}} e^{i\mathbf{k}\cdot\mathbf{r}}$$

the Wannier function may be calculated explicitly. For simplicity, we consider a simple cubic lattice of lattice parameter a, so that the Brillouin zone is a cube $-\pi/a \leqslant k_x, k_y, k_z \leqslant \pi/a$, and we put $\mathbf{r} - \mathbf{R}_n = \hat{\mathbf{i}} X + \hat{\mathbf{j}} Y + \hat{\mathbf{k}} Z$, where $\hat{\mathbf{i}}, \hat{\mathbf{j}}, \hat{\mathbf{k}}$ are the usual unit vectors along the x, y, and z axes. We obtain:

$$a(\mathbf{r} - \mathbf{R}_n) = \frac{(2\pi)^{3/2}}{\Omega^{1/2}} \frac{\sin \pi X/a \sin \pi Y/a \sin \pi Z/a}{\pi XYZ/a^3} \quad (2.171)$$

This function oscillates and decreases relatively slowly. For wave functions less localized in momentum space, the Wannier function decays more rapidly for large values of its argument. Kohn (1959) and Blount (1962a) have shown that in the case of a periodic potential, the Wannier function falls off exponentially at large distances.

The Wannier functions are not solutions of the Schrödinger equation, but are linear combinations of eigenfunctions with varying energy. It is useful, however, to determine the matrix elements of the Hamiltonian on the basis of Wannier functions.

$$Ha_j(\mathbf{r} - \mathbf{R}_n) = \frac{\Omega^{1/2}}{(2\pi)^{3/2}} \int e^{-i\mathbf{k}\cdot\mathbf{R}_n} H\psi_j(\mathbf{k}, \mathbf{r}) \, d^3k$$

$$= \frac{\Omega^{1/2}}{(2\pi)^{3/2}} \int e^{-i\mathbf{k}\cdot\mathbf{R}_n} E_j(\mathbf{k})\psi_j(\mathbf{k}, \mathbf{r}) \, d^3k$$

$$= \frac{\Omega}{(2\pi)^3} \sum_l \left(\int e^{-i\mathbf{k}\cdot(\mathbf{R}_n - \mathbf{R}_l)} E_j(\mathbf{k}) \, d^3k \right) a_j(\mathbf{r} - \mathbf{R}_l)$$

We define a Fourier coefficient of the energy $\mathscr{E}_j(\mathbf{R}_n)$ by

$$\mathscr{E}_j(\mathbf{R}_n) = \frac{\Omega}{(2\pi)^3} \int e^{i\mathbf{k}\cdot\mathbf{R}_n} E_j(\mathbf{k}) \, d^3k \quad (2.172)$$

(the integration includes a Brillouin zone). The Wannier function is seen to satisfy the equation

$$Ha_j(\mathbf{r} - \mathbf{R}_n) = \sum_l \mathscr{E}_j(\mathbf{R}_l - \mathbf{R}_n)a_j(\mathbf{r} - \mathbf{R}_l) \qquad (2.173)$$

The matrix elements of the Hamiltonian can be obtained from (2.173) with the help of the orthonormality relation for the Wannier functions.

$$H_{j'm,jn} = \int a_{j'}^*(\mathbf{r} - \mathbf{R}_m)Ha_j(\mathbf{r} - \mathbf{R}_n)\,d^3r = \mathscr{E}_j(\mathbf{R}_m - \mathbf{R}_n)\,\delta_{jj'} \qquad (2.174)$$

The significance of the quantities $\mathscr{E}_j(\mathbf{R}_n)$ is made apparent when Eq. (2.172) is inverted by multiplying by $e^{-i\mathbf{k'}\cdot\mathbf{R}_n}$ and summing over n. With the use of (A 9.b), we find

$$E_j(\mathbf{k'}) = \sum_n \mathscr{E}_j(\mathbf{R}_n)\,e^{-i\mathbf{k'}\cdot\mathbf{R}_n} \qquad (2.175)$$

The energy is seen to possess a Fourier expansion in which the wave vectors of the plane waves are direct lattice vectors.

It is somewhat difficult to employ Wannier functions directly in energy band calculations because of the multiplicity of terms on the right-hand side of (2.173). Variational principles for these functions have, however, been derived by Koster (1953) and by Parzen (1953). Wainwright and Parzen (1953) attempted to calculate energy bands in lithium in this way; unfortunately, their work contains a serious error.

2.12 The Hartree-Fock Equations

Our discussion of the methods calculating the energy levels of electrons in crystals has been based on a one-particle Schrödinger equation with a periodic potential $V(\mathbf{r})$. It is now desirable to inquire how this potential is obtained.

One must begin by considering the many-electron system. The electrons interact with the atomic nuclei and with each other. The Hamiltonian (N electrons, atomic units) can be expressed as:

$$H = \sum_{i=1}^{N}\left[-\nabla_i^2 + V(\mathbf{r}_i) + \sum_{j(j>i)}^{N}\frac{2}{r_{ij}}\right] \qquad (2.176)$$

The first term represents the kinetic energy of the electrons; the second, their potential energy in the field of all the nuclei; the third, the electrostatic interaction of the electrons with each other: ($\mathbf{r}_i$ is the position of electron i, and r_{ij} is the distance between electrons i and j). The full Schrödinger equation with this Hamiltonian is not separable on account of the interaction term, so that approximate methods of solution must be employed.

The principal approximation procedure for reducing the complexities of the many-body problem is the Hartree-Fock method. Most energy band calculations involve this approximation, although it is usually not possible to solve the Hartree-Fock equations directly. These equations will be derived in this section, and some of limitations of the procedure will be mentioned.

First, let us consider some general properties of the wave function of a many-electron system. A total spin operator may be defined:

$$\mathbf{S} = \sum_i \mathbf{s}_i \qquad (2.177)$$

The following commutation rules hold for the Hamiltonian of Eq. (2.176)

$$[H, \mathbf{S}^2] = 0 \qquad [H, S_z] = 0 \qquad (2.178)$$
$$[H, \mathbf{s}_i^2] = 0$$

Since the spin operators $\mathbf{S}^2$, S_z, $\mathbf{s}_i^2$ also commute with each other, the stationary states of the system will be eigenfunctions of these operators as well as of the energy.[12] Also, if the potential $V(\mathbf{r})$ is periodic with respect to translations by lattice vectors $\mathbf{R}_n$, then we also have

$$[H, \mathscr{T}(\mathbf{R}_n)] = 0 \qquad (2.179)$$

where $\mathscr{T}(\mathbf{R}_n)$ displaces the coordinate of each electron by $\mathbf{R}_n$. The wave function of the entire system satisfies Bloch's theorem and thus can be characterized by a total wave vector $\mathbf{K}$. A further property of the wave function is antisymmetry: The wave function must change sign when the coordinates and spins of any two electrons are interchanged.

In the Hartree-Fock method the wave function of the system is approximated by a determinant of orthonormal one-electron functions. In general, these functions, which we denote as $\psi_q(\mathbf{r}_i)$, are two-component

---

[12] In the case of a single atom, H commutes with $\mathbf{L}^2$, the total orbital angular momentum, L_z, and the angular momentum for each electron, $\mathbf{l}_i^2$.

spinors (q stands for all the quantum numbers required to specify the one electron functions). The determinantal wave function

$$\Psi(1\ldots N) = \frac{1}{\sqrt{N!}} \begin{vmatrix} \psi_1(\mathbf{r}_1) & \ldots & \psi_N(\mathbf{r}_1) \\ \vdots & & \vdots \\ \vdots & \ldots \psi_q(r_i) \ldots & \vdots \\ \vdots & & \vdots \\ \psi_1(\mathbf{r}_N) & \ldots & \psi_N(\mathbf{r}_N) \end{vmatrix} \qquad (2.180)$$

satisfies the antisymmetry requirement, since the interchange of two particles corresponds to the interchange of two rows of the determinant. Interchange of a pair of rows changes the algebraic sign of a determinant. The determinantal wave function is not, in general, an eigenfunction of either $\mathbf{S}^2$ or S_z, and thus does not correspond to a state of definite multiplicity. (Two very important exceptions are: (1) the state of complete spin alignment: all spins "up" along a "z" axis, and (2) the singlet state in which the space part of ψ_q is the same for two states with opposite spins.) It is generally necessary to form linear combinations of determinantal wave functions to obtain eigenfunctions of $\mathbf{S}^2$. This process has been discussed in detail by Löwdin (1955).

The best determinantal wave function Ψ which can be constructed is the one which minimizes the expectation value of the energy, subject to the orthonormality of the one-electron functions. The variation principle can be used to determine the equations satisfied by the one electron functions if the average energy is to be minimized. The expectation value of the Hamiltonian (2.176) with the wave function (2.180) can be expressed as:

$$E = \sum_{s=1}^{N} \left[\int \psi_s^*(\mathbf{r}_1)(-V_1^2 + V(\mathbf{r}_1))\psi_s(\mathbf{r}_1)\,d^3r_1 \right. \qquad (2.181)$$

$$+ \frac{1}{2}\sum_{t=1}^{N} \iint |\psi_s(\mathbf{r}_1)|^2 |\psi_t(\mathbf{r}_2)|^2 \frac{2}{r_{12}} d^2r_1\,d^3r_2$$

$$\left. - \frac{1}{2}\sum_{t=1}^{N} \iint \psi_s^*(\mathbf{r}_1)\psi_t^*(\mathbf{r}_2)\frac{2}{r_{12}}\psi_s(\mathbf{r}_2)\psi_t(\mathbf{r}_1)\,d^3r_1\,d^3r_2 \right]$$

2.12 THE HARTREE-FOCK EQUATIONS

The integrations include summation over the spinor indices implicit in the one-electron wave functions. Application of the variational principle leads an equation for the one-electron functions. (Details of the derivation of the Hartree-Fock equations may be found in standard texts, for instance, Seitz, 1940.) It is convenient in the following to exhibit the spinor indices explicitly, which we do by writing $\psi_{s\alpha}(\mathbf{r}_1)$, etc., to designate the αth component of the Pauli spinor $\psi_s(\mathbf{r}_1)$. (α may have the values 1 or 2, and we will adopt the summation convention that a repeated spin index implies a sum over both possible values.) We find:

$$\left[-\nabla_1^2 + V(\mathbf{r}_1) + \sum_t \int \psi_{t\alpha}^*(\mathbf{r}_2) \psi_{t\alpha}(\mathbf{r}_2) \frac{2}{r_{12}} d^3r_2 \right] \psi_{s\beta}(\mathbf{r}_1) \tag{2.182}$$

$$- \sum_t \left[\int \psi_{t\alpha}^*(\mathbf{r}_2) \psi_{s\alpha}(\mathbf{r}_2) \frac{2}{r_{12}} d^3r_2 \right] \psi_{t\beta}(\mathbf{r}_1) = \varepsilon_s \psi_{s\beta}(\mathbf{r}_1)$$

These are the Hartree-Fock equations (Fock, 1930a, b) in the form given by Thompson (1960). For a system of N electrons, one evidently has a set of $2N$ coupled integro-differential equations.

These equations may be rewritten in a form in which the spin indices are suppressed, ψ_t, etc., being regarded as a column matrix with two rows, and the Pauli spin operators appear explicitly.

$$\left[-\nabla_1^2 + V(\mathbf{r}_1) + \sum_t \int |\psi_t(\mathbf{r}_2)|^2 \frac{2}{r_{12}} d^3r_2 \right] \psi_s(\mathbf{r}_1) \tag{2.183}$$

$$- \sum_t \int \psi_t^*(\mathbf{r}_2) \frac{(\mathbf{I} + \boldsymbol{\sigma}_1 \cdot \boldsymbol{\sigma}_2)}{2} \psi_t(\mathbf{r}_1) \frac{2}{r_{12}} \psi_s(\mathbf{r}_2) d^3r_2 = \varepsilon_s \psi_s(\mathbf{r}_1)$$

In the exchange term of this equation (second integral), $\boldsymbol{\sigma}_1$ acts on the spinor ψ_t; $\boldsymbol{\sigma}_2$ acts on ψ_s, and **I** is a unit operator on both ψ_s and ψ_t. A sum on the spin indices associated with ψ_t is understood. The equality of the exchange terms in (2.182) and (2.183) may be verified by explicit comparison of the components.

Slater (1951a) and Thompson (1960) have discussed the interpretation of these equations in detail. The third term on the left evidently represents the electrostatic potential energy of an electron in state s in the field of other electrons. The exchange term reduces the electrostatic interaction of the electrons, the reduction being greatest when the spins are parallel. This occurs because the probability of close encounters between electrons of parallel spin is reduced by the Pauli principle. Contributions to the sums in (2.183) from terms with $s = t$ cancel: the electron does not act on itself. It must be noted that the exchange operator does not have the form of an ordinary potential: an integral operator is involved. The exchange operator is conventionally defined as follows: The exchange term in (2.183) may be written as

$$\int A(1, 2)\psi_s(\mathbf{r}_2)\, d^3r_2 \qquad (2.184)$$

where

$$A(1, 2) = \sum_t {}'\psi_t^*(\mathbf{r}_2)\frac{(\mathbf{I} + \boldsymbol{\sigma}_1 \cdot \boldsymbol{\sigma}_2)}{2}\frac{2}{r_{12}}\psi_t(\mathbf{r}_1)$$

Nonetheless it is often necessary to approximate this term by a potential: Two such approximations have been discussed by Slater (1951a). This problem will be considered in the next section.

The spinors $\psi_s(\mathbf{r})$ which are solutions of the Hartree-Fock equation will not, in general, be eigenfunctions of the z component of spin. This complication is usually neglected in order to simplify the use of these equations. Such a simplification is probably not justified in the case of materials with a degree of ferromagnetic or antiferromagnetic alignment. The assumption that the wave functions are eigenfunctions of σ_z does not lead to an inconsistency in the Hartree-Fock equations since if one of the components of ψ_s is zero, the equation for this component is trivially satisfied. (The Hartree-Fock equations may, however, have other solutions.) In this approximation, the exchange interaction only occurs between electrons of parallel spin. This assumption does not imply that the position dependence of the wave functions of states with "up" and "down" spin is the same if the number of "up" and "down" spin electrons is different. The difference may be, in fact, quite significant (Pratt, 1956, refers to inclusion of this as the "unrestricted Hartree-Fock method"), and is responsible for the phenomena of spin polarization which will be discussed in Section 3.10.

2.12 THE HARTREE-FOCK EQUATIONS

If wave functions satisfying the Hartree-Fock equations are used in the evaluation of the energy of the N-electron system according to (2.181), the result is

$$E = \sum_s \left[\varepsilon_s - \tfrac{1}{2} \sum_t \iint |\psi_t(\mathbf{r}_2)|^2 |\psi_s(\mathbf{r}_1)|^2 \frac{2}{r_{12}} d^3r_1 \, d^3r_2 \right. \quad (2.185)$$

$$\left. + \tfrac{1}{2} \sum_t \iint \psi_s{}^*(\mathbf{r}_1)\psi_t{}^*(\mathbf{r}_2) \frac{2}{r_{12}} \psi_s(\mathbf{r}_2)\psi_t(\mathbf{r}_1) \, d^3r_1 \, d^3r_2 \right]$$

The reason that the total energy is not just a sum of "one-particle energies" ε_s is that this sum counts each electrostatic interaction twice. It is a general characteristic of the theory of interacting particles that the total energy is not just the sum of the one-particle energies, but rather depends on the distribution of the particles.

The justification for the designation "one-particle energy" comes from Koopmans' theorem (Koopmans, 1934): Let us consider the difference in energy between two systems containing N and $N-1$ particles, respectively, but otherwise identical. In particular, we suppose that the one-electron wave functions ψ_s are the same in each case; however in one, the state ψ_j is unoccupied. The approximation of unchanged wave functions is likely to be reasonably valid in a solid, at least for states belonging to a reasonably wide band, since the electrons are not bound to any particular atom but spread throughout the solid. The possible errors in this assumption have been discussed by J. C. Phillips (1961). From (2.181), we find then that the difference in energy, ΔE, is given by

$$\Delta E = \int \psi_j{}^*(-\nabla^2 + V(\mathbf{r}))\psi_j \, d^3r + \sum_t \iint |\psi_t(\mathbf{r}_1)|^2 |\psi_j{}^*(\mathbf{r}_2)|^2 \frac{2}{r_{12}} d^3r_1 \, d^3r_2$$

$$- \sum_t \iint \psi_j{}^*(\mathbf{r}_1)\psi_t{}^*(\mathbf{r}^2) \frac{2}{r_{12}} \psi_j(\mathbf{r}_2)\psi_t(\mathbf{r}_1) \, d^3r_1 \, d^3r_2 \quad (2.186)$$

The factors of $\tfrac{1}{2}$ in (2.181) disappear because the deleted state, ψ_j, occurs in both summations. If the wave functions satisfy (2.182), then we have just

$$\Delta E = \varepsilon_j \quad (2.187)$$

Hence the one-electron energy parameter ε_j has the significance that it gives the energy required to remove an electron from state j. As a corollary, the energy which has to be added to the system in order to remove an electron from state j and place it in an unoccupied state i is $\varepsilon_i - \varepsilon_j$. Since the energy bands in a solid are determined, in principle, from the Hartree-Fock energy parameters, the physical interpretation of the bands is that afforded by Koopmans' theorem.

In the case of a solid for which the potential $V(\mathbf{r})$ is periodic, it is possible to find wave functions as solutions of the Hartree-Fock equation which satisfy Bloch's theorem. The potential and exchange integrals which are computed with these functions have the proper periodicity. Hence the use of Bloch's theorem as a boundary condition on the wave functions is self-consistent. The Hartree-Fock equations may, however, possess other solutions which do not have this periodicity, and it is a matter of some sublety to determine exactly what the lowest state is.[13] Other solutions of the Hartree-Fock equations have been discussed by Thompson (1960) and by Overhauser (1962). In particular, Overhauser has shown that the ground state of a free electron gas in the high density limit contains static spin density waves. Spin density waves are discussed briefly in Appendix 5.

Except in the case of such simple systems as the free-electron gas, particular solutions of the equations must be found by the method of self-consistent fields (see for instance Hartree, 1957). In this method the potential and exchange integrals are calculated with an assumed set of wave functions. When these integrals are specified, the Hartree-Fock equations are linear. The linear equations are then solved; the potential and exchange integrals are recalculated with the new functions, and the process is repeated until the successive iterations agree within some assigned limit. The process has been carried through with precision for some relatively simple atomic systems, but application to solids is rendered very difficult by the necessity of evaluating the sum over occupied states in (2.182). What is generally done is to assume a set of ordinary and exchange potentials. One then has a one-particle Schrödinger

---

[13] The essential point is that more than one set of wave functions may satisfy the Hartree-Fock equations. In such a case, one must determine which set yields the lowest total energy. This set will describe the Hartree-Fock ground state.

equation. The derivation of such potentials will be considered in the next section. Authors often advance arguments of reasonable plausibility that the resulting solutions are nearly self-consistent.

To conclude this section, we prove a theorem concerning determinantal wave functions. Suppose that the N electrons present just suffice to fill states in a Brillouin zone belonging to a single band at the absolute zero of temperature. We consider the situations for which the determinantal wave function is an eigenfunction of S^2: either complete ferromagnetic alignment, or the singlet state with $N/2$ spatial wave functions and equal numbers of up and down spins. There are N or $N/2$ atomic sites in the system, respectively. The individual wave functions are supposed to be solutions of the Hartree-Fock equations which satisfy Bloch's theorem so that each column of the determinant is characterized by a particular $\mathbf{k}$. It is necessary to consider $\mathbf{k}$ to be a discrete quantity, taking N (or $N/2$) possible values. Let us express the Bloch one-electron wave functions $\psi(\mathbf{k}, \mathbf{r})$ in terms of Wannier functions by the inverse of (2.168b)

$$\psi(\mathbf{k}, \mathbf{r}) = \frac{1}{\sqrt{N}} \sum_n e^{i\mathbf{k}\cdot\mathbf{R}_n} a(\mathbf{r} - \mathbf{R}_n)$$

This substitution is made in the determinantal wave function. Then we have

$$\psi(1\ldots N) = \frac{1}{\sqrt{N!}} \det |\psi(\mathbf{k}, \mathbf{r})| = \frac{1}{\sqrt{N!}} \det \left| \frac{1}{\sqrt{N}} \sum_n e^{i\mathbf{k}\cdot\mathbf{R}_n} a(\mathbf{r} - \mathbf{R}_n) \right|$$

$$= \frac{1}{\sqrt{N!}} \det \left| \frac{e^{i\mathbf{k}\cdot\mathbf{R}_n}}{N^{1/2}} \right| \det |a(\mathbf{r} - \mathbf{R}_n)|$$

$$= \frac{e^{i\theta}}{\sqrt{N!}} \det |a(\mathbf{r} - \mathbf{R}_n)| \qquad (2.188)$$

In the second line of (2.188) we have made use of the rule for multiplying determinants; and then observed that since the transformation between Bloch functions and Wannier functions is unitary, the determinant of the coefficients is a complex number of modulus unity. Since the phase factor can be discarded, we see that the many-electron wave function may be equally well represented as a determinant of Bloch functions or as a determinant of Wannier functions. It must always be observed that this replacement is only valid in the case of a filled band. Conversely, in the case of atoms with filled shells, a determinant of localized atomic wave functions may be transformed to a determinant of tight binding wave functions (Seitz, 1940).

2.13 Determination of the Crystal Potential

The usual approximation to the Hartree-Fock equations is the following: Instead of (2.183) we write:

$$[-\nabla^2 + V(\mathbf{r}) + V_0(\mathbf{r}) + V_{\text{ex}}(\mathbf{r})]\psi_\mathbf{k}(\mathbf{r}) = E(\mathbf{k})\,\psi_\mathbf{k}(\mathbf{r}) \qquad (2.189)$$

in which $V(\mathbf{r})$, as before, is the potential energy of the electron in the field of the nuclei of the system; $V_0(\mathbf{r})$ is the ordinary average electrostatic potential energy of an electron in the field of all the charges of the system; and V_{ex} is an average exchange potential whose explicit spin dependence is neglected. This potential may be defined formally as follows (from 2.184)

$$\int A(1,2)\psi_\mathbf{k}(\mathbf{r}_2)\,d^3r_2 = V_{\text{ex}}(\mathbf{r}_1)\psi_\mathbf{k}(\mathbf{r}_1)$$

or

$$V_{\text{ex}} = \sum_t{}' \frac{\int \psi_t^*(\mathbf{r}_2)\psi_\mathbf{k}(\mathbf{r}_2) 2/r_{12}\, d^3r_2}{\psi_\mathbf{k}(\mathbf{r}_1)}\,\psi_t(\mathbf{r}_1) \qquad (2.190)$$

The wave functions are assumed to be eigenfunctions of the z-component of spin and the sum on t includes only states with the same spin as $\psi_\mathbf{k}$. The exchange potential is singular at those points where $\psi_\mathbf{k}(\mathbf{r}_1)$ has zeros; these singularities make no contribution to the energy, however. Of greater importance is the fact that V_{ex} depends on the state $\mathbf{k}$ whose wave function is being calculated, whereas V_0 is the same for all states.

It is often desired to approximate V_{ex} by a potential which is the same for all states. The most celebrated such approximation is that proposed by Slater (1951a): The exchange potential in an electron system

2.13 DETERMINATION OF THE CRYSTAL POTENTIAL

with a given charge density $\rho = \Sigma_t \psi_t{}^* \psi_t$ should be the same as in a free electron gas of the same density. (The computation of the exchange potential energy of a free electron gas is discussed below.)

$$V_{\text{ex}} \approx -6 \cdot \frac{3}{4\pi} \left[\sum_t \psi_t{}^*(\mathbf{r}) \psi_t(\mathbf{r}) \right]^{1/3} \qquad (2.191)$$

The sum in (2.187) includes only states of the same spin as the one on which V_{ex} acts. Several authors have discussed the adequacy of this approximation, including Herman *et al.* (1954), Callaway (1955), Maslen (1956), and Hartree (1958). The general conclusion would seem to be that, although (2.191) reproduces the general trend of (2.186) fairly well, it may be in error quantitatively by a substantial factor, particularly when the charge density is small.

An approximation which probably is of greater accuracy is to assume that the exchange potential for a given state depends primarily on the angular momentum of the state considered, rather than on the energy; or in the case of a solid, on the predominant angular momentum in the expansion of the wave function in spherical harmonics (Herman, 1954). One may then approximate (2.190) by allowing ψ_k to be a wave function belonging to the particular angular momentum considered. This approach, however, is only pertinant to the problem of calculating the exchange interaction between the band electrons, and electrons tightly bound in inner shells. The latter states are very nearly eigenfunctions of angular momentum.

The problem of calculating the exchange interaction between band electrons is a very difficult one. If this interaction is included in a straight-forward way serious disagreement with experiment is likely to occur. To see why this is so, let us calculate the exchange energy of a free electron gas: Let us consider

$$\varepsilon_1(\mathbf{k}) = \sum_{\mathbf{k}'} \iint \psi_{\mathbf{k}}{}^*(\mathbf{r}_1) \psi_{\mathbf{k}'}^{*}(\mathbf{r}_2) \, (2/|\mathbf{r}_1 - \mathbf{r}_2|) \psi_{\mathbf{k}}(\mathbf{r}_2) \psi_{\mathbf{k}'}(\mathbf{r}_1) \, d^3r_1 \, d^3r_2 \qquad (2.192)$$

Let

$$\psi_{\mathbf{k}}(\mathbf{r}) = \frac{1}{\sqrt{\Omega}} e^{i\mathbf{k}\cdot\mathbf{r}}$$

where Ω is the volume in which the wave functions are normalized. Then we get for the integral in (2.192)

$$\frac{1}{\Omega^2}\int\int e^{i(\mathbf{k'}-\mathbf{k})\cdot(\mathbf{r_1}-\mathbf{r_2})}\frac{2}{|\mathbf{r_1}-\mathbf{r_2}|}d^3r_1\,d^3r_2$$

The variables of integration may be changed to $\mathbf{r_1}$, $\mathbf{r_2}-\mathbf{r_1}$. The integral yields

$$\frac{8\pi}{\Omega}\frac{1}{|\mathbf{k}-\mathbf{k'}|^2} \tag{2.193}$$

This must be summed over $\mathbf{k'}$, but only states of like spin are to be included. We replace $\Sigma_{\mathbf{k'}}$ by $[\Omega/(2\pi)^3]\int d^3k'$.
Then

$$\sum_{\mathbf{k'}}\frac{8\pi}{\Omega}\frac{1}{|\mathbf{k}-\mathbf{k'}|^2}=\frac{1}{\pi^2}\int d^3k'\,\frac{1}{k^2+k'^2-2kk'\cos\theta} \tag{2.194}$$

in which θ is the angle between $\mathbf{k}$ and $\mathbf{k'}$. We suppose the electrons occupy all states with wave vectors $k\leqslant k_F$ where k_F is the radius of the Fermi surface. The integration yields:

$$\varepsilon_1(\mathbf{k})=\frac{2k_F}{\pi}\left[1+\frac{k^2-k_F^2}{2kk_F}\ln\left|\frac{k-k_F}{k+k_F}\right|\right] \tag{2.195}$$

If we look at the Hartree-Fock exchange term (2.184) we see that for the free electron gas with wave functions which are eigenfunctions of σ_z,

$$\int A(\mathbf{r_1}-\mathbf{r_2})\psi_{\mathbf{k}}(\mathbf{r_2})\,d^3r_2=\varepsilon_1(\mathbf{k})\psi_{\mathbf{k}}(\mathbf{r_1}) \tag{2.196}$$

The gas must, of course, be neutralized by a uniform distribution of positive charge in order to prevent catastrophe. Hence the average electrostatic potential in the gas must be zero. Consequently the one-particle energy of the free electron gas is

$$E(\mathbf{k})=\mathbf{k}^2-\varepsilon_1(\mathbf{k}) \tag{2.197}$$

Next, we observe that the derivative of $\varepsilon_1(\mathbf{k})$,

$$\frac{d\varepsilon_1}{dk}=\frac{1}{\pi}\left[\left(1+\frac{k_F^2}{k^2}\right)\ln\left|\frac{k-k_F}{k+k_F}\right|+\frac{2k_F}{k}\right] \tag{2.198}$$

contains a term which is logarithmically infinite on the Fermi surface. From (1.47) we see that the density of states, which depends inversely on dE/dk, would then vanish on the Fermi surface. This result would have great influence on the calculation of the properties of metals, which depend critically on the state of affairs at the Fermi surface and is in violent conflict with experiment.

It is evident, then, that something is fundamentally wrong with the Hartree-Fock theory. In fact, the influence of the electron interaction, when calculated with a more detailed theory, turns out to be much weaker than is predicted by (2.198) for instance. We see that it is undesirable to include the full exchange interaction between nearly free electrons in a band calculation.

The total exchange energy per particle of the system is found by integrating (2.195) over $\mathbf{k}$. If we make use of the relation implied by (1.50) between k_F and the density of electrons, we find

$$\frac{E_{ex}}{N} = -\frac{1}{2} \sum_{\mathbf{k},\mathbf{k'}} \iint \psi_\mathbf{k}^*(\mathbf{r}_1) \psi_{\mathbf{k'}}^*(\mathbf{r}_2) \frac{2}{|r_1 - r_2|} \psi_\mathbf{k}(\mathbf{r}_2) \psi_{\mathbf{k'}}(\mathbf{r}_1) \, d^3r_1 \, d^3r_2$$

$$= -6 \left(\frac{3\rho}{8\pi}\right)^{1/3} \quad (2.199)$$

in which $\rho = N/V$ is the density of electrons. (To obtain agreement with (2.191), we observe that in the present case, both spins are included in ρ; the quantity in the parenthesis becomes $3\rho_\pm/4\pi$ if we include electrons of either ($\pm$) spin.)

Even if the difficulties associated with inclusion of the exchange interaction are neglected, the computation of the effective crystal potential is a most difficult one. It is certainly necessary to include the screening effect of the distribution of band electrons on the average electrostatic potential. This implies, however, that the electron wave functions can be found conveniently throughout the occupied portion of the Brillouin zone. In fact, the best that can be done is to sample the distribution for some selected states, as has been discussed by Kleinman and Phillips (1959), among others. Most authors have not attempted to carry the iterative process to obtain self-consistency very far. Some approximations found in the literature include: (1) self-consistency with respect to states at the center of the zone; (2) the charge density in the crystal is represented as a sum of free atom densities; or (3) the charge density of the band electrons is assumed to be nearly uniform. The latter approximations may be reasonably accurate for the alkali metals (Callaway and Glasser, 1958).

Faced with these problems, Wigner and Seitz (1933) proposed an approximation which appears to be remarkably good for the alkali metals, insofar as can be judged from the very substantial measure of agreement between the calculated and observed values of the cohesive energies of these metals: Each electron moves in the field of a singly charged ion. The argument is that on account of exchange effects, and also the coulomb repulsion of electrons, it is unlikely (in a monovalent material), that two electrons will be found on the same lattice site. Consequently the potential energy of a single electron will be that due to its presence in the field of a single ion, the rest of the lattice being neutral. The polyhedral cells approximate spheres reasonably closely so that multipole components in the potential may also be neglected. The potential within the cell thus does not differ appreciably from what it is in the free atom at the same position. In a certain sense, this approximation transcends the Hartree-Fock method, since the neutrality of the cells required here is not a consequence of the Hartree-Fock equations.[14]

2.14 *The Quantum Defect Method*

If we accept the Wigner-Seitz approximation with respect to the crystal potential, it seems reasonable to try to use spectroscopic data from the free atom as a guide in the construction of the potential. This was realized quite early. Accordingly, in their calculation of the cohesive energy of sodium, Wigner and Seitz (1933) employed a semiempirical potential previously constructed by Prokofjew (1929) to account for the observed spectrum of the free sodium atom. Similarly, Seitz (1935) determined the potential to be used in a calculation of the cohesive energy of metallic lithium from the spectrum of the free atom. Gorin (1936) attempted the same thing for potassium.

---

[14] Cohen (1960) has criticized the Wigner-Seitz approximation on the grounds that the "correlation hole" (the electronic density excluded from the cell by the coulomb repulsion) must have a significantly different position dependence for states of different symmetry.

2.14 THE QUANTUM DEFECT METHOD

The construction of an empirical potential in the fashion of Prokofjew and Seitz is, however, a difficult problem. It is complicated by the fact that the effective potential seen by an electron actually must depend on the energy state, as is seen from (2.190). This dependence is presumably quite weak for lithium and sodium; for substantially heavier atoms, such as potassium, construction of a single empirical potential for all states did not appear to be possible (Gorin, 1936). It was subsequently suggested by Kuhn and Van Vleck (1950) that the spectroscopic information could be used almost directly in the energy level problem, without the necessity of constructing a potential explicitly. This procedure, which has come to be known as the "Quantum Defect Method" has been significantly refined and extended by Brooks (1953, 1958; Brooks and Ham, 1958) and by Ham (1954, 1955).

The essential idea of the procedure is the following. In a free atom of an alkali metal, the valence electron is loosely bound to a compact spherical core of electrons in closed shells. These closed shells will not be significantly modified in the solid, and occupy only a small portion of the volume of an atomic cell. For this reason, the electrostatic field which acts on the valence electron is nearly a pure coulomb field throughout most of the cell. Hence its wave function must be a (negative energy) coulomb wave function in the outer regions of the cell. The same is true in the free atom. (It is not implied that the wave function is hydrogenic.) There are two linearly independent solutions of Schrödinger's equation with a coulomb potential. These are confluent hypergeometric functions. For a given energy, there is a unique "coupling constant" which determines (up to a multiplicative factor) the combination of these two functions which will vanish exponentially at infinity. This behavior of the wave functions of course occurs at the energy eigenvalues of the free atom. The coupling constant is thus determined for those energies. If it is assumed that the coupling constant depends smoothly on energy, it is possible to extrapolate it as a function of energy, and hence to determine it for any energy. This coupling constant is, in reality, determined by solving the wave equation with the actual potential energy inside the core; it is possible, however, to determine this constant empirically without taking explicit account of interactions in the core.

In the solid, the wave functions must satisfy certain prescribed boundary conditions (Section 2.6). These conditions may all be stated in a fashion independent of the normalization of the wave function.[15] These boundary conditions also determine, in principle, a unique value of the coupling constant. Since this coupling constant now is known as a function of energy, the energies of interest in the solid state problem are also determined. Brooks and Ham (1958) have also shown that much information about the wave function in the interior can be obtained. The method in its present form has been applied to the alkali metals by the papers to which reference has already been made; it has also been applied to the computation of the cohesive energies of the noble metals by Kambe (1955). The difficulty of interpreting spectroscopic data for multivalent atoms to effectively "determine" a potential has so far prevented application to such materials.

To formalize this discussion, we consider the radial wave equation for the cellular method, (2.70), and make the following substitutions: $U_l = rR_l$, $V = -2/r$ (since we are concerned with the outer regions of the cell), and $E = -1/n^2$ (n is an integer in the hydrogenic problem). Then we have

$$\frac{d^2 U_l}{dr^2} + \left[-\frac{1}{n^2} + \frac{2}{r} - \frac{l(l+1)}{r^2} \right] U_l = 0 \qquad (2.200)$$

We need two linearly independent solutions of (2.200). It is convenient to choose them so that one of them, $U_{l,0}$, vanishes at the origin and the other, $U_{l,1}$, is singular there. Consequently $U_{l,1}$ could not appear in the usual hydrogenic problem. These functions may be expressed in terms of Bessel functions (Wannier, 1943; Kuhn, 1951; Ham, 1957)

$$U_{l,0}(n,r) = \frac{z}{2} J^n_{2l+1}(z) = \frac{n^{l+1}}{\Gamma(2l+2)} M_{n,l+1/2}\left(\frac{2r}{n}\right) \qquad (2.201)$$

$$U_{l,1}(n,r) = \frac{z}{2} N^n_{2l+1}(z)$$

---

[15] This is most easily seen for the lowest state (Γ_1) for which the logarithmic derivative of the wave function is required to vanish on the atomic sphere (in the spherical approximation). We have also determined the band parameters E_2, E_4 independently of the normalizations of the wave function.

2.14 THE QUANTUM DEFECT METHOD

in which

$$z = (8r)^{1/2} \tag{2.202}$$

The functions J^n_{2L+1}, N^n_{2L+1} are combinations of Bessel functions, and $M_{n,l+1/2}$ is a regular Whittaker function. The relation between the coulomb functions and Bessel functions is obtained in the following way: The change of independent variable (2.202), coupled with the substitution $U_l = (z/2)V_l$ converts (2.200) into the form

$$z^2 \frac{d^2 V_l}{dz^2} + z \frac{dV_l}{dz} + [z^2 - (2l+1)^2]V_l = \frac{1}{n^2}\left(\frac{z}{2}\right)^4 V_l$$

If $n = \infty$ ($E = 0$), V_l is evidently a Bessel function of order $(2l+1)$. Further, for interesting states in the alkali metals, one has $E < 1$ so that $n^2 > 1$. The following expansion is suggested

$$V_l(z) = \sum_k \frac{1}{n^{2k}} V_{l,k}(z) \tag{2.203}$$

The functions $V_{l,k}$ satisfy:

$$\bar{V}_l V_{l,0} = 0 \tag{2.204}$$

$$\bar{V}_l V_{l,k} = (z/2)^4 V_{l,k-1}(z)$$

in which

$$\bar{V}_l = z^2 \frac{d^2}{dz^2} + z \frac{d}{dz} + z^2 - (2l+1)^2 \tag{2.205}$$

Kuhn (1951) has shown that the functions $V_{l,k}$ may be generated from cylindrical functions. Let C_{2l+1} be an arbitrary linear combination of the ordinary Bessel functions $J_{2l+1}(z)$ and the Weber function $Y_{2l+1}(z)$

$$C_{2l+1}(z) = A J_{2l+1}(z) + B Y_{2l+1}(z)$$

(To generate J^n_{2L+1}, we take $A = 1$, $B = 0$.) It follows from the recurrence relations obeyed by the cylindrical functions that

$$\bar{V}_l \left\{ \frac{2l+2+q}{4(2+q)} \left(\frac{z}{2}\right)^{q+2} C_{2l+3+q}(z) - \frac{1}{4(3+q)} \left(\frac{z}{2}\right)^{q+3} C_{2l+4+q}(z) \right\} \tag{2.206}$$

$$= \left(\frac{z}{2}\right)^{q+4} C_{2l+1+q}$$

It is evident that $V_{l,0} = C_{2l+1}$. Next, set $q = 0$ in (2.206). The equation is then identical with that satisfied by $V_{l,1}$ provided that $V_{l,1}$ is given by the bracket on the left of (2.206) for $q = 0$, namely,

$$V_{l,1} = \frac{l+1}{4}\left(\frac{z}{2}\right)^2 C_{2l+3} - \frac{1}{12}\left(\frac{z}{2}\right)^3 C_{2l+4} \qquad (2.207)$$

Higher terms may be generated. The details of the series for the regular function J^n_{2l+1} and the irregular function N^n_{2l+1} are given by Kuhn (1951) and, more completely, by Ham (1955). The functions $U_{l,0}$ and $U_{l,1}$ may be obtained from tables computed by Ham (1955) and (more completely) by Blume et al. (1959).

We now express the general solution of (2.200) as

$$U_l(r) = \alpha(n)U_{l,0}(n,r) + \gamma(n)U_{l,1}(n,r) \qquad (2.208)$$

Both $U_{l,0}$ and $U_{l,1}$ are present because the potential has a core region in which it is not coulombic.

The problem is now to determine the ratio $\alpha(n)/\gamma(n)$. At an eigenvalue of the free atom, the wave function goes to zero exponentially at ∞ and, consequently, must be represented by the function $W_{n,l+1/2}(2r/n)$, which has this property Whittaker and Watson (1952). Wannier (1943) has given the relation between the quasi-Bessel functions J^n_{2l+1}, N^n_{2l+1}, and the Whittaker function $W_{n,l+1/2}$. It is

$$W_{n,l+1/2}(2r/n) = \Gamma(n+l+1)n^{-l-1}(z/2)J^n_{2l+1}(z)\cos\pi(n-l-1) +$$
$$\Gamma(n-l)n^l(z/2)N^n_{2l+1}(z)\sin\pi(n-l-1) \qquad (2.209)$$

Hence, at eigenvalues of the free atom the ratio $\alpha(n)/\gamma(n)$ is determined to be:

$$\frac{\alpha(n)}{\gamma(n)} = \frac{\Gamma(n+l+1)}{n^{2l+1}\Gamma(n-l)\tan\pi(n-l-1)} \qquad (2.210)$$

For any energy, the ratio $\alpha(n)/\gamma(n)$ is defined as

$$\frac{\alpha(n)}{\gamma(n)} = -\frac{\Gamma(n+l+1)}{n^{2l+1}\Gamma(n-l)\tan\pi\nu(n)} \qquad (2.211)$$

When the two previous equations are compared we see that

$$\nu(n) = l + m' + 1 - n \qquad (2.212)$$

where m' is an arbitrary integer. The energy at an eigenvalue is

$$\varepsilon = -\frac{1}{n^2} = -\frac{1}{(m-\nu)^2} \qquad (2.213)$$

in which $m = l + m' + 1$ is an integer. Consequently, ν differs from the experimental quantum defect by an integer at most. The fundamental procedure of the quantum defect method is to set ν equal to the observed quantum defect at energies corresponding to the eigenvalues of the free atom, and to determine it at other energies by putting a smooth curve through the experimental points. The justification for this procedure has been given by Ham (1955) and Brooks and Ham (1958), in terms of the WKB approximation. Of course, once ν is determined, the radial wave function is known as a function of energy in the outer part of an atomic cell (except for a constant factor), and the standard techniques of the cellular method may be applied. The quantum defect procedure is also easily adapted for use with the Kohn-Rostoker method of band structure calculations (Section 2.8), since the required values of the logarithmic derivative on the inscribed sphere can be obtained once the ratio $\alpha(n)/\gamma(n)$ has been found for all energies.[16]

Also on the basis of the WKB approximation, Brooks and Ham obtain the following formula for the ratio of the amplitude of the wave function, $\psi_0(r) = U_0(r)/r$ at a nucleus of atomic number Z to its value at r_s:

$$\frac{\psi_0(0)}{\psi_0(r_s)} = \frac{2Z^{1/2} r_s}{\cos \pi\nu(n) \left[U_{0,0}(n,r_s) - \tan \pi\nu(n) U_{0,1}(n,r_s)\right]} \qquad (2.214)$$

The use of this result, in conjunction with the Kohn-Rostoker method, to determine values for the amplitude at a nucleus of a wave function of an electron on the Fermi surface, has been discussed (and applied to potassium) by Milford and Gager (1961).

---

[16] Brooks and Ham find that a more satisfactory extrapolation procedure is obtained by defining $\alpha(n)/\gamma(n) = -1/\tan \pi\eta(n)$. $\eta(n)$ is a smoother function of n than ν, since if n is an integer ν must also be an integer.

From a theoretical point of view, the Quantum Defect Method is extremely attractive. Many of the difficulties of the Hartree-Fock procedure are avoided. Brooks and Ham show that interaction effects concerning core electrons on a lattice site are included: It is necessary to assume that the valence electron may be described by a one-electron wave function, but the wave function of the core may be quite complicated. Further, if proper account is taken of the spin orbit splitting of energy levels of nonzero angular momentum, relativistic effects may also be included (Callaway et al., 1957).

As a practical matter, success of the quantum defect method depends on the extrapolation of the quantity ν (or η) obtained from spectral data for energies corresponding to eigenvalues of the free atom. The energies of states at the bottom of the valence electron band in the alkali metals will be, however, 2 or 3 ev below the lowest free atom level: some uncertainty in the extrapolation procedure is evidently possible. It is also necessary to average out the spin orbit splitting for levels of nonzero angular momentum. A second difficulty arises from the fact that the potential energy of a valence electron in the outer part of the cell is not strictly coulombic, but contains a term $-\alpha/r^4$, where α is the polarizability of the ion. The ion deforms in the field of the external electron, and the induced dipole moment produces the additional contribution to the energy. This effect is actually a manifestation of the correlation between core and valence electrons (Callaway, 1957a). The significance in the present instance is that the wave functions in the exterior are not precisely coulomb functions. The necessary corrections to the procedure have been discussed by Ham (1955) and Brooks and Ham (1958). They become important for the heavier alkali metals. These topics will be considered in more detail in Chapter III.

2.15 The Pseudopotential in Relation to the Quantum Defect Method

In Section 2.5, it was shown, within the general framework of the OPW method, that the requirement that the valence electron wave functions be orthogonal to core states tends to produce an effective repulsive potential acting on the valence electron. The general correctness of this reasoning, and its independence from the specifics of the OPW method may be inferred from the example of the free sodium atom, for instance. Beyond the core, the potential energy is $-2/r$, while inside the core, the potential decreases, becoming $-2Z/r$ at the center. Nevertheless, the energy of the valence electron (-0.3777 rydbergs) is larger than that of the lowest state in a pure coulomb potential (-1 rydberg).

This suggests that one might introduce a potential energy for the valence electron which would be $-2/r$ for large r, but contain a repulsive part for small r. This pseudopotential would include the effects of orthogonality of the valence electrons to core electrons, so that the lowest valence electron state is also the lowest bound state in this potential. A Yukawa form, $Ae^{-\beta r}/r$, was suggested by the crude considerations of Section 2.5. This idea was introduced by Hellmann and Kassatotschkin (1936a, b) who called it the "combined approximation." They applied the method to the determination of the cohesive energies of the alkali metals. It has also been applied to the study of molecules.

There is, in addition, a certain similarity to the quantum defect method. Spectroscopic data may be used to determine the pseudopotential parameters (A and β above, or a more complicated expression could be employed). A different potential for each l may also be determined. The procedure is as follows: It was seen in the previous section that, for eigenvalues of the free atom, the wave function is uniquely determined in the coulomb region if the energy is known. One may then numerically integrate the wave equation with the repulsive potential, adjusting the pseudopotential parameters until the logarithmic derivative of the wave function in the coulomb region agrees with that computed from the coulomb wave functions. An iterative procedure has been given which facilitates this determination[17] (Callaway, 1958b).

There is, moreover, a certain advantage over the quantum defect method. The quantum defect method cannot be used directly in band calculation procedures based on plane wave expansions. All these require an explicit potential. The required potential can be furnished by the pseudopotential procedure. There is also the further advantage that additional terms may be added to the pseudopotential to take account of differences between the crystal potential and that of the free atom.

---

[17] The variational procedure employed by Hellmann and Kassatotschkin does not seem to give accurate results if simple trial functions are employed.

Chapter 3

Band Structure of Materials

In this chapter the results of experimental and theoretical determinations of band structures in some materials will be surveyed. No attempt at completeness is intended. Previous reviews of calculations prior to 1958 are those of Callaway (1958a), Herman (1958), Slater (1956), and Raynor (1952). Experimental information concerning band structures has been surveyed by Lax (1958), and with emphasis on the theory underlying experimental procedures by Pippard (1960).

3.1 *The Alkali Metals: Cohesive Energy*

The calculation of the cohesive energy of the alkali metals was, historically, the first problem in which band theory was applied to real materials. The physical principles underlying the cohesion of the alkali metals were first determined by Wigner and Seitz (1933, 1934) in their classic study of the binding of metallic sodium. More recently, they have reviewed the general theory of the cohesion of metals (Wigner and Seitz, 1955). Subsequent studies have extended their work, but have not caused a modification of the basic ideas.

The cohesive energy is essentially the difference between the average energy of a valence electron in the solid, and the energy of the valence electron in the lowest state in the free atom. The latter may be obtained directly from spectroscopic data, or it may be found from an atomic self-consistent field calculation. In the solid, there are two principal effects to be considered. The energy of the lowest valence electron state in the solid is lower than in the free atom because the boundary condition for this state (Γ_1, see Section 2.6) reduces the kinetic energy by forcing the wave function to have zero normal derivative on the surface of the atomic polyhedron. The difference in energy between these states is called the boundary correction by Wigner and Seitz (1955). The lowest state in the solid is predominately s-like, and the second radial derivative of the wave function tends to be quite small near the boundary of cell: If r_s is the radius of the sphere whose volume equals the volume of the atomic polyhedron, one generally has $V(r_s)$ nearly equal to the energy of the lowest state. Consequently, the wave function tends to be quite flat over much of the volume of the cell. If ψ_0 is the wave function, for this state, the average kinetic energy is, to a good approximation,

$$- 4\pi \int_0^{r_s} \psi^* \frac{1}{r^2} \frac{d}{dr}\left(r^2 \frac{d\psi}{dr}\right) r^2\, dr = 4\pi \int_0^{r_s} r^2 \left|\frac{d\psi}{dr}\right|^2 dr \qquad (3.1)$$

(The result follows on integration by parts. The integrated part vanishes on account of the boundary condition.) In the free atom, the derivative of the wave function is negative at the typical r_s. There is a further contribution to the binding resulting from a compression of the valence charge distribution into a region of more negative potential energy. The boundary correction amounts to 73 kcal/mol in the case of sodium.[1]

Opposing the boundary correction is the Fermi energy. The electrons occupy a Fermi distribution with only two electrons in each state. It is customary to expand the energy of a state of wave vector **k** as a power series in **k**, in conformity with Eq. (2.73):

$$E(\mathbf{k}) = E_0 + E_2 k^2 + E_4 k^4 + \ldots$$

The average energy per particle is, in atomic units,

$$E_b = \frac{\int E(\mathbf{k})\, d^3k}{\int d^3k} = E_0 + E_K \qquad (3.2)$$

$$E_K = \frac{2.21\, E_2}{r_s^2} + \frac{5.81\, E_4}{r_s^4}$$

---

[1] Energies relating to the calculation of cohesive energies of solids are traditionally measured in units of kilocalories per mole. One kcal/mol is equal to 0.04336 ev/atom or to 0.003187 rydberg/atom.

3.1 THE ALKALI METALS: COHESIVE ENERGY

in which E_K may be considered to be the average kinetic energy, or the average Fermi energy. Terms of sixth order and higher in $E(\mathbf{k})$ are usually neglected as also is the fourth order cubic term. E_K amounts to about 46 kcal/mol in the case of sodium ($r_s = 3.93$). The cohesive energy, estimated on this basis, is 27 kcal/mol for sodium, already in good agreement with the experimental value of 26.0 kcal/mol.

It is also necessary to take account of the electrostatic interaction of the electron distribution. If one stays within the framework of the Hartree-Fock approximation, but uses the Wigner-Seitz approximation to compute the crystal potential, the total energy per particle is determined from (2.191) and (3.2) to be

$$\langle E \rangle = E_b + \frac{1}{N} \sum_{\mathbf{kk'}} \left[\iint |\psi_\mathbf{k}(\mathbf{r}_1)|^2 |\psi_\mathbf{k}(\mathbf{r}_2)|^2 \frac{1}{r_{12}} d^3r_1\, d^3r_2 - \right. \tag{3.3}$$

$$\left. \iint \psi_\mathbf{k}^*(\mathbf{r}_1)\psi_{\mathbf{k'}}^*(\mathbf{r}_2) \frac{1}{r_{12}} \psi_\mathbf{k}(\mathbf{r}_2)\psi_{\mathbf{k'}}(\mathbf{r}_1) d^3r_1\, d^3r_2 \right]$$

These integrals may be evaluated with good accuracy in the case of the alkali metals under the assumption that the wave functions are plane waves. (The accuracy of this approximation has been demonstrated by Wigner and Seitz, 1934.) The first term, which represents the "self-energy" of the charge distribution is easily determined to be $1.2/r_s$. The second term, which is the exchange energy of the electrons, can be determined from (2.199) to be $0.916/r_s$. The difference of these contributions, $+ 0.284/r_s$, amounts to 23 kcal/mol in the case of sodium. Clearly, if the electron interaction is treated in the Hartree-Fock approximation, the computed electrostatic repulsion is large enough to destroy the good agreement between the simple theory and experiment.

It is evidently necessary to treat the electron interaction more accurately than is permitted by the Hartree-Fock approximation. In a lengthy and difficult calculation, Wigner (1934, 1938) attempred to estimate this "correlation energy." The designation is applied because the problem arises from the necessity to consider the way in which the probability of close encounters of electrons, particularly those of unlike spin, is reduced by more detailed consideration of their electrostatic repulsion.

Wigner obtained[2]

$$E_c = -\frac{0.88}{r_s + 7.79} \qquad (3.4)$$

The electrostatic interaction terms then yield

$$E_I = \frac{0.284}{r_s} - \frac{0.88}{r_s + 7.79} \qquad (3.5)$$

With the inclusion of the correlation energy, the electrostatic interaction is very small for sodium (about -1 kcal/mol), and actually increases the binding. However, this result is certainly smaller than the theoretical uncertainties involved in the calculation. In the other alkali metals, the same general result is obtained: the cohesive energy is primarily the difference of the boundary correction and the average kinetic energy; the contribution from the electrostatic interaction with the inclusion of the correlation energy is small (and binding for all except lithium). The comparison between theory and experiment will be given in the next section.

Wigner obtained the expression (3.4) for the correlation energy in the following way. He estimated a constant (independent of r_s) correlation energy -0.11 ry/electron for small r_s. At large r_s ($r_s > 10$), Wigner introduced the hypothesis that the electrons would localize themselves on lattice sites (a continuum of positive charge to neutralize the system is, of course, assumed). Assuming further that the lattice would be body-centered cubic, the correlation energy may be obtained by subtracting from the electrostatic energy of such a system (Fuchs, 1935) the exchange and kinetic energies. The correlation energy approaches $-0.88/r_s$ in the limit of large r_s. The formula (3.4) is nothing more than an interpolation between these limits.

Recently, it has been possible to make considerable improvements in the determination of the correlation energy in both limits. The correlation energy in the high density (small r_s) limit has been obtained more exactly by Gell-Mann and Bruckner (1957). They found, for $r_s < 1$

$$E_c = 0.0622 \, ln \, r_s - 0.096 \qquad (3.6)$$

---

[2] Pines (1955) pointed out an error in Wigner's formula. The corrected expression is given in (3.4).

The calculation of the energy of the electron lattice has also been improved (Coldwell-Horsfall and Maradudin, 1960, Carr, 1961; Carr et al., 1961) with the inclusion of contributions from the zero point motions ($+ 2.65/r_s^{3/2}$) and the contribution from the anharmonic interactions ($- 0.73/r_s^2$).

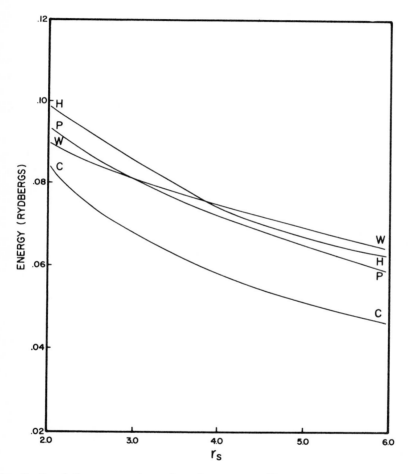

FIG. 8. Correlation energy for a free electron gas. The negative of the correlation energy, $- E_c$, is plotted as a function of the sphere radius r_s in the region $2.0 \leqslant r_s \leqslant 6.0$ according to the calculations of Hubbard (1958) (curve H), Pines (1958) (curve P), Wigner (1934) (curve W), and Carr et al. (1961) (curve C).

$$E_c = -\frac{0.876}{r_s} + \frac{2.65}{r_s^{3/2}} - \frac{2.94}{r_s^2} \tag{3.7}$$

(the term in r_s^{-2} includes the kinetic energy term in the free electron limit, $2.21/r_s^2$, which must be subtracted). Terms of order $r_s^{-5/2}$ and $\exp(-r_s^{1/2})$ have been ignored. No calculations have as yet been performed successfully which are accurate in the region of electron densities appropriate to real materials ($2 \geqslant r_s \geqslant 6$).

Several authors have discussed the problem of interpolating the correlation energy between the low density and high density limits (Pines, 1958; Nozieres and Pines, 1958; Hubbard, 1958; Carr et al., 1961) from different viewpoints. The agreement between the various interpolations (including that of Wigner) is only fair. The results are illustrated graphically in Fig. 8.

The situation with respect to the correlation is evidently still quite unsatisfactory. Not only must one resort to interpolation procedures for densities of interest (and there does not appear to be any general agreement on a "best" interpolation) but the effect of considering more realistic single particle wave functions (Bloch waves) has not been studied. It is to be hoped that there will be significant progress on these most difficult questions in the near future.

With the inclusion of the electrostatic interaction energy

$$E_I = 0.284/r_s + E_c,$$

the final expression for the cohesive energy (E_{coh}) becomes

$$E_{\text{coh}} = E_b - E_a + E_I \tag{3.8}$$

in which E_a is the energy of the lowest valence electron state in the free atom. Results for the alkali metals are given in the next section.

Aside from the question of the correlation energy of the valence electron distribution, the most serious uncertainty in the calculations of the cohesive energy are concerned with the core polarization effect, which is a manifestation of the correlation between core and valence electrons. The polarization effect, which was briefly discussed in Section 2.14, is probably the principal cause of the discrepancy between the experimental values of the ionization energy of free alkali metal atoms and the results of self-consistent field calculations.

3.1 THE ALKALI METALS: COHESIVE ENERGY

In Section 2.14 it was asserted that, outside the ion core, the effect of polarization is to add a term to the effective potential seen by the valence electron which is proportional to r^{-4}. Except in calculations using the quantum defect method, it is also necessary to determine the change in the effective potential due to core polarization, inside the core, for small values of r. It is evident that the r^{-4} dependence of this quantity cannot persist to $r = 0$, since an infinite contribution to the energy would result. A more elaborate treatment of the polarization potential has been given (Callaway, 1957a; Reeh, 1960). Under the influence of the field of the valence electron, one-electron wave functions for the core electrons depend (parametrically) on the coordinate of the valence electron. This effect is calculated by perturbation theory. The energy of the core electrons also must depend on the position of the valence electron. This coordinate dependent energy then serves as a potential energy function for the valence electron. After a number of approximations, a polarization potential is derived:

$$V_p(r_v) = \frac{2}{r_v^2} \sum_i \int_0^{r_v} u_i^{(0)}(\mathbf{r}_1) r_1 \cos \theta u_i^{(1)}(\mathbf{r}_1, \mathbf{r}_v) \, d^3 r_1 \tag{3.9}$$

The function $u_i^{(0)}$ is the unperturbed core function for the state i, whereas $u_i^{(1)}$ is the wave function for the core state perturbed by the field of an external electron at $\mathbf{r}_v$, as calculated, for instance, by Sternheimer (1954) Explicit polarization potentials have been obtained for lithium, sodium, and potassium.

If the same polarization potential existed in the solid as in the free atom, there would be a substantial contribution to the cohesive energy, particularly for the heavier alkali metals, since the valence charge distribution is compressed in the solid. (This effect has been estimated at somewhat more than one-third the observed cohesive energy in the case of potassium.) Such a large contribution would spoil the rather good agreement between observed and calculated values of the cohesive energy, which is obtained by both the quantum defect method and the self-consistent field approach when polarization is neglected. It has been suggested, however, that the polarization potential is greatly reduced in the solid compared to the free atom.[3] The fluctuating dipole moment

---

[3] H. Brooks, private communication.

140   CHAPTER 3. BAND STRUCTURE OF MATERIALS

in the core would be screened by a compensating deformation of the valence electron distribution so that the effective polarization interaction would be nearly zero on the boundary of the atomic cell. The question of this screening of the core polarization potential is one of the most important problems remaining in the study of the cohesive energies of the alkali metals.

3.2 Cohesive Energies of the Alkali Metals. Results: Lattice Constant and Compressibility

The results of some cohesive energy calculations for the alkali metals are given in Table XV. Calculations based on the quantum defect method (QDM) are compared with those based on empirical potentials (EP) (in the case of lithium and sodium) and on potentials derived from self-consistent fields (SCF) (potassium, rubidium, and cesium). Explicit polarization corrections are not included: these would certainly increase the magnitude of the cohesive energy in the case of calculations based on self-consistent fields. The two values given for the cohesive energy for each calculation are obtained by including the extreme values of the correlation energy from Fig. 8.

The effect of inclusion of core polarization has been discussed in detail for the case of potassium by Brooks (1958) and Callaway (1958a). In the self-consistent field calculation, the polarization potential may be treated in perturbation theory. When the potential computed from (3.9) is used, and the change in energy is found to second order in the wave vector $\mathbf{k}$, the cohesive energy is apparently increased by 8.7 kcal/mol. The change in energy of the lowest state of the valence electron in the free atom has been included, but the screening of the polarization potential in the solid has been neglected. In default of a better way to estimate this screening, we may suppose that the polarization potential is reduced to zero on the surface of the atomic sphere, and subtract from the computed polarization correction a constant amount equal to the apparent value of the polarization potential on the atomic sphere. Secondly, the value for the free atom polarizability obtained from the calculations of Sternheimer (1954) is probably too large, and one might better scale the polarization potential to agree with the estimates of Van Vleck (1934). If these crude

3.2 COHESIVE ENERGIES OF THE ALKALI METALS

arguments are accepted the polarization correction to the cohesive energy is reduced to 2.6 kcal/mol, which would not destroy the reasonably good agreement between theory and experiment. This question must, however, be reconsidered in a more fundamental manner.

There are two other quantities which can, in principle, be obtained from cohesive energy calculations of the sort previously described: the equilibrium lattice constant and the compressibility. The crystal structure cannot be predicted if the spherical approximation to the cellular method is used, since the actual polyhedral cell is replaced by a sphere of equal volume. The lattice constant is found by determining the minimum of the cohesive energy as a function of r_s; the compressibility is determined from the second derivative of the energy at the minimum. At absolute zero, the compressibility, K, is given by

$$\frac{1}{K} \equiv - V \frac{dp}{dV} = V \frac{d^2 E}{dV^2} = \frac{1}{12\pi r_s} \frac{d^2 E}{dr_s^2} \qquad (3.10)$$

(Here we have used the relation $p = - dE/dV$ in which V is the volume per atom and E is the cohesive energy per atom.) In spite of the uncertainties which exist in the calculation of the cohesive energy, the prediction of the lattice constant and compressibility is interesting because the electrostatic interaction energy is believed to be a slowly varying function of r_s, and hence the uncertainty in this quantity does not seriously affect the computation.

The determination of the cohesive energy for sufficiently many values of r_s to enable accurate location of the minimum and to permit numerical differentiation in its vicinity would be a very laborious task indeed. On the basis of a suggestion of Frohlich (1937) concerning the variation of the energy of the lowest valence state in the metal as a function of r_s, Bardeen (1938b) proposed that the dependence of the cohesive energy on r_s should be

$$E = A/r_s + B/r_s^2 + C/r_s^3 \qquad (3.11)$$

It would then suffice to know the energy for three values of r_s.

The second term in (3.11) represents the average kinetic or Fermi energy of the material (in the approximation that E_4 can be neglected.) The other terms are of the form proposed by Frohlich, who showed that

TABLE XV

COHESIVE ENERGY AND BAND PARAMETERS FOR THE ALKALI METALS

	Lithium		Sodium		Potassium		Rubidium		Cesium	
	EP[c]	QDM[a]	EP[d]	QDM[a]	SCF[e]	QDM[a]	SCF[f]	QDM[b]	SCF[g,m]	QDM[b]
E_0	−0.6832	−0.6865	−0.6113	−0.6011	−0.452	−0.4876	−0.444	−0.462	−0.4156	−0.4274
E_2	0.727	0.7305[a] 0.752[n]	1.066	1.022[a] 1.036[n]	1.160	1.149[a] 1.161[n]	1.181	1.10[b] 1.284[n]	1.393	1.197[b] 1.376[n]
E_4	−0.03	−0.0303 +0.004[n]	−0.153	−0.0096 −0.069[n]	−0.85	−0.3 −0.58[n]	−1.28	−1.01[n]	−3.64	−1.85[n]
E_a		−0.3963		−0.3777	−0.2915m	−0.3190	−0.305	−0.3070		−0.2862
r_s at which values were calculated	3.21	3.20 3.25[n]	3.94	4.0[a] 3.93[n]	4.84	4.86[a,n]	5.21	5.21[b] 5.20[n]	5.74	5.74[b] 5.62[n]
$-E_b$ (rydbergs)	0.1384	0.1399	0.0856	0.0824	0.0601	0.0648	0.053	0.065	0.0557	0.058
$-E_b$ (kcal/mol)	43.3	43.8	26.8	25.8	18.8	20.3	16.6	20.3	17.4	18.2
E_I^k (kcal/mol)	(+7.2 → +1.7)		(+4.2 → −1.0)		(+2.1 → −3.4)		(+1.4 → −4.1)		(+0.7 → −4.9)	
$(-E)_{coh}^k$	41.6 36.1	42.1 36.6	27.8 22.2	26.8 21.2	22.2 16.7	23.7 18.2	20.7 15.2	24.5 18.9	22.3 16.7	23.1 17.5
r_s (exp) 4° K	3.25[h]		3.93[h]		4.86[h]		5.20[h]		5.62[h]	
$(-E)_{coh}$ exp[m]	36.5		26.0		22.6		18.9		18.8	
r_s (pred)[m]	3.16		3.96		4.80		5.07		5.52	
P (pred)[j,m]	10,500		9,000		10,500		7,200		13,100	

3.2 COHESIVE ENERGIES OF THE ALKALI METALS

NOTES FOR TABLE

[a] H. Brooks, *Nuovo cimento* **7**, Suppl., 165 (1958).
[b] H. Brooks, *Phys. Rev.* **91**, 1027 (1953). The values here attributed to Brooks were determined by interpolation between the published values.
[c] R. A. Silverman and W. Kohn, *Phys. Rev.* **80**, 912 (1950); **82**, 283 (1951); R. A. Silverman, *Phys. Rev.* **85**, 227 (1952).
[d] J. Callaway, *Phys. Rev.* **123**, 1255 (1961b).
[e] J. Callaway, *Phys. Rev.* **119**, 1012 (1960a).
[f] J. Callaway and D. F. Morgan, Jr., *Phys. Rev.* **112**, 334 (1958).
[g] J. Callaway, *Phys. Rev.* **112**, 1061 (1958c).
[h] C. S. Barrett, *Acta Cryst.* **9**, 671 (1956. The values for lithium and sodium refer to the bcc phase, and in the case of lithium, to a temperature of 77°K.
[i] The quantity P (pred) is the pressure predicted to produce the same compression as is experimentally observed at a pressure of 10,000 atm. The compressions are given by C. A. Swenson, Phys. Rev. **99**, 423 (1955).
[j] The values given are the highest and lowest (for the experimental r_s) as determined graphically from Fig. 8. (See text for discussion of correlation energies.)
[k] The two values of the cohesive energy given for each calculation differ in the correlation energy. The "extreme" correlation energies (k, above) are used. *QDM* values refer to calculations a and b.
[l] D. R. Hartree and W. Hartree, *Proc. Cambridge Phil. Soc.* **34**, 550 (1938).
[m] F. S. Ham, *Solid State Phys.* **1**, 127 (1955).
[n] F. S. Ham, *Phys. Rev.* **128**, 82, 2524 (1962a, b).

the energy of the lowest state should be given approximately, in the vicinity of its minimum, by

$$E_0 = -\frac{3}{r_s} + \frac{r_0^2}{r_s^3} \qquad (3.12)$$

The quantity r_0 appears as a constant of integration in the derivation of (3.12); one finds by differentiation that the minimum of E_0 as a function of r_s occurs at r_0.

The approximate derivation of (3.12) has been given by Bardeen (1938b) whose discussion is followed here. The fundamental assumption is that the quantity γ defined in Eq. (2.87) (or equivalently below) is equal to unity

$$\gamma = \frac{r_s}{3} R_0^2(r_s) = 1 \qquad (3.13)$$

where R_0 is r times the radial wave function for the lowest valence state in the solid. R_0 is normalized so that

$$\int_0^{r_s} R_0^2 \, dr = 1$$

This assumption is fairly well justified for the alkali metals at the observed spacings. For instance, one has $\gamma = 1.073$ (Li); 1.0062 (Na); 1.122 (K); 1.082 (Rb); 1.145 (Cs).

The equation satisfied by R_0 is

$$\frac{d^2 R_0}{dr^2} = [V(\mathbf{r}) - E_0] R_0 \tag{3.14}$$

The boundary condition for this state is

$$\left(\frac{dR_0}{dr}\right)_{r=r_s} = \left(\frac{R_0}{r}\right)_{r_s} \tag{3.15}$$

We may regard both R_0 and E_0 as functions of r_s through the boundary conditions. Equation (3.14) may be differentiated with respect to r_s. [Let (′) denote the derivative with respect to r_s.] Then

$$\frac{d^2 R_0'}{dr^2} = [V(\mathbf{r}) - E_0] R_0' - E_0' R_0$$

We multiply this equation by R_0 and integrate. Equation (3.14) is used to eliminate $V(\mathbf{r}) - E_0$. We find, on using the normalization of R_0:

$$E_0' = \left[R_0' \frac{dR_0}{dr} - R_0 \frac{dR_0'}{dr}\right]_{r=r_s} \tag{3.16}$$

R_0' may now be eliminated from (3.16): The boundary condition (3.15) is differentiated with respect to r_s, yielding

$$\left[\frac{d^2 R_0}{dr^2} + \frac{dR_0'}{dr}\right]_{r_s} = \left(\frac{R_0'}{r}\right)_{r_s} \tag{3.17}$$

3.2 COHESIVE ENERGIES OF THE ALKALI METALS

This equation is substituted into (3.16). We obtain

$$E_0' = \frac{dE_0}{dr_s} = R_0(r_s)\left(\frac{d^2R_0}{dr^2}\right)_{r_s} = [V(r_s) - E_0]R_0^2(r_s) \qquad (3.18)$$

The assumption that $\gamma = 1$ enables us to reduce this to

$$\frac{dE_0}{dr_s} = \frac{3}{r_s}[V(r_s) - E_0] \qquad (3.19)$$

The potential may reasonably be assumed to be coulombic in the region beyond the core so that $V(r_s) = -2/r_s$. When this is inserted in (3.19), the equation may be integrated immediately to yield (3.12).

The chief weakness of this approach is the assumption that $\gamma = 1$ for the range of values of r_s of interest. Equation (3.11) is probably a suitable interpolation formula for the cohesive energy as a function of r_s provided that the constants A, B, C are determined from cohesive energies calculated for three different values of r_s.

Brooks (1953) applied this procedure to the alkali metals. The values of r_s he predicted are listed in Table XV. Instead of the compressibility, K, given by Eq. (3.10), Brooks and Ham calculated the theoretical pressure corresponding to the compression which is observed experimentally at 10,000 atm (Swenson, 1955). These results are also given in Table XV.

If we use Eq. (3.12) to give the variation of E_0 with r_s, assume that the effective mass is that of a free electron, and neglect E_4, we can give an approximate expression for the cohesive energy which depends only on the one parameter, r_0. This is, with Wigner's expression for the correlation energy:

$$E_{coh} = -\frac{3}{r_s} + \frac{r_0^2}{r_s^3} + \frac{2.21}{r_s^2} + \frac{0.284}{r_s} - \frac{0.88}{r_s + 7.79} \qquad (3.20)$$

If the parameter r_0 is known [it can be determined by finding a single value of $E_0(r_s)$], a simplified theory of the cohesive energy is possible. Kuhn and Van Vleck (1950) and Brooks (1958) have discussed the determination of the minimum value of E_0 by the quantum defect method.

Raimes (1952) has proposed an extension (3.20) to apply to the computation of the cohesive energies of the polyvalent metals. Let there be n valence electrons per atom, all of which are treated as free for the purpose of calculating the interaction energy. In Eq. (3.19), one inserts $V(r_s) = -2n/r_s$, and we obtain, instead of (3.12),

$$E_0 = n\left(-\frac{3}{r_s} + \frac{r_0^2}{r_s^3}\right)$$

To compute the total energy per atom of the valence electrons in the solid, E_t, we observe that the average kinetic energy per electron is proportional to $n^{2/3}$; the exchange energy to $n^{1/3}$. The self-energy of the charge distribution is proportional to n (per electron) and if the correlation energy is a function only of the average electron density, we merely replace r_s (as determined from the lattice structure only) by $r_s/n^{1/3}$ in the correlation energy formula [such as (3.4)]. Since there are n electrons per atom, another factor of n enters each term. Hence

$$E_t = n^2\left(\frac{r_0^2}{r^2} - \frac{3}{r_s}\right) + \frac{1.2n^2}{r_s} + \frac{2.21n^{5/3}}{r_s^2} - \frac{0.92n^{4/3}}{r_s} - nE_c\left(\frac{r_s}{n^{1/3}}\right) \quad (3.21)$$

(in which E_c is the free electron correlation energy). It is not so simple in this case to specify the theoretical cohesive energy because of problems concerning exchange and correlation effects in the free atom; however, the lattice spacing and compressibility can be computed from (3.21).

In order to apply this expression, it is necessary to determine r_0. Raimes (1952) has given an appropriate expression for computing this quantity from the appropriate ionization potential of the free atom. He found that this extremely simplified theory is capable of accounting for the experimental results of Bridgman on the compression of divalent metals as a function of pressure and could give, with somewhat less accuracy, the lattice constant and cohesive energy. He has also applied this treatment to aluminum (Raimes, 1953).

There are, of course, ample grounds for criticism of this very simple theory. It is thus very interesting that it has a significant degree of success.

3.3 Band Calculations in the Alkali Metals

A simple discussion of the general features of the band structure of the alkali metals has been given by Cohen and Heine (1958). There have been many detailed calculations of energy bands in these metals. Calculations prior to 1958 have been reviewed elsewhere (Callaway, 1958a), and will not be discussed here. The most complete set of calculations at present are

3.3 BAND CALCULATIONS IN THE ALKALI METALS

those of Ham (1960, 1962a,b) who has applied the quantum defect method in conjunction with calculational procedure of Kohn and Rostoker (Section 2.8). Rather careful calculations of energy bands in lithium have also been made by Callaway (1961c) and by Schlosser (1960) (who has also studied sodium in detail). These calculations were based on empirical potentials and employed the OPW and augmented plane wave methods respectively. The degree of numerical agreement between the OPW and the APW calculations in the case of lithium is impressive, and suggests that both methods, if carefully applied, will yield accurate solutions of the periodic potential problem.

The band structures of the alkali metals may be classified in terms of relative degree of departure from the free electron approximation. In that approximation, the Fermi surface is spherical, and the effective mass ratio is unity. All the alkali metals will depart from this idealization to some extent. The Fermi surface in the free electron approximation for an electron concentration of one per atom, would approach most closely to the Brillouin zone face for the body-centered cubic lattice along the 110 axis near the point N. (The surface would be approximately 7/8 of the distance from Γ to N.) However, symmetry considerations require that $\nabla_{\mathbf{k}} E$ vanish at N. Further, there is a degeneracy in the free electron approximation of states at the point N; and this degeneracy is necessarily removed when a periodic potential is present. The splitting is in first order equal to $2V(110)$ where $\dot{V}(110)$ is the (110) Fourier coefficient of potential. These considerations suggest that the most likely distortion of the Fermi surface from a spherical shape would be in the form of bulges along the 110 axes, and raise the interesting question whether the surface might be in contact with the zone face at N.[4] This contact would have important consequences for many experiments.

---

[4] A distortion of the Fermi surface of this sort requires important sixth order nonspherical terms in the expansion of $E(\mathbf{k})$. It cannot be described by the fourth order Kubic harmonic only. If the expansion is written as

$$E(\mathbf{k}) = E_0 + E_2 k^2 + k^4 (E_4^{(1)} + E_4^{(2)} K_4) + k^6 (E_6^{(1)} + E_6^{(2)} K_4 + E_6^{(3)} K_6) + \ldots$$

where K_4, K_6 are Kubic harmonics (normalized to 4π), the coefficients found by a least squares fit to the band calculation of Callaway (1961c) concerning lithium are $E_2 = 0.748$, $E_4^{(1)} = 0.105$, $E_4^{(2)} = -0.020$, $E_6^{(1)} = -0.545$, $E_6^{(2)} = 0.128$, and $E_6^{(3)} = 0.175$. The large size of the sixth order terms compared to those of fourth order should be noted.

A convenient indication of the distortion of the Fermi surface from a spherical shape is furnished by the splitting of the lowest states at N: (these are the states N_1 and $N_1{}'$ in the notation of Table VI; N_1 may be considered as approximately s-like although d functions may also be included; $N_1{}'$ is p-like). Unfortunately, the question of contact cannot be decided from knowledge of this parameter alone. In Table XVI, the calculated splitting at N is given according to the work of Callaway and Ham.

TABLE XVI

Energy Gap at Zone Face for Alkali Metals[a]

	Li	Na	K	Rb	Cs
Ham (1962a,b)	+0.209 (0.557)	+0.018 (0.057)	−0.037 (0.176)	−0.063 (0.324)	−0.088 (0.5!
Callaway (1958a) (1961c)	+0.233 (0.598)	+0.053 (0.172)	−0.034 (0.152)		+0.057 (0.2(

[a] The energy gap, $Eg = E(N_1) - E(N_1{}')$ (in rydbergs) at the zone face is given for all the alkali metals at the equilibrium lattice constant. The fractional gap, defined as $E_g\{\tfrac{1}{2}[E(N_1) + E(N_1{}')] - E(\Gamma_1)\}^{-1}$ is given in the parentheses.

It is evident that the Fermi surfaces in lithium and cesium are the most distorted; sodium and potassium the least. The question of contact has not yet been definitely settled as it is first necessary to compute the Fermi energy (at absolute zero); this in turn requires knowledge of the density of states. Estimates by several authors for the case of lithium: Ham (1962a,b), Schlosser (1960), Cornwell (1961a), Callaway (1961c), predict a close approach of the Fermi surface to the Brillouin zone, but not contact. The calculations are sufficiently precise (within of course, the framework of the Wigner-Seitz approximation discussed earlier) so that the prediction of no contact is unambiguous. Moreover, a more complete inclusion of electron interaction effects (the calculations mentioned are all based on the one-electron picture) will alter the predictions significantly only if the correlation effects are anisotropic or strongly k-dependent. Probably one would have to go beyond the free electron gas treatment of the correlation energy to change the one-electron results in this respect. The calculated band structure of lithium is shown in Fig. 9.[5]

Some of the experimental evidence concerning the shape of the Fermi surface has been summarized by Cohen and Heine (1958) and is in support of the general trend as discussed above. At the time of their review

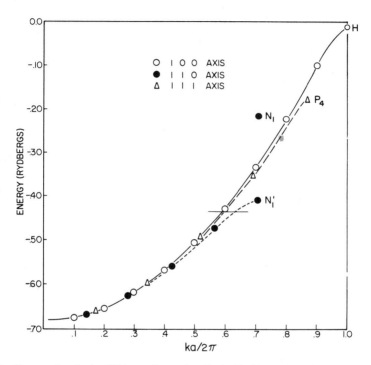

FIG. 9. Energy bands in lithium along the (1, 0, 0), (1, 1, 0) and (1, 1, 0) axes in the Brillouin zone. The horizontal line at -0.433 rydberg represents the Fermi energy. Only a relatively small distortion of the Fermi surface is predicted.

conclusive experiments had not been performed. Recently the De Haas-Van Alphen effect has been observed in potassium and rubidium (Thorsen

---

[5] Experimental evidence supporting the assumption of contact between the Fermi Surface and the Brillouin Zone in lithium comes from measurements of the soft x-ray emission spectrum (Crisp and Williams, 1960). This has the shape that would be expected if contact existed.

and Berlincourt, 1961a,b), and cyclotron resonance has been observed in sodium and potassium (Grimes, 1962). The Fermi surfaces in potassium and rubidium do not seem to be significantly distorted (in sodium, single crystals cannot be obtained at low temperatures, owing to the martensitic phase transformation, so that anisotropy measurements have not been made) The effective masses on the Fermi surface in sodium and potassium are 1.24 ± 0.02 and 1.21 ± 0.02, respectively, from the cyclotron resonance measurements.

The states at symmetry points closest to the Fermi surface in lithium are p-like whereas in potassium, and in rubidium and cesium (according to Ham's calculation) the states are of mixed s and d character. The increasing importance of d states in the heavier alkali metals is naturally to be expected from their position in the periodic table.

Information concerning certain average properties of the Fermi surface can be obtained from thermal and optical measurements. In particular, low temperature specific heat measurements, which reveal the electronic contribution, make possible a determination of a "thermal" effective mass simply related to the area of the Fermi surface.

Luttinger and Ward (1960) have shown that the specific heat of a system of interacting electrons at low temperatures in the presence of a periodic potential is given by

$$C_v = \frac{\pi^2}{3} K^2 T \sum_{\mathbf{k}, \varrho} \delta(\mu - E_{\mathbf{k}\varrho}) \tag{3.22}$$

in which ρ is the band index, K is Boltzmann's constant, and μ is the chemical potential (or Fermi energy). The summation can be replaced by an integral

$$\sum_{\mathbf{k}} \rightarrow \frac{2}{(2\pi)^3} \int d^3k \,.$$

The procedures of Section 1.8 enable us to write (per unit volume of material)

$$C_v = \gamma\, T, \quad \text{where} \quad \gamma = \frac{\pi^2}{3} K^2 n(\mu) \tag{3.23}$$

3.3 BAND CALCULATIONS IN THE ALKALI METALS

$n(\mu)$ is the density of states at the Fermi surface, which is given by

$$n(\mu) = \frac{2}{(2\pi)^3} \int \frac{dS_F}{|V_k E_k|} \qquad (3.24a)$$

Evidently a measurement of the electron specific heat determines the density of states on the Fermi surface. If the energy bands were parabolic, characterized by an effective mass ratio m^*, we would have (in atomic units):

$$n(\mu) = m^* k_F / 2\pi^2, \quad \text{or} \quad m^* = 2\pi^2 n(\mu)/k_F \qquad (3.24b)$$

where k_F is the radius of the (spherical) Fermi surface. It is convenient to *define* a "thermal effective mass, m_t" through (3.24b) for arbitrary band structures (k_F still being given by the usual expression for free electrons). Then (in atomic units)

$$m_t = 6\gamma / k_F K^2 \qquad (3.25)$$

To obtain a calculated m_t to compare with the experimental value deduced from (3.25) requires a detailed calculation of the Fermi surface, which has been performed by Ham (1962b). A comparison of theoretical and experimental values for m_t is presented in Table XVII.

For a spherical Fermi surface, the thermal effective mass should agree with that determined from cyclotron resonance.

TABLE XVII

Thermal Effective Masses for the Alkali Metals

	Li	Na	K	Rb	Cs
m_t (exp)	2.19[b]	1.27[c]	1.25[a]	1.25[a]	1.47[a]
m_t (theor)	1.64[d]	1.00[d]	1.07[d]	1.18[d]	1.75[d]
	1.62[e]				

[a] Lien and Phillips (1960). See also Ham (1962b).
[b] Martin (1961a).
[c] Martin (1961b).
[d] Ham (1962b).
[e] Calculated using the band parameters of Callaway (1961c). The sixth order terms in $E(\mathbf{k})$, including those of Kubic, rather than spherical, symmetry, have been included.

Except in the case of cesium, the theoretical values of m_t are consistently too low. There are two effects which could account for this: (a) warping of the Fermi surface: if there are patches of low velocity regions on the Fermi surface the effective mass would be enhanced, or (b) the discrepancy might be the result of the neglect of electron-electron and electron-phonon interactions. The correction to the effective mass due to electron-electron interactions effects can be calculated exactly only for a free-electron gas in the limit of very high densities (Gell-Mann and Bruckner, 1957). In that limit, the effective mass is reduced. Estimates have been made (DuBois, 1959) which suggest that the correction changes sign so that the effective mass is increased at higher densities. An additional increase in the effective mass do to electron-phonon interactions has been predicted by Quinn (1960).

There have been some calculations of energy levels in the alkali metals which are considerably above the Fermi level (see Callaway, 1958a; Ham, 1962a). This work is rather remote, at present, from possible experimental investigations. There is one feature of interest, however, which has significance for the discussion of the effective crystal potential.

We have seen in Chapter I that it is possible to make a rough classification of states at symmetry points of the Brillouin zone as s-like, p-like, etc. It is interesting to consider the relative order of levels of a given type. It turns out that this is generally in accord with the order of the corresponding levels in a free-electron system (the empty lattice): If we expand the wave function for the state concerned in plane waves, the relative order of states of a given type is rather independent of the assumed crystal potential, and is determined primarily by the kinetic energy of the lowest plane wave appearing in the expansion. Thus, in the body-centered cubic lattice, if we consider the principal symmetry points Γ, H, P, and N, (see Fig. 1) the s-like levels are arranged in order of increasing energy as Γ_1, N_1, P_1, H_1: the P levels are arranged as N_1', P_4, H_{15}, (N_3', N_4'), Γ_{15}. (The states N_3' and N_4' are degenerate in the empty lattice.) This ordering of the levels may be considered as a normal level order. Exceptions to it may generally be explained as due to admixtures of significant amounts of differing angular momentum states. The normal level order is given in Table XVIII for s, p, d, and some f levels in the body-centered and face-centered cubic lattices. The reader should be aware, however, that there is some ambiguity in this classification of levels due to this admixture.

TABLE XVIII

NORMAL LEVEL ORDER FOR ENERGY BANDS[a]

I. s band	bcc		fcc		
	Γ_1		Γ_1		
	N_1		L_1		
	P_1		X_1		
	H_1		K_1		
			W_1		
II. p band	bcc		fcc		
	N_1'		L_2'		
	P_4		X_4'		
	H_{15}		K_3		
	N_3'	degenerate in	(W_2', W_3)	degenerate in	
	N_4'	empty lattice	W_5'	empty lattice	
	Γ_{15}		L_3		
			Γ_{15}		
			K_4		
III. d band	bcc		fcc		
	H_{12}		X_3		
	N_2		L_3		
	Γ_{25}'	degenerate in	Γ_{25}'		
	Γ_{12}	empty lattice	K_2		
	N_4		Γ_{12}		
	P_3		(X_2, X_5)	degenerate in	
	H_{25}'		W_1'	empty lattice	
	N_3				
IV. f band	Γ_{25}		Γ_2'		
	P_5		X_3'		
	H_2'		W_2		
	N_2'		X_2'		
	H_{25}		Γ_{25}		
	Γ_2'		L_1'		

[a] Levels are listed in order of increasing energy

Levels are assigned a type on the basis of the lowest angular momentum state in the decomposition of the wave function in spherical harmonics.

In contrast to the relatively invariant ordering of levels of a given type, the position of levels of different types does depend critically on the detailed crystal potential, and varies in a systematic way from element to element. This fact may be easily explained in terms of the pseudopotential concept previously discussed (Sections 2.5 and 2.15), provided that a different pseudopotential is considered for each angular momentum.

3.4 Wave Functions for the Alkali Metals

Wave functions of the valence electrons in the alkali metals have been studied by several authors (Callaway and Morgan, 1958; Callaway, 1958c, 1960a, 1961b; Callaway and Kohn, 1962; Kohn, 1954; Kjeldaas and Kohn, 1956; Brooks and Ham, 1958; Milford and Gager, 1961). Much attention has been given to the problem of calculating the average over the Fermi surface of the square of the wave function evaluated at the nucleus (conveniently denoted as P_F). This quantity is important in the simple theory of the Knight Shift (see Knight, 1956): The position of the nuclear magnetic resonance line in the metallic state is shifted with respect to nonmetallic compounds by an amount proportional (in the simple theory) to P_F. Let $\Delta H/H$ be the fractional shift in the applied field for which resonance occurs. It can be shown that

$$\frac{\Delta H}{H} = \frac{8\pi}{3} X_p \Omega P_F \qquad (3.26)$$

where

$$P_F = [4\pi^3 n(\mu)]^{-1} \int |\psi(\mathbf{k}, 0)|^2 (|\nabla_\mathbf{k} E|)^{-1} dS_F \qquad (3.27)$$

In these equations, X_p is the spin paramagnetic susceptibility (per unit volume), Ω is the atomic volume in which ψ is normalized, $n(\mu)$ is the density of states on the Fermi surface, and the integral goes over the Fermi surface. This formula is derived under the assumption that electrons in closed shells do not contribute to the effective magnetic field at a nucleus. This assumption will be discussed more fully below.

The spin paramagnetic susceptibility is known from experiment for lithium and sodium; for the heavier alkali metals, it can be obtained only from the theoretical estimates such as those of Pines (1955). When

3.4 WAVE FUNCTIONS FOR THE ALKALI METALS

TABLE XIX
Knight Shift Results for the Alkali Metals

	Li		Na		K		Rb		Cs	
P_F (theor)	0.110^a	0.110^b	0.555^c	0.664^b	0.862^e	0.91^k	2.16^j	1.76^b	2.89^f	2.47^b
P_A (theor)	0.223^a	0.242^l	0.685^c	0.840^l	0.743^e	1.16^l	1.96^j	2.18^l	2.6^g	2.97^l
ξ (theor)	0.49^a	0.455^b	0.81^c	0.790^b	1.16^e	0.78^k	1.10^j	0.81^b	1.1^f	0.832^b
P_F (exp)	0.10 ± 0.005^d		0.53 ± 0.05^d		$0.95^{k,h}$	$1.035^{n,h}$	$2.32^{d,h}$		$4.39^{d,h}$	
P_A (exp)	0.231^i		0.751^i		$1.11^{i,n}$		2.34^i		3.88^i	
ξ (exp)	0.442 ± 0.015^m		0.705 ± 0.07^d		$0.86^{k,h}$	0.93^n	$0.993^{d,h}$		$1.13^{d,h}$	

Notes to Table

a W. Kohn, *Phys. Rev.* **96**, 590 (1954).
b H. Brooks (unpublished), quoted by G. B. Benedek and T. Kushida, *J. Phys. Chem. Solids* **5**, 241 (1958).
c T. Kjeldaas and W. Kohn, *Phys. Rev.* **101**, 66 (1956).
d G. B. Benedek and T. Kushida, *J. Phys. Solids* **5**, 241 (1958).
e J. Callaway, *Phys. Rev.* **119**, 1012 (1960a).
f J. Callaway, *Phys. Rev.* **112**, 1061 (1958c).
g R. M. Sternheimer, Private communication.
h Deduced from experiment using theoretical values of the paramagnetic susceptibility according to the work of D. Pines, *Solid State Phys.* **1**, 367 (1955).
i Determined from hyperfine structure measurements listed by W. D. Knight, *Solid State Phys.* **2**, 93 (1956).
j J. Callaway and D. F. Morgan, *Phys. Rev.* **112**, 334 (1958c).
k E. J. Milford and W. B. Gager, *Phys. Rev.* **121**, 716 (1961).
l H. Brooks and F. S. Ham, *Phys. Rev.* **112**, 344 (1958).
m C. H. Ryter, *Phys. Rev. Letters* **5**, 10 (1960).
n W. H. Jones, T. P. Graham, and R. G. Barnes, *Acta Metallurgica* **8**, 663 (1960).

calculations of P_F are made using the quantum defect method or the empirical potentials of Seitz and Prokofjew, it is possible to make a reasonably accurate prediction of P_F. In calculations using potentials obtained from self-consistent fields for free atoms, better agreement with experiment is obtained if one compares theoretical and experimental values of the ratio $\xi = P_F/P_A$, where P_A is the square of the magnitude of the wave function of the valence electron in the free atom evaluated at the nucleus. An experimental value of P_A may be obtained from measurements of the hyperfine splitting (ΔW) in energy units of the valence electron level $^2S_{1/2}$ the free atom through the Fermi formula (Fermi, 1930):

$$\Delta W = \frac{8\pi}{3} \frac{2I+1}{I} \mu \mu_N \mu_B P_A \qquad (3.28)$$

where $P_A = |\psi_A(0)|^2$. In this equation, μ is the nuclear magnetic moment in units of the nuclear magneton, μ_N is the nuclear magneton, μ_B is the Bohr magneton, and I is the nuclear spin. The reason for emphasizing the comparison of ξ with experiment is that many of the approximations of self-consistent field calculations (neglect of correlation and relativistic effects) will be nearly the same for P_F and P_A and will tend to cancel when the ratio is computed. Theoretical and experimental values for P_F, P_A, and ξ are compared with experiment in Table XIX.

The evident disagreement between the values of P_A as determined from experimental measurements of hyperfine structure according to (3.28) and those obtained from the quantum defect method is in large part due to the inadequacy of the Fermi formula (3.28). Several corrections have been derived, which, when applied to (3.28) give

$$\Delta W = \frac{8\pi}{3} \frac{2I+1}{I} \mu \mu_N \mu_B P_A F_r(j,Z)(1-\delta)(1-\varepsilon) \qquad (3.29)$$

The most important correction is the function

$$F_r(j,Z) = \frac{4j(j+\tfrac{1}{2})(j+1)}{\rho(4\rho^2-1)^{1/2}} \ ; \qquad \rho = [(j+\tfrac{1}{2})^2 - \alpha^2 Z^2]^{1/2} \qquad (3.30)$$

derived by Breit (1930) and Racah (1931) to include relativistic effects in a coulomb field. In Eq. (3.30), Z is the nuclear charge, j the total electronic angular momentum, and α is the fine structure constant $e^2/\hbar c$. The remaining factors of $(1-\delta)$ and $(1-\varepsilon)$ are included to an account of the finite size of the nucleus (Rosenthal and Breit, 1932; Crawford and Schawlow, 1949) and the distribution of the magnetic dipole moment over the nucleus (Bohr and Weisskopf, 1950). With these corrections, the experimental values of P_A become 0.231 (Li), 0.741 (Na), 1.07 (K),

2.06 (Rb) and 2.92 (Cs), in much better agreement with the calculated values. A similar correction evidently must be applied to the Knight shift formula (3.27).

In spite of the relatively good agreement between theory and experiment in many of the calculations summarized in the Table XIX, particularly in respect to the parameter ξ, there is an important, unanswered question: Since the atom has a net spin, the exchange interaction between the valence electron and the core electrons leads to a polarization of the closed shells. This effect causes the charge distribution of the electrons in closed shells, whose spin is parallel to that of the valence electron, to be different from that of the electrons of opposite spin. Clearly the net spin density at the nucleus is altered by this interaction; and it is not evident whether the spin density will be increased or diminished by the interaction. This exchange polarization is different from the core polarization previously described, and can be described in the "Unrestricted Hartree Fock Method" mentioned in Section 2.12.

Exchange polarization effects are apparently of extreme importance for transition metals: measurements based on the Mössbauer effect have shown that the sign of the net spin density may even be changed from the expected value (Hanna *et al.*, 1960). The effect was discussed for the alkali metals by Cohen *et al.*, (1959). They found an additional contribution to P_F amounting to 25% in the case of Li and 5% for Na. The agreement between theory and experiment is lessened by these considerations. The relatively small correction for Na is a consequence of cancellation of opposite polarizations of the 2s and 1s electrons. Calculations of this effect have not been reported for K, Rb, or Cs. A substantial correction may be present in any or all of these cases.

3.5 Valence Crystals: Diamond, Germanium, and Silicon

The rapid development of semiconductor technology which followed the invention of the transistor gave great impetus to programs of research into the fundamental characteristics of the most important semiconductor materials: germanium and silicon. A very large amount of information has been obtained about the band structure of these materials from both experimental and theoretical sources. Particularly detailed experimental information is available concerning the band structure in the immediate vicinity of the maximum of the valence band and the lowest minimum of the conduction band. In general, the success of band theory in correlating the results of numerous experiments concerning semiconductors is perhaps the most striking verification of applicability of the general principles and language of band theory. Attempts at quantitative prediction of the basic parameters have not been so successful.

There have been many calculations of energy bands in these materials: [Diamond: Kimball (1935); Hund and Mrowka (1935a); Morita (1949a, 1958); Herman (1952); Herman (1954c); Hall (1952); Schmid (1953); Zehler (1953); Slater and Koster (1954); Kleinman and Phillips (1959, 1962); Bassani and Celli (1961); Nran'yan (1961). Silicon: Mullaney (1944); Holmes (1952); Yamaka and Sugita (1953); Bell et al. (1954); Jenkins (1956), Kane (1956a); Woodruff (1956); Bassani (1957), J. C. Phillips (1958); Kleinman and Phillips (1960b); Quelle (1962); Brust, Cohen, and Phillips (1962); Kleinman (1962); Bassani and Brust (1963). Germanium: Herman and Callaway (1953); Herman (1954b, c); Segall (1958), Gashimzade and Khartsiev (1961), Liu (1962), Brust, Phillips, and Bassani (1962); Bassani and Yoshimine (1963).][6]

Before the detailed discussion of the band structure is undertaken, it is desirable to consider some of the qualitative aspects. Germanium and silicon have the same lattice structure as diamond. The diamond lattice is composed of two interpenetrating face-centered cubic lattices: the structure is face-centered cubic but with two atoms in each unit cell: If a is the lattice constant of one of the face-centered cubic lattices, atoms are placed not only at the sites of this lattice $\frac{1}{2}a(l\hat{i} + m\hat{j} + n\hat{k})$ where l, m, and n are integers whose sum is an even integer, but also at points displaced from the first set by $\frac{1}{4}a(\hat{i} + \hat{j} + \hat{k})$. The theory of the symmetry properties of wave functions in the diamond lattice is necessarily more complicated than that for the ordinary face-centered cubic structure because the space group is not simple: it contains a screw motion. The

---

[6] Some of the calculations listed above concern more than one of the crystals mentioned. In cases in which two or more papers concern the same calculation, the more detailed account is included here.

3.5 VALENCE CRYSTALS: DIAMOND, GERMANIUM, AND SILICON

Brillouin zone is the same as for the ordinary face-centered cubic structure. The representations at symmetry points have been worked out by Herring (1942) and for the double group by Elliott (1954).

It is simplest to consider the wave functions in the language of the tight binding approximation: We consider Bloch functions formed from the s and p_x, p_y, and p_z orbitals belonging to atoms on each of the lattices. One then obtains an 8×8 secular equation. It is found that the solutions fall into two groups of four bands each; the lowest group or valence band, capable of holding four electrons from each atom, is separated by a gap from the higher group, or conduction band. The wave functions may be roughly characterized by saying that the valence band is formed from bonding combinations of the orbitals on the two sublattices, the conduction band from antibonding combinations of functions. This description is exact in the tight binding approximation only at $\mathbf{k} = 0$, however, where the solutions are: (1) a combination of s orbitals symmetric about the midpoint of a line joining two atoms (state Γ_1, bottom of the valence band); (2) a similar symmetric combination of p orbitals (this state (Γ_{25}') is triply degenerate (spin degeneracy is not included) since there are three p orbitals from each atom, and is at the top of the valence band); (3) a triply degenerate state (Γ_{15}) which is an antisymmetric combination of p orbitals; and (4) an antisymmetric combination of s orbitals (Γ_2'). In diamond and silicon, Γ_{15} lies below Γ_2', but this order is reversed in germanium. When spin orbit coupling is included the states Γ_{25}' and Γ_{15} are split into a doubly degenerate and a nondegenerate level (fourfold and twofold degenerate if the spin degeneracy is included); the nondegenerate level is lowest. Although the valence band maximum is located at $\mathbf{k} = 0$, the conduction band minimum is elsewhere: along the 100 axis (but not at the square face center X, since the bands need not have zero slope there) in diamond and silicon; at the hexagonal face center (L) in germanium.

Away from the center of the zone, the wave function in the tight binding approximation mixes s and p functions on the different sublattices It is characteristic of lattices with more than one atom in the unit cell that a wave function may possess different symmetries about each of the atoms. On the 100 axis (direction k_x) the wave function of Δ_1 symmetry combines the s and p_x functions (separately) on the two atoms in a symmetric way, whereas for Δ_2', the same functions are combined in an antisymmetric manner. The ratio of the coefficients of the s and p combinations is, of course, a function of **k** and has to be determined by solving the band problem. The other band that is formed on this axis is $\varDelta_5$, which is doubly degenerate. This function contains a combination of the p_y and p_z orbitals ($p_y + p_z$ or $p_y - p_z$) mixed with a similar function obtained from its neighbor in the unit cell. At the face center X(100), all wave functions are doubly degenerate. In X_1, an s function on one atom is combined with a p_x function on another. The $\varDelta_5$ band goes into X_4. At the point L the wave function for the state L_1 is composed of a combination of s functions with p functions having the symmetry $x + y + z$ which is symmetric between the two atoms. In the case of L_2', the combination is similar but antisymmetric. There are also the doubly degenerate states L_3 and L_3' for which the wave function is (respectively) a symmetric or antisymmetric combination of functions on the two atoms having symmetries $x - y$ and $z - \frac{1}{2}(x + y)$.

To obtain quantitative results for the energy band structure, even in an interpolation scheme, it is necessary to use procedures other then the tight binding approximation. J. C. Phillips (1958) observed that the energy bands may be characterized readily in terms of small number of parameters if a plane wave approach based on a pseudopotential is used (see Section 2.5). The disposable parameters are then the Fourier coefficients of the pseudopotential.

Before proceding further, it is desirable to consider the available experimental information concerning the band structures We will discuss silicon and germanium in detail. The information has been summarized and interpreted by Phillips (1962). Much less is known concerning diamond, but the energy bands in this material are believed to resemble those of silicon.

The lowest minimum in the conduction band in silicon lies on the 100 axis, approximately 0.85 ($\pm$ 0.03) (Feher, 1959) of the distance from the center of the zone to the square face center X. The surfaces of constant energy in the conduction band are a set of six (one on each axis) ellipsoids of revolution. These ellipsoids are characterized by a transverse and a longitudinal effective mass: $m_t/m_0 = 0.192 \pm 0.001$; $m_1/m_0 = 0.90 \pm 0.02$ (Rauch et al., 1960). The surfaces are thus somewhat cigar shaped. The energy gap between the valence band maximum at **k** = 0 and these (100) conduction band minima is about 1.17 ev at absolute zero (Macfarlane

3.5 VALENCE CRYSTALS: DIAMOND, GERMANIUM, AND SILICON

et al., 1958). The (vertical) band gap at $\mathbf{k} = 0$ is not known with great accuracy: it seems to be about 3.5 ev.

The description of the valence band is somewhat more complex. If we neglect spin orbit coupling, the valence bands at $\mathbf{k} = 0$ are determined by second order degenerate perturbation theory (see Section 1.7) as solutions of the secular equation:

$$\begin{vmatrix} Lk_x^2 + M(k_y^2 + k_z^2) - E & Nk_x k_y & Nk_x k_z \\ Nk_x k_y & Lk_y^2 + M(k_x^2 + k_z^2) - E & Nk_y k_z \\ Nk_x k_z & Nk_y k_z & Lk_z^2 + M(k_x^2 + k_y^2) - E \end{vmatrix} = 0 \quad (3.31)$$

The quantities L, M, and N are sums of squares of certain matrix elements (these will be given below). If spin orbit coupling is taken into account, the $E(\mathbf{k})$ expressions become (sufficiently close to $k = 0$):

$$E_{1,2}(\mathbf{k}) = A\mathbf{k}^2 \pm [B^2 \mathbf{k}^4 + C^2(k_x^2 k_z^2 + k_y^2 k_z^2 + k_x^2 k_z^2)]^{1/2} \quad (3.32)$$

$$E_3(\mathbf{k}) = -\Delta + A\mathbf{k}^2$$

Here Δ is the spin orbit splitting (about 0.04 ev). The cyclotron resonance measurements give (Hensel and Feher, 1963; Phillips, 1962)

$$A = -4.28 \pm 0.02, \quad B = -0.75 \pm 0.05, \quad |C| = 5.25 \pm 0.08$$

The constants A, B, and C of Eq. (3.32) are related to the parameters L, M, and N of (3.31) by the equations (Dresselhaus, 1954; Dresselhaus *et al.*, 1955).

$$A = 1 + (L + 2M)/3$$
$$B^2 = [(L - M)/3]^2 \quad (3.33)$$
$$C^2 = \frac{1}{3}[N^2 - (L - M)^2]$$

The relation between L, M, and N and the basic matrix elements is as follows: Let $\langle \Gamma_{25'}^{(1)}(0)|$ stand for a state in Γ_{25}' at the top of the valence band (basis functions are given by Dresselhaus, 1954). Define

$$F = \frac{\hbar^2}{m^2} \sum_l \frac{|\langle \Gamma_{25'}^{(1)}(0)|p_x|\Gamma_2'(l)\rangle|^2}{E(0) - E(l)} \quad (3.34a)$$

The sum runs over all states belonging to Γ_2', these states being characterized by the index l; $E(0)$ is the energy of $\Gamma_{25}'(0)$, etc. Similarly, define

$$G = \frac{\hbar^2}{m^2} \sum_l \frac{|\langle \Gamma_{25'}^{(1)}(0)|p_x|\Gamma_{12}(l)\rangle|^2}{E(0) - E(l)} \tag{3.34b}$$

$$H_1 = \frac{\hbar^2}{m^2} \sum_l \frac{|\langle \Gamma_{25'}^{(1)}(0)|p_x|\Gamma_{15}(l)\rangle|^2}{E(0) - E(l)} \tag{3.34c}$$

$$H_2 = \frac{\hbar^2}{m^2} \sum_l \frac{|\langle \Gamma_{25'}^{(1)}(0)|p_x|\Gamma_{25}(l)\rangle|^2}{E(0) - E(l)} \tag{3.34d}$$

The relations between the quantities L, M, and N and the sums of matrix elements of (3.34) is:

$$\begin{aligned} L &= F + 2G \\ M &= H_1 + H_2 \\ N &= F - G + H_1 - H_2 \end{aligned} \tag{3.35}$$

Evidently, it is not possible to determine F, G, H_1, and H_2 unambiguously from the experimental data. One might try to neglect H_2 on account of the supposed remoteness of the nearest Γ_{25} level. Then one deduces (Phillips, 1962) $F = -5.5$, $G = -0.7$, $H_1 = -4.5$.

The valence band in silicon can be crudely described, with regard to gross structure, as composed of the superposition of a wide band formed from combinations of s and p states, capable of holding two electrons per atom and a more narrow band based on the remaining p functions. (This description is exact in the tight approximation along the 100 and 111 axes.) The over-all width of the valence band has been estimated as 16.7 ev from x-ray measurements of Tomboulian and Bedo (1956), and as in the range from 13 ev to 16 ev by Kern (1960). The lack of precision of x-ray measurements in these respects is notorious. The width of narrower p subband has been estimated at 5.1 ev by Hagstrum (1961) from measurements of the energy distribution of electrons emitted in the Auger-type electron transitions which occur when an ion is neutralized at a solid surface. He also finds that the over-all width of the valence bands is in the range of 14 to 16 ev.

3.5 VALENCE CRYSTALS: DIAMOND, GERMANIUM, AND SILICON

The band structure of germanium is similar in many qualitative respects to that of silicon. The lowest minimum of the conduction band occurs, however, at the extremities of the 111 axes (hexagonal face centers) L (representation L_1). Since these points are equivalent in pairs, the energy surfaces of electrons consist of four ellipsoids of revolution along the 111 axes. These ellipsoids are characterized by effective mass ratios: $m_l/m_0 = 1.588 \pm 0.005$, $m_t/m_0 = 0.08152 \pm 0.00008$ (Levinger and Frankl, 1961). The L_1 minima are located 0.744 ± 0.001 ev above the Γ_{25}' state. The lowest energy level in the conduction band at $\mathbf{k} = 0$ has Γ_2' symmetry and lies 0.897 ± 0.001 ev. above Γ_{25}' at 4°K. The constant energy surfaces near this minimum are spherical, characterized by an effective mass $m_s/m_0 = 0.041 \pm 0.002$ (Zwerdling et al., 1957). The small effective mass near this minimum is due to the proximity of the Γ_{25}' level and, conversely, implies the existence of light holes in the valence band. In contrast, the vertical energy separation between the conduction and valence bands at L is larger (about 2.2 ev). Evidence has also been obtained from measurements of the pressure dependence of the resistivity, and from optical measurements for the existence of a third set of minima in the conduction band lying 0.15 to 0.20 ev above the L_1 minima, along the 100 axis (Paul, 1959). These minima are analogous to those found in silicon, probably with a similar ratio of $m_l/m_t \approx 5$ (Glicksman and Christian, 1956).

The valence band in germanium can be characterized in a fashion similar to that in silicon. The parameters involved are, according to Levinger and Frankl (1961; see also Stickler, 1962),[7]

$$A = -13.27 \pm 0.025; \quad |B| = 8.63 \pm 0.12; \quad |C| = 12.4 \pm 0.25$$

If one neglects the matrix element sums of the type H_2, one obtains
$F = -29.2, \quad G = -1.2, \quad H_1 = -5.6$

The relative magnitudes of F and H_1 are consistent with the Γ_2' state being lowest in the conduction band at the origin. The spin orbit splitting of the valence band is substantially larger than in silicon: about

---

[7] Neither in the case of germanium or silicon is there complete agreement concerning the valence band parameters. For example, Stickler, Zeiger, and Heller (1962) give $A = 13.2$, $|B| = 8.2$ and $|C| = 13.3$ for germanium; and $A = 4.22$, $|B| = 1.0$, and $|C| = 4.34$ for silicon.

0.29 ev (Kahn, 1955). The over-all width of the valence band is given as 7.0 ev by Tomboulian and Bedo (1956); however, Hagstrum (1961) obtains a width in the range 14–16 ev, as in silicon, with the upper p bands occupying 4.3 ev.

Alloys of germanium and silicon have band structures intermediate between those of the elements composing them (Johnson and Christian, 1954; Herman, 1954a; Dresselhaus et al., 1955b). Conduction occurs by migration of electrons in the 111 (L_1) minima of the conduction band

TABLE XX

ENERGY LEVEL SEPARATIONS IN GERMANIUM AND SILICON[a]

	$L_3' - L_1$	$L_3' - L_3$	$\Gamma_{25}' - \Gamma_{15}$	$X_4 - X_1$
Silicon	3.1	5.4	3.4	4.3
Germanium	2.1	5.7	3.2	4.3

[a] All energies are in electron volts.

in germanium. Although the 100 minima exist, they are too high above the 111 minima to be populated. The addition of silicon can be considered, in the language of Phillips' pseudopotential theory and the virtual crystal approximation, to introduce a weak, essentially repulsive potential. This perturbation causes the bands to rise, and the band gap increases. In particular, s states rise faster than p states as the concentration of silicon is increased, so that Γ_2' crosses Γ_{15}. The 111 minima rise faster than the 100 minima. In the range of composition in which 8 to 20% silicon is present, conduction takes place as a result of carriers present in both the 100 and 111 minima. For higher concentrations (20 to 100% Si), only the 100 minima are populated. Thus, there is a continuous transition between the energy bands in germanium and those in silicon. It is no doubt quite significant for the theory of alloys that these simple ideas seem to be successful.

It has recently become possible to determine the approximate positions of levels at symmetry points in the Brillouin zone which are a few electron

3.5 VALENCE CRYSTALS: DIAMOND, GERMANIUM, AND SILICON

volts away from the lowest band edges. This is accomplished through the association of peaks observed in the optical absorption of the semiconductor in the visible and ultraviolet spectral regions with interband transitions between these states. The relevant measurements are

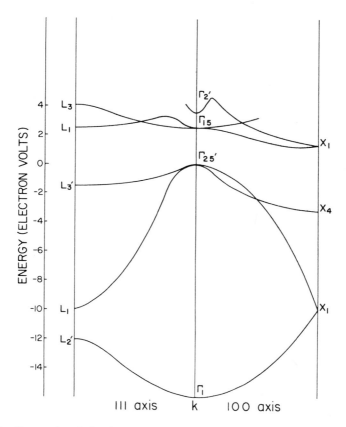

FIG. 10. Energy bands in silicon are shown along the (1, 0, 0) and (1, 1, 1) axes. Spin orbit coupling is neglected.

those of Philipp and Taft (1959, 1960), Tauc and Abraham (1961), and Cardona and Sommers (1961) relating to germanium, silicon, and germanium-silicon alloys. Similar information can be obtained from studies of the photoemission of electrons (Spicer and Simon, 1962).

The observation of spin orbit splitting of some of the peaks has been particularly helpful in identifying the transitions. The energy separation of some important levels in silicon and germanium is given in Table XX

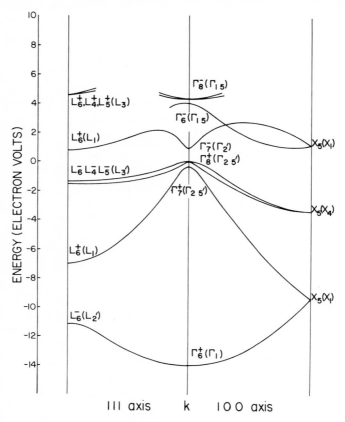

FIG. 11. Energy bands in germanium are shown along the (1, 0, 0) and (1, 1, 1) axes with spin orbit coupling included. Representations are labeled in the notation appropriate to the double group (Elliott, 1954); the labels in parentheses indicate the single group designation which would apply if spin orbit coupling were neglected.

according to Ehrenreich, Philipp, and Phillips (1962), Brust, Phillips and Bassani (1962), and Brust, Cohen, and Phillips (1962). The spin orbit splitting of the transitions at L is about 0.2 ev. Similar measurements have been made on some 3–5 and 2–6 semiconductors.

The energy band structures of silicon and germanium are shown in Figs. 10 and 11, as deduced from the available experimental and theoretical information.

3.6 Valence Crystals: Results of Calculations

The complicated shape of the Wigner-Seitz cell for the diamond lattice, and the necessity to consider functions with different symmetries about each of the two atoms in the unit cell, have favored the use of calculational procedures based on plane wave expansions for energy band calculations in these materials. Some studies have, however, employed the cellular method with modifications (Bell *et al.*, 1954; Jenkins, 1956), or the Kohn-Rostoker method (Segall, 1958).

In a calculation briefly mentioned in the previous section, Phillips (1958) [see also Bassani and Celli (1961); Brust, Cohen, and Phillips (1962); and Brust, Phillips, and Bassani (1962)] has attempted to use experimental information concerning the band structures of silicon and germanium to determine the parameters of an interpolation scheme and thus to characterize the band structure throughout the zone. The interpolation scheme is a plane wave calculation in which the Fourier coefficients of potential are regarded as disposable parameters.

The values of the Fourier coefficients chosen by Brust, Bassani and Phillips (1962) were (in Rydbergs)

$$V(1,1,1) = -0.21; \quad V(2,0,0) = 0.04,$$

$$V(3,1,1) = 0.08 \tag{3.36}$$

$$V(\mathbf{K}) = 0 \quad \text{for} \quad a^2 \mathbf{K}^2/4\pi^2 > 11$$

where $(1, 1, 1)$ stands for any of the eight reciprocal lattice vectors of the type $(2\pi/a)$ $(\pm 1, \pm 1, \pm 1)$ etc. These parameters are determined to match approximately the experimental values of the separation between valence and conduction bands at Γ, X, and L given in Table XX. The parameters given above yield values of 3.4, 4.0, and 3.1 ev. for these quantities. The smallest indirect gap is 0.9 e.v.

When the parameters have been determined, it is possible to compare other features of the computed band structure with experiment. In general, the agreement is reasonably good. Earlier work by Phillips (1958) with somewhat different parameters showed that the minimum of the conduction band occurs for $k_{min}/k_X = 0.88 \pm 0.02$; the effective mass ratios of electrons were found to be: $m_l/m_0 = 0.98$; $m_t/m_0 = 0.30$. The valence band parameters were $A = -3.9$; $|B| = 0.9$, $|C| = 3.6$; $H_1 = -5.8$, $F = -1.7$, and $G = -0.7$. The description of the band structure by the plane wave pseudopotential method is in much better agreement with experiment than is obtained with the use of the tight binding approximation (see Slater and Koster, 1954) and more satisfactory theoretically in that the number of disposable parameters required is much smaller.

Phillips, Bassani and Celli, and Brust, Phillips and Bassani have also applied this approach to germanium. The pseudopotential parameters given below are those of Bassani and Celli (1961). In this case different values of the average crystal potential $V(0)$ were included for s-like and p-like states (these are denoted by $V_s(0)$ and $V_p(0)$ respectively).

$$V_s(0) = -0.60, \quad V_p(0) = -0.64, \quad V(1,1,1) = -0.23, \quad V(2,2,0) = 0.00,$$
$$V(3,1,1) = 0.055 \tag{3.37}$$

The parameters were chosen to give reasonable agreement with the experimentally determined separations of the levels (see Table XX). Values found for these quantities were: $\Gamma_2' - \Gamma_{25}'$, 0.6 ev; $L_1 - L_3'$, 1.8 ev; $\Gamma_{15} - \Gamma_{25}'$, 3.6 ev; $X_4 - X_1$, 3.6 ev; $L_3 - L_3'$, 5.4 ev. In the earlier calculation of Phillips (1958), the effective mass ratios at L were computed to be $m_l/m_0 = 1.27$; $m_t/m_0 = 0.15$. The valence band parameters are also computed to be $A = -7.6$, $|B| = 5.0$, and $|C| = 8.4$. The agreement with experiment is reasonably good considering the small number of parameters employed. Bassani and Brust (1963) have extended the pseudopotential calculations by considering the band structure of Ge-Si alloys and the change in band structure under pressure.

The calculations which have attempted to determine the band structure form first principles have been extremely laborious, and quantitative success has been hard to obtain. We will consider first the work of Herman (see Herman and Callaway, 1953; Herman, 1954b; Herman and Skillman, 1960), who has employed the OPW method in a continuing effort to determine theoretically the band structure of germanium. Several important difficulties were encountered: most important perhaps is that concerned with the inclusion of the exchange interactions of the valence electrons. Problems of self-consistency also arise both in connection with the exchange interaction and in the determination of the average electrostatic potential. These questions are particularly serious since the quantitative characteristics of the band structure seem to be quite sensitive to the crystal potential. Herman has employed the average exchange potentials suggested by Slater (in his 1954c paper an arithmetic average of two of Slater's formulas was used).

Herman has also experienced difficulty with the application of the OPW method: it is essential, as was pointed out in Section 2.4 that the core functions be eigenfunctions of the same Hamiltonian that is used in the calculation of valence electron wave functions. In the early work (Herman, 1954c), the core wave functions were obtained by orthogonalizing Hartree (self-consistent field, without exchange) functions for the free germanium atom. The core energy levels were determined from experimental x-ray data. It is clear that this procedure is not consistent with the requirements of the OPW method. In more recent work (Herman and Skillman, 1960) the situation has been greatly improved by recalculation of the core functions. Unfortunately some ambiguities still remain which may be real limitations on the applicability of the OPW method: these include the question of the nonspherical (cubic) components of the crystal potential; and the determination of the average crystal potential $[V(0)]$. The relative positions of the calculated energy levels appears to be sensitive to the choice of $V(0)$. This sensitivity must result from an erroneous application of the OPW method: A change in the crystal potential by a constant amount also changes the energies of the core states by the same constant. Inspection of the OPW matrix elements as given in (2.52) and (2.53) shows that the energies of the valence states are changed by the same amount, so that the relations of the energy levels to each other are unchanged.

Herman's (1954c) calculation correctly predicted that the lowest conduction band level at the center of the zone should be Γ_2' (instead of Γ_{15} as previously obtained by Herman and Callaway, 1953), but the $(1, 1, 1)$ conduction band minima (L_1) was found to be more than 1 ev above Γ_2'.

In a series of papers, Kleinman and Phillips (1959, 1960a,b, 1962) Phillips and Kleinman (1962b) have studied the band structure of covalently bonded semiconductors. We shall be particularly concerned here with their calculation of energy bands in silicon. This work may be regarded as an extension of the semiempirical calculations previously described in which an attempt is made to calculate the effective repulsive potential from the OPW method rather than to determine coefficients from experimental data. The interpretation of the OPW method in terms of an effective repulsive potential has been discussed in Section 2.5. For the convenience of the reader, the formulas are reproduced here: The energy levels are obtained as the solutions of the equation

$$(-\nabla^2 + V + V_p)v_{ki} = E_{ki}v_{ki} \qquad (2.65)$$

in which v_{ki} transforms according to the ith irreducible representation of wave vector $\mathbf{k}$ and is a smooth function. V is the "real crystal potential," and V_p is given by

$$V_p = \sum_n a^i_{kn}(E_{kn} - E_{ki})\frac{\varphi_{kn}}{v_{ki}} \qquad (2.64)$$

in which φ_{kn} is the core wave function for state n, and energy E_n; and

$$a^i_{kn} = -\int \phi^*_{kn}(\mathbf{r})v_{ki}(\mathbf{r})\, d^3r \qquad (2.62)$$

The determination of the repulsive potential V_p requires knowledge of the final wave function v_{ki} and energies E_{ki}; and this requires a self-consistent calculation. Further, one must note that V_p will, in general, have a significant angular dependence. Kleinman and Phillips argue, however, that it is possible to make a sufficiently accurate estimate of the repulsive potential, and that the angular variation may be neglected. To do this, they assume that in the region where the valence function v overlaps the core functions, it may be represented as a simple hydrogenic function $C_s \exp(-\alpha_s r)$ for the s part and $C_p \exp(-\alpha_p r)$ for the p part. They calculate a repulsive potential for s and p states separately, and then assign a percentage of s and p repulsive potential to be used for each state based on an estimation of the s and p contribution to the diagonal matrix elements of the OPW method for symmetrized combinations of plane waves. The coulomb potential of the core was taken from a Hartree type

3.6 VALENCE CRYSTALS: RESULTS OF CALCULATIONS

(no exchange) calculation for Si^{4+} (McDougall, 1932). The core-valence exchange potential was obtained by extrapolating such a potential previously calculated for aluminium by Heine (1957c). The coulomb potential produced by the valence electrons was calculated from Poisson's equation under the assumption that the valence wave functions in the crystal could be represented as a superposition of the smooth functions above made orthogonal to core electron wave functions. The mutual exchange potential of the valence electrons was included through Slater's free electron approximation (modified to take approximate account of correlation effects: see Kleinman and Phillips, 1959). It was assumed in calculating these potentials that each atom has effectively $17/8\ p$ electrons and $15/8\ s$ electrons.

An attempt was made to obtain a reasonable degree of self-consistency in the calculation of the potential in the following manner: Equation (2.29), which relates the Fourier coefficients of the crystal potential to those of the charge density may be generalized to:

$$V(\mathbf{K}_n) = - \frac{8\pi}{\mathbf{K}_n^2 \tau_k} \int_{BZ} \rho(\mathbf{K}_n, \mathbf{k})\, d^3k \qquad (3.38)$$

in which $\rho(\mathbf{K}_n, \mathbf{k})$ is the amplitude of the $\mathbf{K}_n$th Fourier coefficient of the charge density for all electrons in the Brillouin zone with reduced wave vector $\mathbf{k}$, and the integral goes over the first Brillouin zone (volume τ_k). Kleinman and Phillips show that the integral in (3.38) may be approximated by a sum of contributions from the states Γ_1, Γ_{25}', L_1, L_2', L_3', X_1, and X_4 with the weights $\frac{1}{8}$, $\frac{3}{8}$, $\frac{1}{2}$, $\frac{1}{2}$, 1, $\frac{3}{4}$, and $\frac{3}{4}$, respectively (considering only a single point L and a single point X). The total valence coulomb potential was computed by writing the wave function in the smooth exponential form previously mentioned and adjusting the parameters until $V(1, 1, 1)$ computed from them agreed with $V(1, 1, 1)$ as computed from approximating the integral (3.39), using the leading symmetrized combination of plane waves to obtain $\rho(\mathbf{K}_n, \mathbf{k})$.

The procedure of Phillips and Kleinman has been given in detail here to enable the reader to appreciate the complexities of band calculations. The adequacy of the approximations employed is not self-evident, however, and a very large amount of work remains to be done before any particular procedure for the determination of the crystal potential can be regarded as established. Neither is it evident that the approximate form of the OPW method employed by Kleinman and Phillips is sufficiently accurate. These authors have given arguments, however, to indicate that the errors are small. Further, the fundamental questions of principle involved in the calculation of the exchange interactions of band electrons are still present. Phillips and Kleinman (1962) have considered the screening of the exchange interaction between band electrons through the introduction of a k-dependent dielectric constant.

The indirect energy gap obtained in this calculation was found to be about 1.5 ev. The valence band parameters A, $|B|$, and $|C|$ were computed to be -4.4, 0.84, and 4.1 in reasonable agreement with the experimental values; however, only the "smooth part," φ, of the wave functions were included in the calculation of the matrix elements. The effective mass ratios for electrons were also found to be $m_l^*/m_0 = 0.97$ and $m_t/m_0 = 0.205$, in good agreement with experiment. Kleinman (1962) has extended these calculations to determine the deformation potential in silicon.

We will not discuss the other calculations of the energy band structure from first principles in detail here. It suffices to point out that they have encountered difficulties similar in many respects to those of Herman and of Kleinman and Phillips; the results have generally not been in any better agreement with experiment.

There is, however, another sort of semiempirical study of energy bands which differs from that of Phillips in that emphasis is placed on determining accurately the form of the bands near $\mathbf{k} = 0$. This approach has been pursued by Kane (1956a,b, 1959a) and Dresselhaus (1954; Dresselhaus et al., 1955). The equation for the periodic part of the Bloch function, $u_\mathbf{k}$ which is defined by

$$\psi_\mathbf{k} = e^{i\mathbf{k}\cdot\mathbf{r}} u_\mathbf{k}$$

may be determined from (1.67) to be

$$[H_0 + H_1] u_\mathbf{k} = E_\mathbf{k} u_\mathbf{k} \tag{3.39}$$

where H_0 is the periodic, one-electron, Hamiltonian (relativistic effects other than spin orbit coupling are neglected in both H_0 and H_1)

$$H_0 = \frac{\mathbf{p}^2}{2m} + V(\mathbf{r}) + \frac{\hbar}{4m^2 c^2} \boldsymbol{\sigma} \cdot [\nabla V \times \mathbf{p}] \tag{3.40}$$

3.6 VALENCE CRYSTALS: RESULTS OF CALCULATIONS

and

$$H_1 = \frac{\hbar}{m} \mathbf{k} \cdot \mathbf{p} + \frac{\hbar^2}{4m^2 c^2} \boldsymbol{\sigma} \cdot [\nabla V \times \mathbf{k}] \tag{3.41}$$

Since the spin orbit coupling term in (3.40) is only appreciable near a nucleus where ∇V is large and the potential is nearly spherically symmetric, this term may be written in the more familiar atomic form

$$\frac{\hbar}{4m^2 c^2} \frac{1}{r} \frac{dV}{dr} \mathbf{L} \cdot \boldsymbol{\sigma}$$

In a study of the valence band structure in germanium and silicon, Kane (1956a) chose basis functions u_0 which were eigenfunctions of H_0, (including spin orbit coupling). The $\mathbf{k} \cdot \mathbf{p}$ term in (3.41) was treated as a perturbation, and the $\boldsymbol{\sigma} \cdot [\nabla V \times \mathbf{k}]$ term was neglected. On the basis of these functions, the energy bands near $\mathbf{k} = 0$ are given as the solution of a 6×6 secular equation which replaces (3.31), but contains only one additional parameter: the spin orbit splitting at $\mathbf{k} = 0$ (previously called Δ). (For details, see Kane 1956a.) The resulting energy bands differ from those given by Eq. (3.32) because of the more accurate inclusion of spin orbit coupling. Kane solved the secular equation for the 100, 110, and 111 directions, and found that the energy bands depart significantly from parabolic form. The deviations from parabolic behavior occur in a range in which the energy, measured from the top of the band, is of the order of the spin orbit splitting of the Γ_{25}' level. Departures of the bands from parabolic form are also caused by higher order mixing of valence and conduction band levels. These effects are, however, smaller than those produced by the spin orbit coupling.

Liu (1962) has calculated the spin orbit splitting of the energy bands in germanium and silicon in a more fundamental way. The matrix elements of the spin orbit coupling between the members of a degenerate set of states (such as Γ_{25}') were determined by representing the wave functions for these states as combinations of orthogonalized plane waves. The matrix elements of spin orbit coupling between orthogonalized plane waves are dominated by the core portion of the OPW's and this may be determined if the core functions and the orthogonality coefficients are known. Good agreement with experiment was obtained.

174 CHAPTER 3. BAND STRUCTURE OF MATERIALS

3.7 Zinc Blende Structures

The zinc blende structure is quite similar to the diamond lattice. The difference is that the two atoms in the unit cell are not identical. Band calculations for some materials with this structure have been reported by Kobayasi (1958), Birman (1958), Shakin and Birman (1958), Birman (1959a, b), Kleinman and Phillips (1960a), Gubanov and Nran'yan (1960), Nran'yan (1960, 1961), Bassani and Celli (1961), Gashimzade and Khartsiev (1961), and Bassani and Yoshimine (1963). We will consider only indium antimonide and gallium arsenide in detail. The experimental information relevant to the band structure of several 3-5 compounds has been reviewed by Ehrenreich (1961). In indium antimonide, both the lowest minimum of the conduction band and the maximum of the valence band occur at $\mathbf{k} = 0$. The energy gap at $\mathbf{k} = 0$ (0.2357 ev at 0°K; Zwerdling et al., 1961) is smaller than the spin orbit splitting of the valence band (probably 0.9 ev). The small energy gap means that the electron effective mass is quite small ($m^*/m_0 = 0.013$). Kane (1956b) has analyzed the band structure near $\mathbf{k} = 0$ in a calculation which is similar to his studies of germanium and silicon which were discussed in the previous section. In this case, it is important to treat the interaction of the lowest conduction band and highest valence bands exactly. This may be done by diagonalizing the matrix representing the perturbation $\mathbf{k} \cdot \mathbf{p}$ on the basis of these states, and including also the spin orbit coupling.

The band system in indium antimonide and many other semiconductors of the zinc blende structure resembles that of the group IV elements with the diamond structure in many respects. The principal differences result from the lack of inversion symmetry (Dresselhaus, 1955; Parmenter, 1955). It was shown in Section 1.11 that in the absence of magnetic fields $\psi_{\mathbf{k}}$ and $\sigma_y \psi_{\mathbf{k}}^*$ are eigenfunctions of the Hamiltonian for the same energy. The second solution belongs to wave vector $-\mathbf{k}$, so that we must have $E(\mathbf{k}) = E(-\mathbf{k})$ (Kramer's theorem). If there is inversion symmetry, the periodic part of the Bloch function must satisfy $u_{-\mathbf{k}}(\mathbf{r}) = u_{\mathbf{k}}(-\mathbf{r})$, so that all states are doubly degenerate. If inversion symmetry is lacking, the condition $u_{-\mathbf{k}}(\mathbf{r}) = u_{\mathbf{k}}(-\mathbf{r})$ is not satisfied, so that a double degeneracy is not required. In this case, the degeneracy of states at symmetry points may be removed either in first order or in third order.

3.7 ZINC BLENDE STRUCTURES

The term $\boldsymbol{\sigma} \cdot [\boldsymbol{\nabla} V \times \mathbf{k}]$ in H_1 (Eq. (3.41)), combined with the lack of inversion symmetry destroys the requirement present for the diamond lattice that all the bands be flat at $\mathbf{k} = 0$. In particular, bands formed from Γ_8, which is situated at the top of the valence band, will not have zero slope. Kane finds that the linear terms move the maximum of the valence band away from the zone center to points on the 111 axis about 0.3% of the distance to the zone face. The energy of the maximum was estimated to be about 10^{-4} ev above the energy at $\mathbf{k} = 0$. The linear term may, however, be more important in the band structure of compounds such as InAs in which the difference in atomic number between the constituents is greater (Stern and Talley, 1957). The small band gap implies the existence of a light hole band ($m^* = 0.015$). The heavy holes have a mass about $m^* = 0.25$.

The small effective mass of electrons in indium antimonide implies that the density of states in the conduction band is quite low. Consequently, the conduction band becomes filled quite readily in n-type material, and the conduction electrons soon become degenerate as the electron concentration is increased (Burstein, 1954). The effective mass should decrease with increasing energy since large k^4 terms are present. In third order in k, the spin degeneracy of the conduction band should be split except along the 100 and 111 directions. Subsidiary conduction band minima should exist at the extremities of the 111 and 100 axes (since the double degeneracy at X present in Ge is removed in this structure.

In spite of the differences in band structure produced by the lack of inversion symmetry, there are also important similarities between the band structures of elements of group IV and corresponding compounds with the zinc blende structure. Consider, for instance, the relation between the conduction band in germanium and that in gallium arsenide. If we assume that the effective mass (m^*) of electrons in the $\mathbf{k} = 0$ conduction band is due solely to the interaction of that band with the valence bands at $\mathbf{k} = 0$, and that the momentum matrix elements are the same in the two materials, we can write:

$$\frac{m^*(\text{GaAs})}{m^*(\text{Ge})} = \frac{E_G(\text{GaAs})}{E_G(\text{Ge})} \tag{3.42}$$

where E_G is the band gap at $\mathbf{k} = 0$. If we use $E_G(\text{GaAs}) = 1.51$ ev (Sturge, 1962) $E_G(\text{Ge}) = 0.90$ ev; $m^*(\text{Ge}) = 0.04$, we obtain $m^*(\text{GaAs})$

= 0.067 in fair agreement with the experimental value $m^* = 0.072$ (Ehrenreich, 1960). The calculation can be improved by taking into account the spin orbit splitting of the valence band (Moss and Walton, 1959; Ehrenreich, (1961). The point is that one of the three valence bands with which the conduction band interacts is located at an energy $(E_G + \Delta)$ below the conduction band (where Δ is the spin orbit splitting). If we apply Eq. (1.41b) to this case, and assume that the momentum matrix elements connecting each valence band with the conduction band are the same, we find

$$\frac{m_0}{m^*} = 1 + \frac{A}{Eg}\left(\frac{2}{3} + \frac{1}{3}\frac{Eg}{Eg+\Delta}\right) \tag{3.43}$$

where A is a constant. This relation with the same A for all materials is able to describe the dependence of effective mass on energy gap with considerable accuracy for InSb, InAs, GaSb, GaAs, InP, and germanium. A has the value 20 ev with a spread of about 20%. The success of this simple relation is strong evidence for similarity of band structure at $\mathbf{k} = 0$.

In order to relate the general features of the band structures of these materials, one may consider sequences of compounds formed from elements of the same row of the periodic table, for example, Ge, GaAs, ZnSe, CuBr. The band structure of the compounds of this sequence may be regarded as generated from that of germanium by a perturbing potential. The major portion of this perturbation will be antisymmetric about the midpoint of the line joining the two in the unit cell. Except at certain points of the zone (W, X) where degeneracies are removed in first order, the principal effect of the perturbation is in second order perturbation theory (Herman, 1955). The change in the energy gap between GaAs and ZnSe is three or four times that between GaAs and Ge, which is about what would be expected from this simple reasonaing. It is also possible to pursue the argument in more detail by considering on a group theoretical basis the matrix elements of the perturbing potential. Since the band structure of germanium is reasonably well understood, it is possible to estimate the energy denominators of second order perturbation theory and so to account for the shift of the lowest conduction band minimum from L (in Ge) to Γ (in GaAs) (Callaway, 1957b). A similar analysis can be employed to trace the behavior of some other levels through the 3–5 and 2–6 semiconductors (Ehrenreich et al., 1962). The relation of energy levels in GaAs to those in germanium has also been investigated by Bassani and Celli (1961), using perturbation theory applied to a pseudopotential.

3.8 Aluminum

Studies of energy levels in aluminum have been reported by Matyas (1948), Gaspar (1952), Raimes (1953), Antoncik (1953), Heine (1957a,b,c), Harrison (1959, 1960a), and Segall (1961b). The calculations of principal interest are the detailed studies of Heine and Segall and the (almost) free-electron model of Harrison. Considerable experimental information exists concerning the detailed characteristics of the Fermi surface: de Haas-van Alphen effect measurements (Gunnersen, 1957); anomalous skin effect (Fawcett, 1960), electronic specific heat (Howling et al., 1955); cyclotron resonance (Moore and Spong, 1962; Fawcett, 1960). Both Heine and Harrison have made an effort to construct a model of the Fermi surface which would be in agreement with the experimental observations. A large measure of qualitative success has been obtained, but significant quantitative discrepancies remain.

The starting point of both Heine and Harrison is the determination of the Fermi surface for a trivalent face-centered cubic metal. Heine then made a detailed band calculation using the orthogonalized plane wave method, in which very considerable care was used in the determination of the crystal potential. This calculation will be discussed below. The calculation was used to suggest the probable deviations of the actual Fermi surface from the predictions of the free electron model. Harrison (1960a) has extended Heine's calculations in an approximate manner to facilitate computation of surfaces of constant energy.

It is interesting that the Fermi surface should be quite complex even in the free-electron approximation. The methods of constructing the Fermi surface in this approximation were discussed in Chapter 1, and the results for aluminum are shown in Fig. 6, which is taken from the work of Harrison (1959). In the terminology of the reduced zone scheme, the wave vector of an electron state **k** always lies within the first zone, and the energy is a multivalued function of **k** with branches characterized by the band index. We shall, however, loosely refer to the nth branch $E_n(\mathbf{k})$ as the energy in the nth zone. In this description, the free electron approximation predicts that the first zone should be full, and the Fermi surface should have portions in the second, third, and fourth zones: In the second zone, the Fermi surface consists of caps of spheres with their convex sides facing Γ. This surface should not touch the zone boundary. It is very probable that the major portion of the area of the Fermi surface in aluminum comes from this surface. To determine the occupied regions in the higher bands it should be remembered that in the free electron approximation for a face centered cubic lattice, the lowest levels at X and L are doubly degenerate; at K, U, triply degenerate; at W, fourfold degenerate. Hence in the third zone there will be a portion of the Fermi surface surrounding the zone edges, $W - U - W$ and $W - K - W$. These pieces can be combined to form the surface shown. In the fourth zone, there will be small occupied regions around the corners W.

Next we consider the effect of the crystal potential (or pseudopotential) on these levels. If the potential is sufficiently weak for it to be considered in first order perturbation theory, the energies of the levels at the point W become

$$E(W_1) = E_0 + V(0) + 2V(1,1,1) + V(2,0,0)$$

$$E(W_3) = E_0 + V(0) - V(2,0,0) \quad \text{(doubly degenerate)} \quad (3.44)$$

$$E(W_2') = E_0 + V(0) - 2V(1,1,1) + V(2,0,0)$$

In these expressions, E_0 is the energy of the levels at W in the absence of the potential and the V's are, of course, Fourier coefficients of the potential. In the case of aluminum, we expect to have the p-like states W_2', W_3 below the s-like state W_1 so that the Fourier coefficients of the pseudopotential should be positive. Then we will expect the W_1 level to be considerably raised above $(E_0 + V(0))$ and W_3 to be lowered, while if $2V(1,1,1) > V(2,0,0)$ (which is probable), W_2' will also be lowered. In consequence, the occupied regions in the fourth band should disappear, but levels in the first three bands at W should remain occupied. This is in agreement with the results of Heine's calculation but not with the Fermi surface he proposed which contained pockets of holes in the first zone centered on W. (This proposal would require that all the levels there should rise.)

The more precise of the two band calculations reported by Heine utilized a crystal potential which was based on a self-consistent field calculation (with exchange) for Al^{3+} (Froese, 1957). A correction made for correlation effects among the core electrons was determined by analogy with a calculation carried out by Bernal and Boys (1952) for sodium; separate core-valence exchange potentials were computed for s and p states. The contribution of the valence electrons to the crystal potential was computed assuming that the wave functions for the valence electrons are single OPW's. This potential resembles that of a uniform charge distribution. Comparison of the assumed potential with that computed from some of the wave functions finally obtained indicates that the calculation is very nearly self-consistent. A correction was applied to take account of the nonspherical character of the potential in the metal since the actual charge distribution does not consist of nonoverlapping spheres. (An error in this correction was discovered by Behringer, 1958). Exchange among the valence electrons was included from the work of Bohm and Pines (Pines, 1955). This energy was determined as a function of k, and the variation with r was also determined for states near the Fermi surface. The use of the Bohm-Pines result is a weak point of the calculation since it is not evident that their approximations are valid. With these exceptions, the potential appears to be one of the most carefully constructed for a multivalent atom. Appropriate core functions were found by numerical integration for this potential. The OPW calculation appeared to be convergent within 0.02 ry.

Heine calculated the lowest energy levels of s- and p-like states at the symmetry points, Γ, X, L, W, and K, and along the 100 and 111 axes. (An earlier and more crude calculation included 140 nonequivalent points in the zone.) Except in the vicinity of symmetry points, the band structure is close to the predictions of the free electron approximation for an effective mass ratio $m^*/m_0 = 1.03$. In view of this result, it seems reasonable to extend the results throughout the zone by a pseudopotential interpolation scheme. This calculation was made by Harrison (1960a), who treated the Fourier coefficients appearing in Eqs. (3.45) as disposable parameters to be determined from Heine's results for the point W. He set $E_0 + V(0) = \alpha \mathbf{k}^2$ ($\alpha = 0.8535$), $V(1, 1, 1) = 0.0295$ ry, and $V(2, 0, 0) =$

0.0550 ry. Higher coefficients have been neglected.[8] When these values were used to compute the energy at the other symmetry points of the zone considered by Heine, the agreement was found to be generally quite good (within about 0.03 ry or 3% except for a single level at K).

The principal reason for using a pseudopotential procedure with a small set of basis functions (rather than the full OPW method) is to facilitate determination of the surfaces of constant energy. This was done for surfaces in the second and third band for energies in the vicinity of the Fermi energy. When the surfaces of constant energy were compared with those found in the free-electron (or single OPW) approximation, it was found that the principal effect of mixing in additional waves is to round off-edges of the Fermi surface in regions where they are sharp in the free electron approximation.

The band calculation of Segall (1961b) employed the Kohn-Rostoker method, and was based on the same potential as used by Heine, with the correction due to Behringer, except for the truncation of the potential outside the inscribed sphere required in the Green's function method. The results are generally in reasonable agreement with those of Harrison and Heine.

There are several sorts of experimental measurements which determine characteristics of Fermi surfaces. Some of these are the following. A measurement of the period of oscillation of the diamagnetic susceptibility of a metal as a function of $1/H$ (the de Haas-van Alphen effect), measures the maximum or minimum cross-sectional area of the Fermi surface; the cross section is taken perpendicular to the magnetic field. In a magnetic field the orbit of an electron wave packet is (in a semiclassical sense) the intersection of an energy surface with a plane perpendicular to the field. In cyclotron resonance, an effective mass is determined from the rate of change of the area of these orbits with energy to be

$$\frac{m^*}{m_0} = \frac{1}{\pi} \frac{dA}{dE} \qquad (3.45)$$

---

[8] Harrison's procedure does not include all the matrix elements of the Hamiltonian which contain $V(1, 1, 1)$ or $V(2, 0, 0)$. The matrix is, instead limited to a small size (2×2, 3×3, or 4×4) depending on the point considered.

Measurements of the anomalous skin effect, if made on a polycrystalline sample, give the total area of the Fermi surface. The theory underlying the de Haas-van Alphen effect and cyclotron resonance will be discussed in detail in Chapter 4. We have already seen that the low temperature specific heat gives the density of states at the Fermi energy, and hence a value for $(|V_k E|)$ there.

Harrison has compared the calculated and experimental results for the properties of the Fermi surface mentioned previously, using principally the single OPW approximation. The agreement is quite good qualitatively (we shall not discuss the details here) in that the main features of the experimental data are reproduced. There are, however, significant discrepancies in a quantitative sense, particularly with regard to the cyclotron resonance masses and the low temperature specific heat. In general, however, the comparison is sufficiently favorable to be a striking confirmation of the applicability of the general language of band theory to a multivalent metal.

Harrison (1960b) has also extended the free-electron calculations to other multivalent metals, and has used the resulting model of the Fermi surface as a guide to the interpretation of experimental results. We will not discuss this work here.

3.9 The Noble Metals

The metals copper, silver, and gold are perhaps second in simplicity of electronic structure only to the alkali metals, and are much easier to handle experimentally. For this reason, the experimental determination of the Fermi surface is, at the time of writing, much further advanced for these materials than for the alkali metals. The lattice structure is face-centered cubic. Most of the band structure calculations, which are listed below, pertain to copper (for obvious reasons): (Krutter (1935), Fuchs (1935, 1936), Tibbs (1938), Chodorow (1939), Howarth (1953, 1955), Kambe (1955), Fukuchi (1956), Segall (1961a, 1962), Cornwell (1961b), Burdick (1963).

Band structure calculations in these materials are usually concerned with the d electron states nearest the valence levels as well as the (supposedly) largely s-like valence electron. Slater (1936) used the results of the band calculation of Krutter (1935) to derive a density of states curve for d bands, which has frequently been used in the interpretation of experiments on the properties of the transition elements. Characteristic features of the Slater curve are the high density of states at the bottom and top of the band and a minimum near the middle of the band. Slater used this density of states to estimate the Curie temperature of nickel, but it has been applied to the entire series of transition elements. It is therefore worth emphasizing here that the errors in the basic band calculation resulting from faulty application of the cellular method are so serious that the density of states derived from it has no validity. For example, the fivefold degeneracy of the d bands was not removed at the center of the zone, and this leads to a quite spurious peak in the density of states at the bottom of the band.

TABLE XXI

COHESIVE ENERGIES AND EFFECTIVE MASSES OF THE NOBLE METALS[a]

	$E_2 = (m_0/m^*)$	E_{coh} (theor)	E_{coh} (exp)	r_s (exp)
Copper	0.988	62.4	81.2	2.67
Silver	1.008	59.0	68.0	2.99
Gold	1.006	52.2	92.0	2.99

[a] After Kambe (1955). Kambe's published results have been modified by using the corrected form of Wigner's formula for the correlation energy.

Interest in recent years has been centred on energy levels in the noble metals near the Fermi surface. We will, however, first review the attempts to calculate the cohesive energies of these materials. The most recent calculation is that of Kambe (1955) who applied the quantum defect method. The calculation is quite analogous to those for the alkali metals discussed in Section 3.1. In particular, the Fermi energy was obtained from Eq. 3.2. Because the atomic cell is small, the potential in the vicinity of a cell boundary is not completely hydrogenic. Corrections for the departure of the potential from the simple coulomb form were determined from a self-consistent field in the case of copper. For silver and gold, the corrections were obtained from the results for copper and may not be very accurate. The extrapolation of the quantum defects is more difficult than for the alkali metals because of the presence of configuration interaction in the atomic spectra. The results obtained by Kambe (all at the observed lattice spacing) are given in Table XXI.

The simple calculations of cohesive energies which were so successful for the alkali metals are much less so in this case. There are several effects which would have to be included in a more rigorous calculation. Among these are the strong van der Waals attractions between the ions (Friedel, 1952; Mott, 1953). It has been suggested that these interactions contribute perhaps 25 kcal/mol to the cohesive energy. Their strength has been related to the optical absorption. The coulomb and exchange interactions of the ions is also important, particularly in the computation of the lattice constant and the compressibility. Further, on account of the small cell size, the d electron cores overlap to some extent, and it becomes necessary to consider the interaction between the valence electron on one atom and the d electron core of neighboring atoms (Hsu, 1951). This is at variance with the usual procedure of the Wigner-Seitz approximation in which an electron is considered to move in the field of a single ion. A different and very important criticism of Kambe's procedure comes from the band calculations of Segall (1961a, 1962) and Burdick (1963). These show that the filled d band (about 4 ev in width and containing ten electrons) lies above the lowest s level (Γ_1), whose energy is calculated by Kambe. If this is correct, it is quite impossible to estimate the Fermi energy according to (3.2), and it is necessary to consider the contribution of the d electrons — as band electrons — to the cohesive energy.

We next turn to a description of the Fermi surface. A free-electron Fermi sphere (one electron per atom) would, if placed inside the Brillouin zone for the face-centered cubic lattice, approach closest to the zone face in the vicinity of the hexagonal face centers L. The radius of the free-electron sphere amounts to 90% of the Γ-L distance (to be denoted by d) — a slightly closer approach than that in the body-centered cubic lattice. Since $E(\mathbf{k})$ must be flat at L, a distortion of the Fermi surface in the vicinity of L is to be expected. In fact, the experimental evidence indicates that in all three of the noble metals, the Fermi surface is in contact with the zone boundary at L. The surface may be represented pictorially as a sphere drawn out along the 8 (1, 1, 1) axes as shown in Fig. 12. A model of this sort was first proposed for copper by Pippard (1957) to interpret the results of measurements of the anomalous skin effect.[9] Confirmatory evidence has come from measurements of the

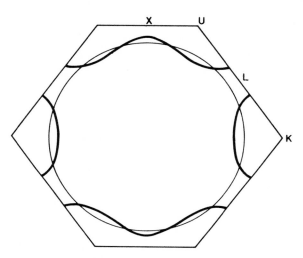

FIG. 12. Cross section of the Fermi surface of copper in a plane perpendicular to a (1, 1, 0) axis.

de Haas-van Alphen effect (Shoenberg, 1962), cyclotron resonance (Kip, 1960; Kip et al., 1961), magnetic field dependence of ultrasonic attenuation (Morse, 1960; Morse et al., 1961); Easterling and Bohm, 1962), and magnetoresistance (Klauder and Kunzler, 1960). A detailed description of the Fermi surface in these materials has emerged. Contact of the Fermi surface with the zone boundary is definitely indicated.

The radius of the contact area (assumed circular) may be deduced from the de Haas-van Alphen (dHvA) measurements and from the

---

[9] A mathematical description of the Fermi surface in copper was given by Pippard (1957), and by Garcia-Moliner (1958), who used a Fourier Series in a fashion similar to the tight binding interpolation scheme of Slater and Koster (1954). The more recent data on the Fermi surfaces of all the noble metals has been treated by Roaf (1962), who has extended the Fourier series approach of Garcia-Moliner.

magnetic field dependence of ultrasonic attenuation (UA). The results are given in Table XXII, in terms of the Γ—L distance, d. The theoretical results of Segall (1962) (using his "l-dependent potential) are also given for the case of copper. The cyclotron effective mass ratios are defined by

$$\frac{m^*}{m_0} = \frac{1}{\pi}\frac{dA}{dE}$$

where A is the area of the orbit.

TABLE XXII

FERMI SURFACE DATA FOR THE NOBLE METALS

	Cu	Ag	Au
Neck radius r			
(dHvA)	$0.18d$	$0.12d$	$0.16d$
(UA)	$0.17d$	$0.13d$	$0.17d$
Theory (Segall)	$0.18d$		
Belly radius R[a]			
(100 axis)	$0.97d$	$0.93d$	$1.02d$
(110 axis)	$0.85d$	$0.87d$	$0.85d$
"Cyclotron" mass ratios			
belly[a]	1.38	0.7	1.2
neck	0.6		0.5
Theory (Segall)			
belly	1.1 ± 0.1		
neck	0.41 ± 0.02		
Γ-L distance d	$1.51 \times 10^8\,\text{cm}^{-1}$	$1.34 \times 10^8\,\text{cm}^{-1}$	$1.34 \times 10^8\,\text{cm}^{-1}$

[a] Since the "belly" is not exactly spherical, there is a spread in the values of the radius and of the effective mass. Most of the data in this table is taken from the discussion of Roaf (1962).

The existence of regions of contact between the Fermi surface and the Brillouin zone implies, when one considers the equivalence of opposite points on the zone surface, that the Fermi surface is actually multiply connected. Such a surface has the property that its intersection with a plane need not be a closed curve. As the orbit of an electron in a magnetic field is, as was pointed out previously, the intersection of a plane perpendicular to the field with the Fermi surface, it is evident that when contact of the surface with the zone boundary occurs, electron orbits may exist which are not closed. This is perhaps easiest to visualize in an extended zone scheme in which the Fermi surface is repeated periodically in a zone structure which fills all space. The possibility then exists for an electron to wander from one zone to another in an open orbit without leaving a plane perpendicular to the applied field.

Ziman (1961) has attempted to develop a simple characterization of the Fermi surface in terms of the nearly free-electron approximation by considering only the interaction of two (orthogonalized) plane waves which would be degenerate on the zone boundary. A single Fourier coefficient of a pseudopotential enters into this calculation $V(1, 1, 1)$. This procedure may be reasonable near the point L, where contact occurs, since in lowest order only the waves $(2\pi/a)$ $(\frac{1}{2}, \frac{1}{2}, \frac{1}{2})$ and $(2\pi/a)$ $(-\frac{1}{2}, -\frac{1}{2}, -\frac{1}{2})$ are connected there; at other symmetry points other waves must be considered, even in first order.

The equations may be easily obtained from (2.23). Let us denote the two interacting waves by $\mathbf{k}$ and $\mathbf{k}'$, where $\mathbf{k}' = \mathbf{k} - (2\pi/a) (\hat{\mathbf{i}} + \hat{\mathbf{j}} + \hat{\mathbf{k}})$, the (1, 1, 1) Fourier coefficient of potential by V, and the quantity $E(\mathbf{k})$ by λ. We may take approximate account of the effect of the crystal potential on the diagonal matrix elements by replacing $\mathbf{k}^2$ in the ordinary free electron approach by $\alpha \mathbf{k}^2$ where α is then a reciprocal effective mass. Then we have the 2 × 2 determinantal equation

$$\begin{vmatrix} \alpha \mathbf{k}^2 - \lambda & V \\ V & \alpha \mathbf{k}'^2 - \lambda \end{vmatrix} = 0 \qquad (3.46)$$

whose solution is:

$$\lambda = \tfrac{1}{2}[\alpha(\mathbf{k}^2 + \mathbf{k}'^2) \pm \sqrt{\alpha^2(\mathbf{k}^2 - \mathbf{k}'^2)^2 + 4V^2}] \qquad (3.47)$$

Since this solution is expected to be useful only in the vicinity of the point L [whose position vector we denote by $\mathbf{k}_0 = (\pi/a) (\hat{\mathbf{i}} + \hat{\mathbf{j}} + \hat{\mathbf{k}})$], it is convenient to put

$$\mathbf{k} = \mathbf{k}_0 + \delta \mathbf{k}; \qquad \mathbf{k}' = \delta \mathbf{k} - \mathbf{k}_0$$

Then (3.47) becomes

$$\frac{\lambda}{\alpha} = \mathbf{k}_0^2 + \delta\mathbf{k}^2 \pm \sqrt{\frac{V^2}{\alpha^2} + 4(\mathbf{k}_0 \cdot \delta\mathbf{k})^2} \tag{3.48}$$

Evidently at the zone boundary where $\delta\mathbf{k} = 0$, the energy gap is just $2V$.

The pseudopotential coefficient V may be related to the neck radius r in the following way. Let the Fermi energy (λ_f) be given. The intersection of the surface of constant energy, $\lambda = \lambda_f$, with the zone face is a circle of radius r, where $r^2 = \delta\mathbf{k}^2$, and (since the face is perpendicular to the 111 axis) $\mathbf{k}_0 \cdot \delta\mathbf{k} = 0$. Further $\mathbf{k}_0^2 = d^2$, where d is the Γ–L distance previously defined. We take the lowest root of (3.48):

$$r = \sqrt{\frac{\lambda_f}{\alpha} + \frac{V}{\alpha} - d^2} \tag{3.49}$$

We define dimensionless measures of the Fermi energy and the pseudopotential as follows: $u = V/\alpha d^2$, $\varepsilon_f = \lambda_f/\alpha d^2$,

$$r = d(\varepsilon_f + u - 1)^{1/2} \tag{3.49a}$$

The catch in this is that it is now necessary to evaluate the Fermi energy, taking into account the distortion of the surfaces of constant energy from the spherical form. If we should use the free electron value for ε_f (with $\alpha = 1$ and $\varepsilon_f = 0.814$), we find that a neck radius of $0.18d$ implies $u = 0.22$, and hence that $V = 1.9$ ev (for copper), which implies a gap of 3.8 ev at the zone face. Use of Segall's (1962) values for the Fermi energy does not change this estimate appreciably since ε_f as deduced from his work is 0.809. This seems to be in reasonable agreement with the indications from optical data (Biondi and Rayne, 1959) that the gap at the point L is perhaps 3.5 ev wide in copper. The value obtained by Segall is considerably larger: 5.9 ev. If we use this gap to determine V, then u comes out to be 0.265.

Ziman has discussed the electron specific heat on the basis of his model. In order to perform the integration in (3.24) it is necessary to have a representation of the Fermi surface as a whole, which he approximates as formed from eight cones around 111 axes, with gaps as given from (3.48). The experimental thermal effective masses for the noble metals are, according to Eq. (3.25) and the measurements of Corak *et al.* (1955): copper, $m_t/m_0 = 1.38$; silver, $m_t/m_0 = 0.96$; gold, $m_t/m_0 = 1.18$. Ziman finds that agreement with experiment can be obtained in his theory by choosing α to be 1.15, 1.61, and 1.37, respectively.

Cohen and Heine (1958) has shown that (see Section 4.10 for a derivation of this result) the real part of the dielectric constant ε of a cubic metal, at frequencies below the main absorption edge but still sufficiently high so that relaxation effects may be neglected, is, in mks units,

$$\frac{\varepsilon}{\varepsilon_0} = 1 + \kappa_a - \frac{\omega_p^2}{\omega^2} \qquad (3.50)$$

and

$$\omega_p^2 = \frac{N_c e^2}{m_{op}^* \varepsilon_0}$$

in which κ_a is the contribution from the atomic polarizability, ω_p is the plasma frequency, N_c is the number of conduction electrons per unit volume, and m_{op}^*, the average "optical" effective mass of the electrons, is given by

$$m_{op}^{*-1} = \frac{1}{12\pi^3 N_c \hbar^2} \int d^3k \, [\nabla_k^2 E(\mathbf{k})]$$

$$= \frac{1}{12\pi^3 N_c \hbar^2} \int d\mathbf{S}_F \cdot \nabla_k E \qquad (3.51)$$

The measured optical effective masses (Schulz, 1957, Ehrenreich and Philipp, 1962) are 1.42, 1.03, and 0.98 in units of m_0 for copper, silver, and gold, respectively. Ziman has calculated these quantities from his theory, and finds rather good agreement with experiment. A more elaborate pseudopotential calculation considering gold and silver as well as copper, and including the (2, 0, 0) Fourier coefficient, has been reported by Cornwell (1961b).

The band calculations which have been performed for these materials disagree with each other in many important respects. However, considerable progress has been made recently in the calculation of energy bands in copper. The work of Segall (1962) and Burdick (1963) utilizing the Kohn-Rostoker and APW methods, respectively, has produced a band structure in reasonable agreement with experiment. We will discuss the former calculation here. Two different crystal potentials were considered: one obtained by Chodorow (1939), which is principally appropriate for d states, and the other calculated by Segall and allowing a different exchange potential for s, p, and d states. An l-dependent potential can be easily accommodated in the Green's function approach. The band structures based on the two potentials are in quite reasonable qualitative agreement with each other. The essential features of the results are that the d bands overlap the s band, but lie below the Fermi level, and that the predicted Fermi surface is in contact with the Brillouin zone in the vicinity of the face center L in the manner required by the experiments previously discussed. Large band gaps are predicted to exist on the zone faces. The top of the d band is apparently located either at W, representation W_1', or X, representation X_5 (the energy difference between these states is very small), approximately 2 ev below the Fermi level. This enables an interpretation of the main absorption edge in copper, which begins at about 2.1 ev and is responsible for the red color of the metal as due to transitions from the filled d bands to the Fermi level (in agreement with Mott, 1953). States near the Fermi surface are generally p-like so that the optical transition is allowed. The alternative explanation of the absorption edge as due to transitions between the Fermi level and higher bands (Suffczynski, 1960) is not supported by this calculation since the relevant band gap is much larger (4 or 5 ev). Portions of the calculated band structure are shown in Fig. 13.

Segall has calculated the parameters describing the Fermi surface. Some of these are given in Table XXII. The "belly" of the Fermi surface is not particularly spherical, being pulled out along the (1, 0, 0) axes, and pushed in along the (1, 1, 0) axes, in general agreement with Morse *et al.* (1961). The distortion may be interpreted as due to the interaction of the conduction band with d bands, since there is a greater admixture of $l = 2$ components into the conduction electron wave function on the (1, 1, 0) axis than along the (1, 0, 0) axis. The thermal and optical effective masses were calculated to be 1.12 ± 0.06 and 1.2 ± 0.10, respectively. The effective masses are seen to lie below the experimental values, as was previously found to be the case for the thermal effective masses in the alkali metals. It is tempting to ascribe the discrepancy to neglect of electron-electron and electron-phonon interactions.

Segall has examined the corrections to the band calculation resulting from the truncation of the potential outside the inscribed sphere, and from the nonspherical components of the actual crystal potential. The corrections are found to be small. This confirms the utility of the Green's function method for band calculations.

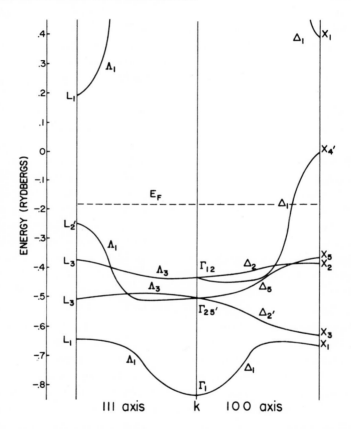

FIG. 13. Energy bands in copper along the (1, 0, 0) and (1, 1, 1) axes according to Segall (1962). The dashed line, E_F, shows the Fermi energy.

A preliminary report of energy band calculations in silver has been given (Segall, 1961a). The band structure appears to be similar to that found for copper.

3.10 d Bands and Transition Metals

Under the title of transition elements, we will consider only those elements of the fourth period of the periodic table in which the free atoms possess incomplete d shells. These metals are particularly interesting because of their magnetic properties: iron, cobalt, and nickel are ferromagnetic; chromium and manganese are antiferromagnetic. Many of the simple compounds of these elements are antiferromagnetic. The magnetic properties are naturally ascribed to the d electrons, and it becomes a principal task of band theory to give a satisfactory quantum-mechanical account of these properties.

Detailed discussions of the band theories of ferromagnetism and antiferromagnetism can be found in papers by Slater (1951b, 1953b, 1956). We will not discuss this work in detail here. It will suffice to state that it has not been possible to give a quantitative account of either ferromagnetism or antiferromagnetism in metals. Some of the difficulties involved should become apparent in the course of this discussion.

Although experimental evidence relating to the band structures of these materials is quite meager in comparison to some other materials we have discussed, what there is (particularly that derived from the magnetic properties) suggests a rather smooth and gradual variation of electronic structure from element to element, regardless of rather substantial changes in crystal structure. Thus it seems reasonable to try to construct a general model of d bands in the transition elements. There have been many attempts to do this: Mott (1935, 1949), Mott and Stevens (1957), Pauling (1938, 1953), Bader et al. (1954), Lomer and Marshall (1958), Goodenough (1960), and Wollan (1960). The basic ideas of the original proposal of Mott were that there is a narrow d band of width less than 1 ev which is overlapped by a wide s band. A narrow d band is necessary in the band theory of ferromagnetism so that the decrease of energy resulting from increased exchange upon spin alignment will outweigh the increase of energy resulting from the promotion of the electrons to higher band states. The magnetic properties are determined by the d band whereas the cohesion and the conductivity are produced by a relatively small number of s electrons: less than one per atom. Scattering of s electrons into vacant d band levels, where the density of states is high, accounts for the relatively large resistivity.

On the other hand, Pauling assumes that the d electrons (two in iron) responsible for the magnetic properties form a band of essentially zero width. The atomic model is adequate for these electrons. The other electrons, s and d in the free atom, go into a broad band which is based on hybridization of s, p, and d orbitals. Ferromagnetism is believed to be produced by the exchange coupling of the atomic d electrons with the conduction band in a manner similar to the proposal of Zener and Heikes (1953). A small number of the conduction electrons contribute to the net magnetic moment.

A basic conflict is evident in the discussion of the transition metals with respect to the question of the adequacy of the band approximation to describe the behavior of the d electrons. Adherents of the band approach believe that a properly self-consistent calculation of energy bands from the Hartree-Fock equation would make possible, at least in principle, reasonably quantitative explanations of magnetic properties. This point of view has been expounded by Slater. Others, including Pauling, Zener, and, in recent years, Mott have contended that a model in which at least some of the d electrons are localized on atomic sites is a closer approximation to reality.

A basic difficulty which arises in the energy band theory of ferromagnetism is this: The theory erroneously predicts ferromagnetism in the limit of infinite atomic separation. This results because the energy of a nonmagnetic state (disordered spins) is overestimated in this limit: polar states of the individual atoms in which some have too many electrons and some have too few are predicted to occur in large number. The energy of the ferromagnetic state does go to the correct limit as the interatomic distance becomes large because the Pauli principle effectively prohibits such polar states. This difficulty is an aspect of the general problem of treating the electron interaction in more detail than is possible in the Hartree-Fock approximation. Its importance in consideration of the transition metals is emphasized by the fact that one seems to be close to the large separation limit for d wave functions in the materials. The problem may be summarized in a different way as follows: if the d band is narrow, neglect of correlations leads to a serious overestimate of the tendency to ferromagnetism; if the band is wide, the promotion energy inhibits ferromagnetism.

3.10 d BANDS AND TRANSITION METALS

Quantitative calculations of energy bands in the transition metals are unusually difficult. The problem of attaining self-consistency is present in an aggravated form: because of the large numbers of d electrons which have to be considered it is not only more difficult to attain self-consistency, but the consequences of departures from self-consistency may be more serious. Thus, d band calculations have not been sufficiently quantitative to lead to a precise theoretical model. However, a number of comments can be made which indicate how improvements must be made in the simple models.

In the first place, none of the numerous band calculations indicates that the d band is extremely narrow: less than 1 ev in width. A figure of 3 ev seems to be a reasonable minimum figure. However, exchange coupling effects involving spin waves may reduce the calculated values somewhat, (Wolfram and Callaway, 1962). Secondly, in spite of a brief flurry of concern in connection with the experiments of Weiss and De Marco (1958; see also Batterman et al., 1961 and Cooper (1962)), the number of "s and p" electrons per atom in iron, cobalt, and nickel is probably small (one per atom or less — particularly in Ni). It is then necessary to assume that the d electrons contribute to the cohesion. Hall effect measurements (Foner and Pugh, 1953; Foner, 1957) indicate predominantly hole conduction in vanadium, chromium, manganese and iron, but electron conduction in cobalt and nickel; this strongly suggests the importance of conduction in d band states.

The question of the general shape of the d band has been particularly serious. It arises in the following way: several authors (Bader et al., 1954; Mott and Stevens, 1957; Wollan, 1960; Goodenough, 1960) have attempted to divide the d electrons into two groups on the basis of crystal field theory. (For a review of crystal field theory, see Moffitt and Ballhausen, 1956). The fivefold degeneracy of free-atom d levels would be split by an electric field with cubic symmetry (such as would arise naturally in a lattice) into a triply degenerate system, based on functions of xy, yz, zx symmetries, usually denoted t_{2g}, and a doubly degenerate set, ε_g, based on functions of $x^2 - y^2$; $3z^2 - r^2$ symmetries. If we approximate the actual metal as a lattice of positive point charges screened by a uniform distribution of negative charge, then in both the body-centered and face-centered cubic lattices, the cubic components of the crystalline field would tend to lower the energies of the t_{2g} states and raise the e_g states. In the limit of large atomic spacing, this effect, which falls off as a^{-5} (a is the lattice spacing), would be larger than effects involving overlap of wave functions on different atoms which tend to widen the bands. The latter must decrease exponentially with increasing a, so that a separation of the d band into t_g and ε_g subbands would be produced. The magnetic electrons in iron might then be put into a half-filled ε_g subband.

Such a splitting produces an ordering of levels which is different from the normal, nearly free-electron, level order. In the body-centered cubic lattice this arrangement places the doubly degenerate level above the triply degenerate one at the center of the zone but the reverse order with a larger splitting is found at the corner H. This ordering is also achieved in the tight binding approximation. Consequently, a crossing of bands on the 100 axis is predicted (something close to this remains even when spin orbit coupling is included).

It is possible to make some simple calculations which indicate the sort of band structure to be expected (Callaway 1959, 1960b, 1961a). In the case of d electrons it is not necessary to distinguish between an ordinary crystal potential and a pseudopotential since the basic d states at symmetry points are orthogonal to the s and p functions (this applies completely, however, only at Γ and H). We can use the simple crystal model of a point charge lattice with atomic number Z, lattice parameter a, and a uniform distribution of negative charge, and treat the crystal potential as a perturbation in accord with the procedure discussed in Section 2.3. When terms of second order in the crystal potential are used, the energies of the four d states at Γ and H are (Γ_{25}' and H_{25}' are the t_{2g} states; Γ_{12} and H_{12} are the e_g states)

$$E(H_{12}) = 39.478/a^2 - 0.2905(Z/a) - 0.00267Z^2$$

$$E(\Gamma_{25}') = 78.957/a^2 - 0.7679(Z/a) - 0.00509Z^2$$

$$E(\Gamma_{12}) = 78.957/a^2 - 0.3435(Z/a) - 0.00499Z^2$$

$$E(H_{25}') = 118.43/a^2 - 0.8740(Z/a) - 0.001165Z^2$$

(3.52)

For small values of the binding parameter Za, the ordering of levels is determined by the free-electron kinetic energy [the first terms in (3.52)], and the band scheme of Fig. 14 is implied. For large values of Za, the perturbation calculation is certainly not reliable numerically, but the qualitative behavior of the expressions is interesting: the t_{2g} levels Γ_{25}' and H_{25}' lie below the ε_g states Γ_{12} and H_{12}, indicating a split of the d bands. A more accurate calculation is required to determine the value of the binding parameter for which this split occurs: It has

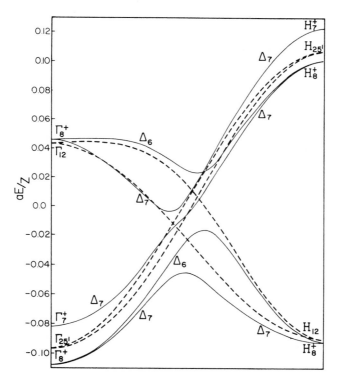

FIG. 14. d Bands along the 100 axis in the body-centered cubic lattice. The dashed curves show the bands with spin orbit coupling neglected; the solid curves show the effect of its inclusion.

been possible to make a reasonable estimate of this using the nearest neighbor tight binding approximation of Suffczynski (1956a,b, 1957) to determine the overlap splitting and a calculation of the crystal field effects in this idealized lattice (Callaway and Edwards, 1960). The crystal field splitting does not dominate for $Za < 72$. It is not possible to make quantitative application to the transition metals from this simple theory, but we see that if $a = 6$, (iron has $a = 5.4$), we would have to have $Z = 12$ to get a split d band. We do not expect the effective charge in this theory to be greater than that of the core of closed shells under the d band (which would imply $Z < 8$ for iron). Hence crystal field effects should not produce a split d band in the case of iron. This conclusion is reinforced by a computation of the crystal field splitting in a point charge lattice using the actual lattice spacing for iron, and self-consistent field wave functions (Watson, 1960a,b). The separation between the t_{2g} and the e_g states due to the crystal field was found to be $-0.1\,Z$ ev. This will be smaller than the d bandwidth for any reasonable value of Z.

There have been a fairly large number of energy band calculations for these metals: Titanium: B. Schiff (1955, 1956), Altmann (1956, 1958a,b), Altmann and Cohan (1958); Chromium: Asdente and Friedel (1961), Asdente (1962), Lomer (1962); Iron: Manning (1943), Greene and Manning (1943), Steinberger and Wick (1949), Callaway (1955), Suffczynski (1956a), Stern (1959), Belding (1959), Wood (1960, 1962); Nickel: Fletcher and Wohlfarth (1951); Fletcher (1952), Koster (1955). The calculations made prior to 1958 have been reviewed previously (Callaway, 1958a). We will emphasize here the calculations concerning iron, on which the most careful work has been done.

The OPW calculation of Callaway (1955) is interesting chiefly insofar as it sets a minimum value for the bandwidth. A potential derived from a self-consistent field calculation for the free iron atom (Manning and Goldberg, 1938) was used. It was assumed that in metallic iron there are six d electrons and two s electrons, and the crystal potential was determined by superposing charge distributions from Manning's calculation placed on lattice sites. An exchange potential computed according to Slater's (1951a) free electron approximation was also included. The calculation determined that this choice of potential was far from being self-consistent in that it yielded a very narrow d band (width 0.1 ry) below the s band. Hence, the predicted configuration in the metal was d^8 instead of d^6s^2. Various corrections were then made to the potential to estimate the position of the bands more reliably on the basis of a potential derived from a d^7s^1 configuration. These corrections need not concern us here. The essential point is that the assumed charge distribution was too diffuse for self-consistency: the potential was too attractive in the region in which the d functions have their maxima, so that the predicted d band was too low and too narrow. We can conclude therefore that the actual d band in iron must be wider than 0.1 ry, and there must be less than two "s electrons" per atom.

Stern (1959) has reported an interesting calculation of the cohesive energy of iron. His method is a modification of the tight binding approximation. The ordinary tight binding wave function $\varphi_\mathbf{k}$ which is formed from an atomic orbital u is

$$\varphi_\mathbf{k}(\mathbf{r}) = \frac{1}{\sqrt{N}} \sum_{\nu}{}' e^{i\mathbf{k}\cdot\mathbf{R}_\nu} u(\mathbf{r} - \mathbf{R}_\nu) \tag{3.53}$$

We saw in Section 2.10 that if the expectation value of the energy is evaluated with this wave function, a complicated series of multicenter integrals must be summed. However, it is possible to avoid multicenter integrals if the wave function is evaluated within one cell, including the contributions from the distant neighbors. It is also necessary to orthogonalize the wave function to the core states. The tight binding function can be written as a sum of the wave function in the central cell plus an overlap contribution, which we denote by

$$\varphi_\mathbf{k}'(\mathbf{r}) = \sum_{\nu \neq 0}{}' e^{i\mathbf{k}\cdot\mathbf{R}_\nu} u(\mathbf{r} - \mathbf{R}_\nu)$$

If $u_b(\mathbf{r})$ is a core wave function in the cell at the origin, we orthogonalize $\varphi_\mathbf{k}'$ to u_b by constructing the function

$$\psi_\mathbf{k}'(\mathbf{r}) = \varphi_\mathbf{k}'(\mathbf{r}) - \sum_b u_b(\mathbf{r}) \int \varphi_\mathbf{k}'(\mathbf{r}) u_b^*(\mathbf{r}) d^3r \tag{3.54}$$

Then, instead of (3.53), Stern uses the wave function (unnormalized)

$$\varphi_\mathbf{k}(\mathbf{r}) = u(\mathbf{r}) + \psi_\mathbf{k}'(\mathbf{r}) \tag{3.55}$$

Stern evaluates the overlap term by a power series expansion in the central cell. He finds that it is necessary to include only first and second neighbors in calculating the $3d$ wave function, but in the case of the $4s$ wave function terms out to sixth neighbors were included, and a qualitative correction was made for those farther out.

Stern uses the approximation of Wigner and Seitz in constructing the crystal potential: each electron moves in the field of a positive ion. Specific exchange effects are not included. In the calculation of the crystal potential, the cells are considered to be spherical and, except for the one in which the wave functions are calculated, neutral. Then the cohesive energy may be expressed as a sum of contributions from individual cells. This differs from the usual formalism in that the expectation value of the energy is not evaluated with a specific determinantal wave function, so that the energy found is not an upper bound to the true energy. But there is a significant advantage which largely compensates for the loss of the applicability of the variational principle: the procedure is correct in the limit of large lattice constants, and avoids the difficulty associated with polar states which was mentioned previously. The number of $4s$ electrons in the solid was treated as a parameter, but the potentials were derived from the wave functions, previously calculated by Stern (1956), for the d^7s^1 configuration of the free iron atom.

Stern calculated energy values for 16 nonequivalent points in the Brillouin zone, distributed along the 100, 110, and 111 axes. He estimated a rough density of states from these and computed the cohesive energy. This calculation was repeated for four values of n_{4s}, the number of $4s$ electrons per atom (0.3, 0.6, 0.9, and 1.0) and three different values of the sphere radius ($r_s = 2.30$, 2.66, and 3.10). He then determined the value of n_{4s} by finding the minimum of the energy with respect to it for fixed r_s. Then the final cohesive energy, lattice constant, and compressibility were determined by finding the minimum of the cohesive energy as a function of r_s. The cohesive energy was found to be 0.43 ry/atom in comparison with the experimental value of 0.32 ry/atom (99.2 kcal/mol). A theoretical uncertainty of $\pm$ 0.2 ry is estimated. The sphere radius comes out to be 2.66 which agrees with experiment, as does the calculated compressibility. No explicit correction for correlation and exchange was made.[10]

---

[10] An interesting auxiliary result of Stern's calculation is the prediction from consideration of the matching of wave functions at the center of the faces of the atomic polyhedron that the normal order of d band levels at N (increasing energy) is N_1, N_2, N_1, N_4, N_3. The normal positions of the two N_1 states are located, which was not possible in Section 3.3.

Stern's calculation has been emphasized here because it is the first significant detailed calculation of the cohesive energy of a metal with more than two valence electrons. It is quite encouraging that reasonable results can be obtained. The calculation can be criticized for neglect of the exchange interaction, both between the valence and core electrons and between the d electrons themselves. Stern has, however, evaluated the exchange energy for a determinantal wave function approximately (which was otherwise included in the assumption that each electron moves in the field of an ion: the coulomb hole), and finds that the binding would have to be reduced by 0.75 ry/atom. However, it would then be necessary to include an explicit correction for correlation, as in Eq. (3.21), which Stern does not consider. Further, the exchange interaction of the d electrons would tend to reduce the d bandwidth. Stern's figure of 0.68 ry (9.2 ev) for this quantity is probably too high. The occupied portion has a width of 0.33 ry (4.5 ev) (not considering magnetic effects, however).

Wood (1960, 1962) has studied the d band in both the body-centered and face-centered cubic forms of iron in detail using the augmented plane wave method. Most of the work has been based on a potential previously obtained by Manning (1943), which does not include exchange effects and consequently may cause an overestimate of the bandwidth. Wood obtained energy values for levels belonging to the lowest six bands at 55 points inside 1/48 of the Brillouin zone for bcc iron, including general points. A smaller number of points were considered for fcc iron. The general structure of the band is in agreement with the considerations presented earlier: in particular, the separation of the d levels at the center of the zone, Γ, is less than one-third the separation at the corner, H, and the order of the levels at these points is reversed. One surprising feature is the presence of the largely p-like level N_1' in the middle of the d band levels at the face center. If this is correct, it suggests that there may be an appreciable mixture of p character in the d band wave functions — a possibility which was not considered by Stern. The total width of the d band was 0.47 ry (6.4 ev).

A density of states was constructed from the results of the band calculation. It has two principal maxima separated by a deep minimum. If allowance is made for the ferromagnetic properties of iron by shifting the bands corresponding to electrons of $+$ and $-$ spin, without inclusion of explicit exchange effects, however, the occupied bandwidth is 5.7 ev for majority spin electrons and 3.8 ev for minority spin electrons. The density of states at the Fermi surface is found to be 1.22 electrons per atom per ev. The density of states calculated by Woods is qualitatively similar to that found in an earlier calculation by Belding (1959), whose work was based on the tight-binding approximation.

The numerical sampling technique employed by Wood in the calculation of the density of states probably tends to obscure some of the important structure of this function. Wohlfarth and Cornwell (1961) have calculated a more detailed function through the use of the Slater-Koster (1954) interpolation scheme to extend Wood's calculation and obtain the energy at a larger number of points than he considered. The density of states they obtained is characterized by several high, sharp peaks.

Wood and Stern have studied the d electron wave functions as well as the band structure. Both have emphasized the change in the relative compactness of wave functions of d electron wave functions on going from the bottom of the d band to the top. Wave functions associated with the lower-lying states tend to be smooth and nearly flat at the boundary of the cell, while those pertaining to higher states are small on the cell boundary, and thus must have a higher peak inside the cell. In a qualitative sense, this result follows from elementary considerations since a rapidly varying wave function implies a high kinetic energy. In terms of the simple perturbation treatment of the crystal potential discussed previously, the low-lying states have large long-wavelength components, while the higher states are based on functions of larger $|\mathbf{k}|$. For instance, at the corner H, the low-lying state H_{12} is based on plane waves of wave vector $\mathbf{k} = (2\pi/a) (1, 0, 0)$ whereas the higher state H_{25}' is based on waves with $\mathbf{k} = (2\pi/a) (1, 1, 1)$. If one assumes that the free atom wave functions correspond roughly to states near the middle of the band, then in iron in which the d band is not full, one would expect that the charge distribution in the metal would be somewhat more diffuse than in the free atom.

In addition the charge distribution around any lattice site need not be spherically symmetric, but only have cubic symmetry. This effect has been discussed by Weiss and Freeman (1959), and Stern (1961) has estimated these effects on the basis of his band structure calculation.

It is found that the net spin-up (parallel to the magnetization) charge density in iron as determined from neutron scattering measurements reveals a slight preferential population of ε_g states: 50% of the net spin up density has ε_g symmetry, whereas 40% would be expected in a spherical distribution.

Experimental information concerning the band structure of iron is quite meager, since none of the experiments which have dramatically clarified our knowledge of Fermi surfaces in some of the materials previously discussed have been performed.[11] Cheng et al. (1960) have measured the electron specific heat of iron (and of many body-centered cubic transition metal binary alloys). From this data, the density of states at the Fermi surface comes out to be 28.7 electrons/ry (or 2.11 electrons/ev), considerably larger than the value obtained by Wood. This large value raises serious problems for theories which assume that some — or all — of the d electrons are in atomic orbitals which do not contribute to the Fermi surface. If the density of states at the Fermi surface were an average for the whole band, a width of 3.8 ev would be implied (for the occupied portion). Stern's calculation (neglecting the important band shift caused by the ferromagnetism) gave a density of states of 25 electrons/ry for this quantity.

Walmsley (1962) has reported a measurement of a linear shift of the Fermi level in iron with applied magnetic field. He concludes that all the electrons on the Fermi surface have spin parallel to the magnetization. This result is in conflict with the band structure calculations, and if true, appears to imply a split of the spin-down band into subbands lying above and below the Fermi surface.

Cheng et al. have attempted to use their data on the specific heats to determine a common density of states for the bcc transition elements. To do so, it is necessary to assume an ultrarigid band model: the effect of alloying (or even of changing the element) is only to alter the occupation of a given density of states (hence, just to change the Fermi energy). The density of states so derived has a deep minimum at a position corresponding to the Fermi level of chromium, and subsequently rises

---

[11] At the time of writing, the only measurement pertaining to the Fermi surface in transition metals is of Fawcett and Reed (1962) concerning the magnetoresistance of Nickel. These authors show that the Fermi surface is in contact with the Brillouin zone in the vicinity of the hexagonal face center, L.

to a very high peak (of the order of 9 electrons/ev) midway between chromium and manganese. This behavior is, of course, extremely interesting, but the justification of the rigid band model is doubtful.

Measurements utilizing the Mossbauer effect have shown that the effective magnetic field at a Fe^{57} nucleus in ferromagnetic iron is large and negative ($-$ 333 kgauss) (Hanna et al. 1960). The negative value implies that the direction of the field is opposite to the direction of magnetization. A number of effects which contribute to this field have been discussed by Marshall (1958). There is an exchange polarization of the $4s$ electrons by the $3d$ electrons which causes an excess of $4s$ electrons whose spin is parallel to that of the majority of the $3d$ electrons (Callaway, 1955; Pratt, 1957). There is some admixture of $4s$ wave functions into the $3d$ band, and a contribution from unquenched orbital motion due to spin orbit coupling. These effects tend, however, to produce a field parallel to the magnetization of between 120 and 370 kgauss, according to Goodings and Heine (Goodings and Heine, 1960). There must be a negative contribution to the field amounting to at least 450 kgauss, and this can come only from the core electrons.

A core electron whose spin is parallel to that of the majority of d electrons will experience a stronger exchange interaction than one whose spin is opposite. Hence the wave functions for core electrons of majority and minority spins will be different, and one may calculate a net spin density at the nucleus coming from these core electrons in closed shells (s electrons but $p_{1/2}$ as well if relativistic effects are included). It is perhaps surprising that the polarization of an inner shell may be opposite to the magnetization. This possibility was first pointed out by Heine (1957d), and occurs because the attractive exchange potential tends to pull the core electrons of parallel spin into a region where the $3d$ functions are large, thus reducing their amplitude at the nucleus. If ρ_n is the unbalanced spin density at a nucleus due to an electron in shell n,

$$\rho_n = |\psi_{n\uparrow}(0)|^2 - |\psi_{n\downarrow}(0)|^2$$

The arrows $\uparrow$ and $\downarrow$ indicate spin parallel and antiparallel to that of the majority. The effective magnetic field due to the core is (Goodings and Heine, 1960)

$$H_{\text{core}} = \frac{8\pi}{3} \mu_B \frac{\bar{s}}{s} \sum \rho_n \qquad (3.56)$$

Here $\bar{s}$ is the effective spin from the measured magnetic moment per atom, and s is the spin of the configuration for which the ρ_n are calculated. This field has been estimated by Goodings and Heine and by Freeman and Watson (1960). These authors show that the net contribution from the core electrons is indeed negative in iron (due principally to the $2s$ shell: the $3s$ shell makes a positive contribution), but it appears somewhat more difficult to make the negative contribution as large as required by experiment. However, these calculations are all nonrelativistic, and one might expect a significant improvement if the Dirac equation were employed since an enlarged contribution from the s electrons would be augmented by a contribution from the $p_{1/2}$ electrons.

3.11 Bismuth

A large amount of work has been done in order to determine the Fermi surface in bismuth. The metal has a rhombohedral structure with

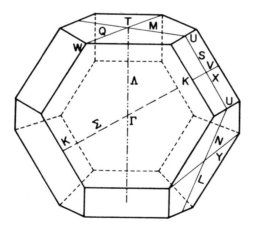

FIG. 15. Brillouin zone for bismuth. Points and lines of symmetry are designated according to the notation of Cohen (1961).

two atoms per unit cell. It is often useful to consider the structure to be a distorted simple cube. Each atom has six neighbors, but the distances are not equal: there are three short bonds and three longer bonds. The

Brillouin zone for this structure is shown in Fig. 15; it may be regarded as a distorted form of the Brillouin zone for the face-centered cubic lattice, and some of the symmetry points have been designated in the same way.

Electronically, bismuth may be characterized as a semimetal. Small numbers of both electrons and holes are present in equilibrium at absolute zero. The concentrations of these are equal but are not known exactly. Estimates range from $2.5 \times 10^{17}/cm^3$ to $6.5 \times 10^{17}/cm^3$ (or from 0.9×10^{-5}/atom to 2.3×10^{-5}/atom) at 0 °K. The stronger evidence favors the smaller number (Zitter, 1962). A number of experiments have suggested that the Fermi surface for the electrons consists of three very prolate ellipsoids (which are not centered at $\mathbf{k} = 0$ — the exact location is not known).[12] If k_x and k_z are chosen along the binary and trigonal axes, one ellipsoid can be represented as:

$$\frac{2m_0}{\hbar^2} E = \alpha_2 k_x^2 + a_2 k_y^2 + a_3 k_z^2 + 2\alpha_4 k_y k_z \qquad (3.57)$$

The values of the α_i are not known exactly. Information concerning these quantities may be obtained from the de Haas-van Alphen effect, cyclotron resonance, ultrasonic attenuation, and far infrared reflectivity and transmissivity. Jain and Koenig (1962) give $\alpha_1 = 119$, $\alpha_2 = 1.31$, $\alpha_3 = 102$, $\alpha_4 = 8.6$; Cohen (1961) quotes $\alpha_1 = 202$, $\alpha_2 = 1.67$, $\alpha_3 = 83.3$, $\alpha_4 = 8.33$. If the latter set is referred to the principal axis system, the effective masses are 0.005, 0.012, and 1.3.[13] Principal axis 3 is tipped 5.8° from the trigonal axis. Estimates of the Fermi energy at 0 °K are in the vicinity of 0.022 ev.

The situation with respect to holes is even less clear. Both light and heavy holes may exist. The existence of light holes occupying one (or possibly two) ellipsoids of revolution about the trigonal axis is established by the work of Galt et al. (1959) and Brandt (1960). The

---

[12] The number of ellipsoids is perhaps not yet definitely determined. There might be six ellipsoids.

[13] The effective masses measured in a cyclotron resonance experiment are not given directly by these numbers. The cyclotron resonance masses must be determined from (3.45), and are of the general form $(m_i m_j)^{1/2}$; depending, of course, on the direction of the magnetic field.

effective masses in this band are $m_1/m_0 = 0.92$; $m_t/m_0 = 0.068$, according to Galt; and 0.7 and 0.05 respectively according to Brandt. The existence of heavy holes is postulated to account for measurements of the electronic specific heat (Kalinkina and Strelkov, 1958; N. E. Phillips, 1960), which appear to show that the density of states at the Fermi energy is larger than can be accounted for by the light electrons and holes previously mentioned. As the two specific heat measurements are in considerable disagreement with each other, the parameters of the heavy hole band cannot be inferred with any certainty. Estimates range from $m^* = 0.66$ to $m^* = 2.5$. Other experiments have not shown heavy hole effects so clearly; however, see Lerner (1962). The Fermi level for the light holes is about 0.012 ev, for the heavy holes, in the range from 0.0006 to 0.002 ev. The overlap between electron and light hole bands is given by the sum of their Fermi energies: this is 0.034 ev.

It is natural to attribute the large curvature of the electron band to the repulsion of some band directly below it. The energy gap between the valence and conduction bands at this point is believed to lie between 0.01 and 0.05 ev. Values of 0.042 (Wolff, 1961b), 0.046 (Weiner, 1962), and 0.047 (Brown *et al.*, 1960) have been quoted; however, Brown, Mavroides, and Lax (1963) find the gap to be 0.015 ± 0.005 ev. The band structure here has been analyzed by Cohen and Blount (1960), Cohen (1961), and Wolff (1961). In bismuth, the electron Fermi energy is not small compared to the band gap, and the parabolic relation between energy and wave vector is only approximate. There is evidence for an energy dependence of the effective mass. The description is based on the work of Kane (1956b) concerning InSb; but in the following, we will largely follow the treatment of Cohen (1961).

In bismuth, it is essential to include spin orbit coupling at all stages of the calculation. Accordingly, we use the Hamiltonian of Eq. (1.64):

$$H = \left(1 - \frac{(E-V)}{2mc^2}\right)\frac{\mathbf{p}^2}{2m} - \frac{\hbar^2}{4m^2c^2}\nabla V \cdot \nabla + \frac{\hbar}{4m^2c^2}\boldsymbol{\sigma} \cdot (\nabla V \times \mathbf{p})$$

We denote the eigenfunctions of this operator as $\psi_{n\rho}(\mathbf{k},\mathbf{r})$, in which $\mathbf{k}$ is the wave vector, n is the band index, and $\rho = 1$ or 2 designates the independent eigenfunctions which are degenerate by time reversal (the bands must be doubly degenerate throughout the zone). The procedure is quite similar to the treatment of a degenerate pair of bands in Section 1.7. Since it is not our intention to make a detailed calculation of the band structure from first principles, we will ignore in the following the terms $(E - V)/2mc^2$ and $(\hbar/4m^2 c^2) \nabla V \cdot \nabla$. We have

$$H\psi_{n\rho}(\mathbf{k}, \mathbf{r}) = E_n(\mathbf{k})\psi_{n\rho}(\mathbf{k}, \mathbf{r}) \tag{3.58}$$

Let the minimum of interest be located at $\mathbf{k}_0$. We expand in terms of the solutions at that point.

$$\psi_{n\rho}(\mathbf{k}, \mathbf{r}) = \sum' A_{n\rho, n'\rho'} e^{i(\mathbf{k} - \mathbf{k}_0) \cdot \mathbf{r}} \psi_{n'\rho'}(\mathbf{k}_0, \mathbf{r})$$

The coefficients $A_{n\rho, n'\rho'}$ satisfy Eq. (1.65). In that equation, we substitute $\hbar \mathbf{s} = \hbar(\mathbf{k} - \mathbf{k}_0) = \mathbf{p}$; and $\mathbf{V} = \mathbf{\Pi}/m$. Then we have

$$\left(E_n(\mathbf{k}_0) + \frac{\mathbf{p}^2}{2m}\right) A_{j\rho, n\tau} + \sum_{m, \mu} A_{j\rho, m\mu} \mathbf{p} \cdot (n\tau|\mathbf{V}|m\mu) = E_j(\mathbf{k}) A_{j\rho, n\tau} \tag{3.59}$$

In (3.60)

$$(n\tau|\mathbf{V}|m\mu) = \int \psi_{n\tau}^*(\mathbf{k}_0, \mathbf{r}) \mathbf{V} \psi_{m\mu}(\mathbf{k}_0, \mathbf{r}) \, d^3r$$

Let us designate the two bands separated by the small gap at $\mathbf{k}_0$ with $n = 0, 1$. We are concerned only with the behavior of these bands. We may eliminate the higher bands by solving for the appropriate coefficients to first order in p, since only $A_{j\rho, 0\tau}$ and $A_{j\rho, 1\tau}$ ($n = 0, 1$) will be large:

$$A_{j\rho, n\tau} = \mathbf{p} \cdot \sum_\mu \left[\frac{(n\tau|\mathbf{V}|0\mu)}{E_j(\mathbf{k}_0) - E_n(\mathbf{k}_0)} A_{j\rho, 0\mu} + \frac{(n\tau|\mathbf{V}|1\mu)}{E_j(\mathbf{k}_0) - E_n(\mathbf{k}_0)} A_{j\rho, 1\mu} \right] \tag{3.60}$$

$$(j = 0, 1)$$

When this expression is substituted into (3.59) the result implies a coupled set of four equations as n, τ run through the values (0, 1) and (1, 2) respectively. Second order off-diagonal terms are neglected. It is most convenient to state these equations in matrix form, and it is evident that we then have the problem of diagonalizing an effective Hamiltonian H' which is a 4×4 matrix. In order to write this down, it is convenient to introduce the abbreviations:

$$K_0 = \mathbf{p} \cdot (0\,1|V|0\,1) + \frac{\mathbf{p}^2}{2m} + \mathbf{p} \cdot \sum_{n\rho} \frac{(0\,1|V|n\,\rho)(n\,\rho|V|0\,1)}{E_0(\mathbf{k}_0) - E_n(\mathbf{k}_0)} \cdot \mathbf{p}$$

$$K_1 = \frac{\mathbf{p}^2}{2m} + \mathbf{p} \cdot \sum_{n\rho} \frac{(1\,1|V|n\,\rho)(n\,\rho|V|1\,1)}{E_1(\mathbf{k}_0) - E_n(\mathbf{k}_0)} \cdot \mathbf{p} \qquad (3.61)$$

$$t = \mathbf{p} \cdot (0\,1|V|1\,1), \qquad u = \mathbf{p} \cdot (0\,1|V|1\,2)$$

$$E_g = E_1(\mathbf{k}) - E_0(\mathbf{k}) \qquad (3.62)$$

The zero of energy is chosen to be the minimum of band 1: $E_1(\mathbf{k}_0) = 0$. Cohen and Blount have shown that time reversal symmetry requires

$$\mathbf{p} \cdot (0\,2|V|1\,1) = -u^*; \qquad \mathbf{p} \cdot (0\,2|V|1\,2) = t^*. \qquad (3.63)$$

The order of the rows and columns in the matrix H' which follows is: 00, 01, 11, 12. Note that a linear term has been included in the first but not the second of Eqs. (3.61) because the lower band (0) at $\mathbf{k}_0$ need not have a minimum there. Band 1, is, however, flat at $\mathbf{k}_0$. The zero of energy is set at the minimum of band 1, i.e., we put $E_1(\mathbf{k}_0) = 0$. Then we have:

$$H' = \begin{pmatrix} K_0 - E_g & 0 & t & u \\ 0 & K_0 - E_g & -u^* & t^* \\ t^* & -u & K_1 & 0 \\ u^* & t & 0 & K_1 \end{pmatrix} \qquad (3.64)$$

The eigenvalues of this matrix are doubly degenerate as required by symmetry, and satisfy the equation:

$$E^2 - (K_0 + K_1 - E_g)E + [K_1(K_0 - E_g) - |t|^2 - |u|^2] = 0 \qquad (3.65)$$

The solutions of (3.65) are:

$$E_{1,0} = \tfrac{1}{2}(K_0 + K_1 - E_g) \pm \sqrt{[\tfrac{1}{2}(K_1 - K_0 + E_g)]^2 + |t|^2 + |u|^2} \qquad (3.66)$$

The minus sign goes with band 0. In the limit of very small $\mathbf{p}$, we get:

$$E_1 = K_1 + \frac{|t|^2 + |u|^2}{E_g} \qquad (3.67\text{a})$$

$$E_0 = K_0 - E_g - \frac{|t|^2 + |u|^2}{E_g} \qquad (3.67\text{b})$$

The contribution of band 0 to the effective mass in band 1 is given by the second term of Eq. (3.67a). The first term contains the contributions from the higher bands. For energies near the Fermi energy, the approximations of (3.67) do not apply, and we must use (3.66) instead.

This model of the band structure contains a larger number of parameters (matrix elements of vector quantities) than was necessary in some of the other models discussed. If the location of the minimum were accurately known, the number of parameters could be reduced. The general formula (3.65) is adequate to account for the highly elongated ellipses that are found: the effective masses imply that the coefficient $p_2{}^2$ in $|t|^2 + |u|^2$ is quite small, possibly zero, leaving the other bands to give the small departure of the effective mass from unity in this direction. The advantage of using the more complicated $E(\mathbf{k})$ expression (3.66) rather than the simple empirical formula (3.57) is that (3.66) includes deviations from parabolic behavior which, because of the small gap, is important at the Fermi energies. A very large (and negative) g factor is also predicted (Cohen and Blount, 1960), which has been observed experimentally (Smith et al., 1960; Everett, 1962).

The location of the band extrema are not accurately known. If the three ellipsoid model is correct, the conduction band minimum could be located at either X or L. A single ellipsoid for holes would imply Γ or T; a pair of ellipsoids would be located on the trigonal axis Λ.

There have been no complete band calculations for bismuth. Two tight binding studies and one nearly free electron calculation have been reported. Morita (1949b) considered the bands in the tight binding approximation, including only nearest neighbor potential integrals and neglecting overlap integrals. Mase (1958) performed a similar study, but included spin orbit coupling. He constructed character tables for the double group of bismuth. The atomic configuration was considered to be $s^2 p^3$, and only the p electrons were included. The two-center approximation was made with nearest neighbors alone considered. The slight overlapping of bands obtaining small numbers of free electrons and holes characterized by small effective masses can be regarded as due to the removal of degeneracies which would be present in a cubic structure. The effect of this rhombohedral distortion is probably smaller than the spin orbit splittings. Harrison (1960c) has studied the Fermi surface in the nearly free electron approximation.

3.12 Graphite

Graphite has a layer structure in which planes of atoms arranged in a hexagonal pattern are stacked and weakly bound to each other. The atoms in a single layer are covalently bonded while the binding between planes is of van der Waals character. In fact, a single layer may be regarded as a large aromatic molecule. In view of these structural characteristics, it is not surprising that many of the electrical properties, including the conductivity, exhibit pronounced anisotropy.

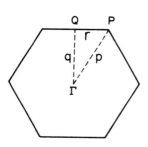
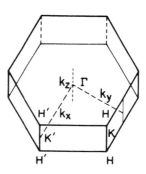

FIG. 16. Brillouin zone for graphite. The left hand drawing shows the zone for a single layer according to Lomer (1955); the right hand drawing is the zone for the three-dimensional crystal according to Slonczewski and Weiss (1958).

In a first attempt to understand the electronic structure of graphite, one might investigate the energy bands pertaining to a single layer. The unit cell for such a layer contains two nonequivalent atoms, and the Brillouin zone is the hexagon shown in Fig. 16. Points and lines of symmetry are indicated according to the notation of Lomer (1955) who has analyzed the symmetry properties of the wave functions in detail.

A simple study of a single graphite layer was performed by Wallace (1947a) who used the tight binding approximation. The carbon atoms are considered to be in the configuration sp^3. The orbitals which are formed from functions of symmetry s, p_x, and p_y lie in the plane of the layer (σ-orbitals). Such orbitals are fully occupied and do not contribute to the conduction process. The fourth electron has a wave function with symmetry p_z; these orbitals are perpendicular to the layer (π orbitals). The π electrons are responsible for conduction; it will be seen that both holes and electrons exist in the band. Wallace considered only potential integrals between π electron wave functions on nearest neighbors of both kinds. He treated these as undetermined parameters, and neglected the lack of orthogonality of functions on different atoms. In this way, he obtained a parameterization of the band structure of graphite which, although crude, indicated some of the most important features. The basic point is that there is degeneracy at the point P between the filled and the empty π bands. This is actually a consequence of the symmetry of the layer model. At absolute zero, the lowest π band would be full, the upper one empty; however there is no energy gap. The energy depends linearly on $|\mathbf{k} - \mathbf{k_0}|$ in the vicinity of P.

Corbato (1956) has made a very detailed calculation on the basis of the single layer model and the tight binding approximation. Bloch functions were constructed including all the $1s$, $2s$, and $2p$ electrons. Two-center overlap and potential integrals were included through ninth neighbors; three-center potential integrals were included through fourth neighbors.

It is worth remarking that the large static diamagnetic susceptibility of graphite can be explained in terms of the layer model as originating in the large gradient of the energy in the vicinity of P.

Calculations which take account of the real three-dimensional nature of the graphite crystal have not been performed in so much detail as the work on a single layer. In this case there are four atoms in the unit cell, associated with two layers. The Brillouin zone is a thin, hexagonal prism shown in Fig. 17. A group theoretical analysis of the three-dimensional lattice has been given by Carter (1953). It has been standard to consider the four bands formed from the p_z orbitals on the four atoms in the unit cell. The interaction between π and σ orbitals has usually been neglected (see, however, Johnston, 1956). The interaction between layers removes some of the degeneracy predicted by the layer model for the edges HKH and H' K' H'. The splittings are small, of the order of 0.1 ev; but this is large compared to kT under normal circumstances and consequently vital in any detailed consideration of transport properties. There are, in the three-dimensional case, two conduction bands and two valence bands of which two are required by symmetry to be degenerate along the edge. Wallace's (1947a) three-dimensional calculation which included only nearest neighbor interactions revealed this degeneracy, but showed no overlap of conduction and valence bands. Calculations by Johnston (1955, 1956) included more distant neighbors, revealed additional degeneracies near the zone edge, a small overlap between valence and conduction bands near the edge, and determined the dependence of E on k_z along the edge.

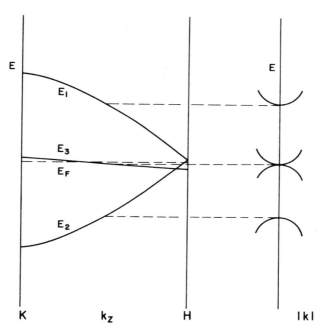

FIG. 17. Energy bands in graphite along the k_z axis according to Nozieres (1958). The drawing on the right shows a small portion of the band structure in a plane perpendicular to this axis.

Recent studies of the magnetic susceptibility of graphite (McClure 1957, 1960) have been based on an analysis of the band structure developed by Slonczewski and Weiss (1958). Since the interesting part of the zone extends only 1% of the distance from the edge of the zone to the center, the **k · p** perturbation theory is useful in determining the variation of the energy perpendicular to the edge. The dependence of the energy on k_z (measured along the zone edge) can be considered in the tight binding approximation — which is equivalent to taking the leading term in a Fourier series expansion of the energy — since the binding between planes is weak. The matrix of $\mathbf{K} \cdot \mathbf{p}$ (where $\mathbf{K}$ is a vector perpendicular to the edge) was constructed on the basis of the states on the edge. The effective Hamiltonian for this problem is given by Slonczewski and Weiss and by McClure (1957; also see Nozieres, 1958). Diagonalization yields the energy levels in the vicinity of the edge. The effective Hamiltonian contains six parameters corresponding to the various momentum matrix elements — which can also be interpreted as integrals from the tight binding approximation.

Possible forms of the energy bands along the edge and perpendicular to it are illustrated in Fig. 17. The surfaces of constant energy in k-space are hyperbolas of revolution. It can be seen from the figure that small numbers of holes and electrons can coexist in the band at 0 °K. The concentrations of holes and electrons are of the order of 10^{-4}/atom. It seems possible to obtain numerical values of the parameters in the Slonczewski-Weiss Hamiltonian which give reasonable agreement with experimental results concerning the diamagnetic susceptibility, de Haas-van Alphen effect, and cyclotron resonance (McClure, 1960).

3.13 Summary

The survey of band structure studies given in this chapter has not been intended to be exhaustive. It should, however, indicate how it is possible to combine experimental and theoretical results to obtain a reasonable knowledge of the electronic structure of particular materials.

Our ability to make quantitative calculations of energy levels from first principles is by no means as complete as is desired. Great progress has, however, been made in recent years. It is now possible to make reasonably accurate calculations of the band structure associated with a given potential, with levels determined at a fairly large number of points in the Brillouin zone. Much additional effort is required before the self-consistency problem is under adequate control, and beyond that lies the problem of the effects of electron-electron and electron-phonon interactions.

When energy band calculations are used qualitatively to describe and interpret the results of experiments, or when simplified Hamiltonians

3.13 SUMMARY

with disposable parameters are used for similar purposes, a very considerable measure of success can often be attained. There is little doubt that the general concepts and language of band theory are adequate to describe a large body of experimental information. This semiempirical usage of band theory is currently quite popular, and probably has many years of life ahead. It remains an unfortunate fact that sometimes the most precise and careful calculations have the least quantitative success.

Chapter 4

Point Impurities and External Fields

4.1 General Discussion

In this chapter, we are concerned with solutions of a Schrödinger equation of the form

$$(H_0 + U)\Psi = i\hbar\, \partial\Psi/\partial t \tag{4.1}$$

where H_0 is a Hamiltonian which contains a periodic potential, and U is a perturbation. We will consider three basic types of perturbation: a point impurity, an external magnetic field, and an external electric field. We will first study uniform fields, but will subsequently discuss the response of band electrons to electromagnetic radiation.

The eigenfunctions of the unperturbed Hamiltonian are Bloch functions:

$$H_0\, \psi_n(\mathbf{k}, \mathbf{r}) = E_n\, \psi_n(\mathbf{k}, \mathbf{r}) \tag{4.2}$$

where $\mathbf{k}$ is the wave vector and n is the band index. The functions ψ have the property that:

$$\psi_n(\mathbf{k}, \mathbf{r}) = e^{i\mathbf{k}\cdot\mathbf{r}}\, u_n(\mathbf{k}, \mathbf{r}) \tag{4.3}$$

in which $u_n(\mathbf{k}, \mathbf{r})$ has the same periodicity as the potential. We recall from Section 2.11 that we can form Wannier functions, which are localized around lattice sites $\mathbf{R}_\nu$ by making a linear combination of the $\psi_n(\mathbf{k}, \mathbf{r})$ which belong to a single band

$$a_n(\mathbf{r} - \mathbf{R}_\nu) = \frac{\Omega^{1/2}}{(2\pi)^{3/2}} \int e^{-i\mathbf{k}\cdot\mathbf{R}_\nu}\, \psi_n(\mathbf{k}, \mathbf{r})\, d^3k \tag{2.168a}$$

in which Ω is the volume of the unit cell. The integral in (2.168a) includes all **k** inside or on the surface of the Brillouin zone. The inverse transformation to (2.168a) is

$$\psi_n(\mathbf{k}, \mathbf{r}) = \frac{\Omega^{1/2}}{(2\pi)^{3/2}} \sum_{\nu} e^{i\mathbf{k}\cdot\mathbf{R}_\nu} a_n(\mathbf{r} - \mathbf{R}_\nu) \tag{2.170}$$

Equation (2.168a) is a special case of a general expression for the expansion of an arbitrary function in the Bloch functions $\psi_n(\mathbf{k}, \mathbf{r})$ which we assume to be a complete set of functions with the following orthonormality properties:

$$\int_{\substack{\text{entire}\\\text{crystal}}} \psi_n^*(\mathbf{k}, \mathbf{r}) \psi_{n'}(\mathbf{k}', \mathbf{r}) \, d^3r = \delta_{nn'} \, \delta(\mathbf{k} - \mathbf{k}') \tag{4.4}$$

The Bloch functions are natural choices for basis functions in which to expand the solutions of (4.1). One writes

$$\Psi(\mathbf{r}) = \sum_n \int \varphi_n(\mathbf{k}) \psi_n(\mathbf{k}, \mathbf{r}) \, d^3k \tag{4.5}$$

It is easily verified that Eq. (4.1) leads to the following equation for the expansion coefficients $\varphi_n(\mathbf{k})$:

$$\left[E_n(\mathbf{k}) - i\hbar \frac{\partial}{\partial t} \right] \varphi_n(\mathbf{k}) + \sum_{n'} \int d^3k' \, \langle n\mathbf{k}|U|n' \, \mathbf{k}' \rangle \, \varphi_{n'}(\mathbf{k}') = 0 \tag{4.6}$$

in which the matrix element is given by

$$\langle n\mathbf{k}|U|n'\mathbf{k}' \rangle = \int \psi_n^*(\mathbf{k}, \mathbf{r}) U \psi_{n'}(\mathbf{k}', \mathbf{r}) \, d^3r \tag{4.7}$$

The representation which is obtained in this way has been called the "crystal momentum representation" by Adams, who has used it extensively (Adams, 1952, 1953, 1957).

A variant of the crystal momentum representation has been introduced by Luttinger and Kohn (1955). They choose a set of basis functions related to Bloch functions for one particular point in the band. Let these functions be denoted by $\chi_n(\mathbf{k}, \mathbf{r})$, and let $\mathbf{k}_0$ be some particular point in the band (usually a maximum or a minimum):

$$\chi_n(\mathbf{k}, \mathbf{r}) = e^{i(\mathbf{k}-\mathbf{k}_0)\cdot \mathbf{r}} \psi_n(\mathbf{k}_0, \mathbf{r}) \tag{4.8}$$

These functions were introduced in Section 1.7. The properties of this set will be discussed in more detail here. The $\chi_n(\mathbf{k}, \mathbf{r})$ may easily be shown to form a complete orthonormal set if the $\psi_n(\mathbf{k}, \mathbf{r})$ do. There are two points to consider: Let $f(\mathbf{r})$ be an arbitrary function of position which can be expanded in the Bloch functions

$$f(\mathbf{r}) = \sum_n \int g_n(\mathbf{k}) \psi_n(\mathbf{k}, \mathbf{r}) \, d^3k$$

$$= \sum_n \int g_n(\mathbf{k}) \, e^{i\mathbf{k}\cdot\mathbf{r}} u_n(\mathbf{k}, \mathbf{r}) \, d^3k \tag{4.9}$$

Since $u_n(\mathbf{k}, \mathbf{r})$ and $u_n(\mathbf{k}_0, \mathbf{r})$ have the same periodicity, we can write

$$u_n(\mathbf{k}, \mathbf{r}) = \sum_{n'} b_{n,n'}(\mathbf{k}) \, u_{n'}(\mathbf{k}_0, \mathbf{r}) \tag{4.10}$$

If we substitute this into (4.9), we have

$$f(\mathbf{r}) = \sum_n \int \tilde{g}_n(\mathbf{k}) \, \chi_n(\mathbf{k}, \mathbf{r}) \, d^3k \tag{4.11}$$

where

$$\tilde{g}_n(\mathbf{k}) = e^{i\mathbf{k}_0\cdot\mathbf{r}} \sum_{n'} b_{n'n}(\mathbf{k}) g_{n'}(\mathbf{k}) \tag{4.12}$$

Hence, neglecting questions of convergence, if $f(\mathbf{r})$ can be expanded in Bloch functions, it can also be expanded in the χ_n. To prove the orthonormality of the χ_n, consider the integral

$$\int_{\substack{\text{entire}\\\text{crystal}}} \chi_{n'}^*(\mathbf{k}', \mathbf{r}) \chi_n(\mathbf{k}, \mathbf{r}) \, d^3r = \int e^{i(\mathbf{k}-\mathbf{k}')\cdot\mathbf{r}} \, u_{n'}^*(\mathbf{k}_0, \mathbf{r}) \, u_n(\mathbf{k}_0, \mathbf{r}) d^3r \tag{4.13}$$

The function $u_{n'}^*(\mathbf{k}_0, \mathbf{r})u_n(\mathbf{k}_0, \mathbf{r})$ is periodic in the crystal and so may be expressed as a Fourier series

$$u_{n'}^*(\mathbf{k}_0, \mathbf{r})u_n(\mathbf{k}_0, \mathbf{r}) = \sum_m B_m^{n'n} e^{-i\mathbf{K}_m \cdot \mathbf{r}} \tag{4.14}$$

On substitution into (4.13), we obtain

$$(2\pi)^3 \sum_m \delta(\mathbf{k} - \mathbf{K}_m - \mathbf{k}') B_m^{n'n} \tag{4.15}$$

We suppose $\mathbf{k}$ and $\mathbf{k}'$ are inside the zone so that $\mathbf{k} - \mathbf{k}'$ is not a reciprocal lattice vector. Then only the term with $m = 0$ in (4.15) can contribute. However,

$$B_m^{n'n} = \frac{1}{\Omega_0} \int_{\text{cell}} u_{n'}^*(\mathbf{k}_0, \mathbf{r}) u_n(\mathbf{k}, \mathbf{r}) e^{i\mathbf{K}_m \cdot \mathbf{r}} \, d^3r$$

The u_n are orthogonal in a unit cell, and are normalized[1] so that

$$B_0^{n'n} = \frac{1}{\Omega_0} \int_{\text{cell}} u_{n'}^*(\mathbf{k}_0, \mathbf{r}) u_n(\mathbf{k}_0, \mathbf{r}) \, d^3r = \frac{1}{(2\pi)^3} \delta_{nn'} \tag{4.16}$$

It then follows that

$$\int \chi_{n'}^*(\mathbf{k}', \mathbf{r}) \chi_n(\mathbf{k}, \mathbf{r}) \, d^3r = \delta_{nn'} \delta(\mathbf{k} - \mathbf{k}') \tag{4.17}$$

In accord with (4.11), we expand the solution of (4.1) in the Kohn-Luttinger functions $\chi_n(\mathbf{k}, \mathbf{r})$

$$\Psi(\mathbf{r}) = \sum_n \int A_n(\mathbf{k}) \chi_n(\mathbf{k}, \mathbf{r}) \, d^3k \tag{4.18}$$

---

[1] This choice of normalization for the u_n can be seen to be consistent with the normalization of the Bloch functions [Eq. (4.4)] when the summation relations of Appendix 2 are used.

The representation of the Hamiltonian on the basis of the $\chi_n(\mathbf{k}, \mathbf{r})$ is somewhat more complicated than in the case of the $\psi_n(\mathbf{k}, \mathbf{r})$ since the χ's are not eigenfunctions of H_0. For simplicity in the calculation of matrix elements, we shall set $\mathbf{k}_0 = 0$ and ignore any possible degeneracy of the states at $\mathbf{k}_0$. Afterwards, these restrictions will be removed. The matrix elements of H_0 are then determined in the following manner:

$$(n\mathbf{k}|H_0|n'\,\mathbf{k}') \equiv \int_{\text{entire crystal}} \chi_n^*(\mathbf{k}, \mathbf{r}) H_0 \, \chi_n(\mathbf{k}', \mathbf{r}) \, d^3r \tag{4.19}$$

$$= \int e^{i(\mathbf{k}' - \mathbf{k}) \cdot \mathbf{r}} u_n^*(\mathbf{r}) \left[E_{n'} + \hbar \frac{\mathbf{k}' \cdot \mathbf{p}}{m} + \frac{\hbar^2 \mathbf{k}'^2}{2m} \right] u_{n'}(\mathbf{r}) \, d^3r$$

We will use parentheses to indicate the matrix elements in the Kohn-Luttinger formalism, and reserve the angular brackets $\langle \rangle$ for the CMR. Also, we have set $E_{n'}(\mathbf{k}_0) = E_{n'}$ and $u_n(\mathbf{k}_0, \mathbf{r}) = u_n(\mathbf{r})$. Since $\mathbf{p} = (\hbar/i)\nabla$, the function multiplying the exponential in Eq. (4.19) is periodic, so that the argument that leads to (4.17) also applies in this case. We can now write

$$(n\mathbf{k}|H_0|n'\,\mathbf{k}') = \frac{(2\pi)^3}{\Omega_0} \delta(\mathbf{k} - \mathbf{k}') \int_{\text{cell}} u_n^*(\mathbf{r}) \left[E_{n'} + \hbar \frac{\mathbf{k} \cdot \mathbf{p}}{m} + \frac{\hbar^2 \mathbf{k}^2}{2m} \right] u_{n'}(\mathbf{r}) \, d^3r$$

$$= \delta(\mathbf{k} - \mathbf{k}') \left[\delta_{nn'} \left(E_n + \frac{\hbar^2 \mathbf{k}^2}{2m} \right) + \hbar \frac{\mathbf{k} \cdot \mathbf{p}_{nn'}}{m} \right] \tag{4.20}$$

In the last line of (4.20) we have introduced

$$\mathbf{p}_{nn'} = \frac{(2\pi)^3}{\Omega_0} \int u_n^*(\mathbf{r}) \mathbf{p} u_{n'}(\mathbf{r}) \, d^3r \tag{4.21}$$

Consequently, the equation satisfied by the expansion coefficients $A_n(\mathbf{k})$ in Eq. (4.18) is

$$\left[E_n + \frac{\hbar^2 \mathbf{k}^2}{2m} - i\hbar \frac{\partial}{\partial t} \right] A_n(\mathbf{k}) + \frac{\hbar \mathbf{k}}{m} \cdot \sum_{n'} \mathbf{p}_{nn'} A_{n'}(\mathbf{k}) \tag{4.22}$$

$$+ \sum_{n'} \int d^3 k' (n\mathbf{k}|U|n'\,\mathbf{k}') A_{n'}(\mathbf{k}') = 0$$

220 CHAPTER 4. POINT IMPURITIES AND EXTERNAL FIELDS

The Wannier functions (Wannier, 1937; Slater, 1949; Adams, 1952; Koster, 1954; Koster and Slater, 1954a,b; des Cloizeaux, 1963) are a third possible set of functions for the expansion of the wave function of Eq. (4.1). Although these functions are not solutions of the Schrödinger equation, they are orthonormal and localized so that the set is particularly useful in the case of localized perturbations. In Section 2.11 we verified their orthonormality, and saw that they satisfied the equation

$$H_0 \, a_n(\mathbf{r} - \mathbf{R}_\nu) = \sum_j{}' \mathscr{E}_n(\mathbf{R}_j - \mathbf{R}_\nu) a_n(\mathbf{r} - \mathbf{R}_j)$$

where $\mathscr{E}_n(\mathbf{R}_i)$ is a Fourier coefficient of the energy (note that the band index n is the same on both sides):

$$E_n(\mathbf{k}) = \sum_l{}' \mathscr{E}_n(\mathbf{R}_l) \, e^{-i\mathbf{k} \cdot \mathbf{R}_l}$$

The wave function is expanded in the Wannier function by writing

$$\Psi = \sum_{\nu, n}{}' B_n(\mathbf{R}_\nu) a_n(\mathbf{r} - \mathbf{R}_\nu) \qquad (4.23)$$

We find that the B_n satisfy the equation:

$$\sum_{\nu, l}{}' \left[\left(\mathscr{E}_n(\mathbf{R}_\mu - \mathbf{R}_\nu) - i\hbar \delta_{\mu\nu} \frac{\partial}{\partial t} \right) \delta_{nl} + U_{nl}(\mathbf{R}_\mu, \mathbf{R}_\nu) \right] B_l(\mathbf{R}_\nu) = 0 \qquad (4.24)$$

In equation (4.24), $U(\mathbf{R}_\mu, \mathbf{R}_\nu)$ is the matrix element of the perturbation between two Wannier functions:

$$U_{nl}(\mathbf{R}_\mu, \mathbf{R}_\nu) = \int a_n^*(\mathbf{r} - \mathbf{R}_\mu) U(\mathbf{r}) a_l(\mathbf{r} - \mathbf{R}_\nu) \, d^3r \qquad (4.25)$$

A discrete set of functions has been employed for the expansion of the wave function in the present case (whereas in both the CMR and the Kohn-Luttinger representations, the functions are characterized by a continuous variable $\mathbf{k}$). Equations (4.24) are difference equations. However, for slowly varying potentials, they may be replaced by a differential equation.

4.2 Point Impurities

As an aid in understanding impurity states, we will study a simple example first discussed by Koster and Slater (Koster and Slater, 1954a). We suppose that the matrix element $\langle n\mathbf{k}|U|n'\,\mathbf{k}'\rangle$ which appears in Eq. (4.6) has the value

$$\langle n\mathbf{k}|U|n'\,\mathbf{k}'\rangle = [V_0\Omega/(2\pi)^3]\,\delta_{nn'} \qquad (4.26)$$

in which Ω is the volume of the atomic cell and $(2\pi)^3/\Omega$ is the volume of the Brillouin zone. Except for diagonality in the band index, the matrix element corresponds to a delta function potential of strength $V_0\Omega/(2\pi)^3$ located at the origin. The solution is stationary with energy E. Equation (4.6) becomes

$$(E_n(\mathbf{k}) - E)\varphi_n(\mathbf{k}) + \frac{V_0\Omega}{(2\pi)^3}\int \varphi_n(\mathbf{k}')\,d^3k' = 0 \qquad (4.27)$$

An equation for the energy is obtained by dividing (4.27) by $E_n(\mathbf{k}) - E$ and integrating over the Brillouin zone. We assume for the present that $E_n(\mathbf{k}) - E$ does not vanish:

$$\frac{\Omega V_0}{(2\pi)^3}\int d^3k\,\frac{1}{E_n(\mathbf{k}) - E} = V_0 \int \frac{G(E')}{E' - E}\,dE' = -1 \qquad (4.28)$$

We have used the definition of the density of states per volume Ω and a single spin direction. The nature of the solutions of Eq. (4.28) may be more readily appreciated if we imagine that the unperturbed levels are discrete (N of them, say). In this case we consider a function

$$f(E) = \frac{1}{N}\sum_n \frac{1}{E - E_n}$$

The function $f(E)$ has poles at each of the unperturbed energies, and in between varies from $-\infty$ to $+\infty$ (see Fig. 18). The energies of the perturbed states are the energies for which the function has the value $1/V_0$. Each eigenvalue is shifted by the perturbation, but not so strongly as to cross the energy of an unperturbed solution. For energies lower than the lowest eigenvalues, the function is negative, going to zero at $E = -\infty$. If V_0 is negative, there will also be a root in this region, representing a state which is split off from the band. (If V_0 is positive, the split-off state is at the top of the band.) The total number of states, N, is not changed by the perturbation. In the limit of a continuous distribution of energies, the energies of the states in the band are unaltered,

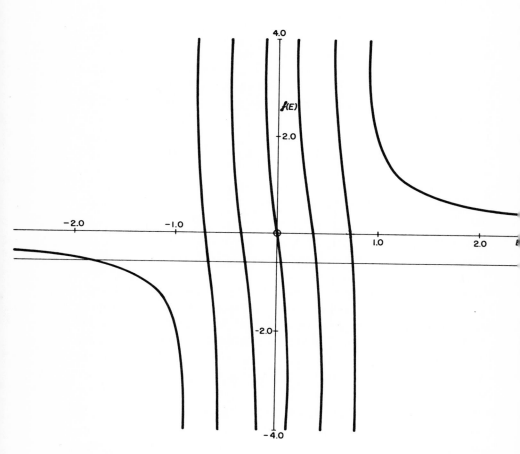

FIG. 18. The determination of the energies of impurity states in the Slater-Koster model is illustrated in the case of a "band" of six levels. The function $f(E) = (1/N) \Sigma_n 1/(E - E_n)$ is shown. The energies of the perturbed states are found from the intersections of $f(E)$ with the horizontal line which corresponds to $V_0 = -5/3$.

except that a single state drops below the band for negative V_0 or rises above it for positive V_0.

We must know $E(\mathbf{k})$ to determine the energy of the split-off state. The nearest neighbor tight binding formula for a one-dimensional crystal furnishes a simple example:

$$E(k) = E_0 + 2E_1 \cos kd \qquad (4.29)$$

where d is the distance between atoms (we suppose E_1 is negative). Then (4.28) becomes

$$\frac{dV_0}{2\pi} \int_{-\pi/d}^{\pi/d} \frac{dk}{E_0 - E + 2E_1 \cos kd} = -1 \qquad (4.30)$$

(since in the one-dimensional case, the Brillouin zone has a "volume" of $2\pi/d$). The integral in (4.30) is standard, and we find after solving the resulting equation

$$E = E_0 + 2E_1 \left[1 + \frac{V_0^2}{4E_1^2} \right]^{1/2} \qquad (4.31)$$

This result was also derived by Koster and Slater from the difference equation (4.24) in the Wannier function representation.

The energy of the lowest state in the original band was $E_0 + 2E_1$. The energy of the split-off state depends quadratically on V_0 if $V_0 \ll 2E_1$, but if V_0 is large, the dependence on V_0 becomes linear, and the split-off state goes to an energy $E_0 + V_0$.

In order to study the wave function of the split-off state, it is desirable to transform from the representation involving Bloch functions to that based on Wannier functions. The transformation may be determined as follows: We multiply Eq. (4.23) by $a_{n'}^*(\mathbf{r} - \mathbf{R}_\nu)$ and integrate over $\mathbf{r}$. Since the Wannier functions are orthonormal, we have

$$B_n(\mathbf{R}_\nu) = \int \Psi(\mathbf{r}) a_n^*(\mathbf{r} - \mathbf{R}_\nu) d^3r \qquad (4.32)$$

We now substitute Eq. (4.5) for $\Psi(\mathbf{r})$ into (4.32), and expand the Bloch functions in Wannier functions, and then make use of the orthonormality of the Wannier functions. The result is

$$B_n(\mathbf{R}_\nu) = \frac{\Omega^{1/2}}{(2\pi)^{3/2}} \int e^{i\mathbf{k}\cdot\mathbf{R}_\nu} \varphi_n(\mathbf{k}) \, d^3k \qquad (4.33)$$

It follows from (4.27) that

$$\phi_n(\mathbf{k}) = -\frac{\Omega^{1/2}}{(2\pi)^{3/2}} V_0 \frac{B_n(0)}{E_n(\mathbf{k}) - E} \qquad (4.34)$$

This is substituted into (4.33). We find

$$B_n(\mathbf{R}_\nu) = \frac{V_0 \Omega}{(2\pi)^3} B(0) \int \frac{e^{i\mathbf{k}\cdot\mathbf{R}_\nu}}{E - E_n(\mathbf{k})} d^3k \qquad (4.35a)$$

In the one-dimensional model, we put $\mathbf{k}\cdot\mathbf{R}_\nu = k\nu d$ where ν is an integer and d is the distance between atoms. Then

$$B_n(\nu) \propto \int_{-\pi/d}^{\pi/d} e^{ik\nu d} \frac{1}{E_n(k) - E} dk \qquad (4.35b)$$

It is necessary to substitute Eqs. (4.29) and (4.31) into (4.35), and then carry out the integration. The work is simplified if we define a quantity γ by

$$\sinh \gamma d = \frac{V_0}{2E_1} \qquad (4.36)$$

(we are assuming that V_0 and E_1 have the same sign). Using this (and dropping the band index, n)

$$B(\nu) \propto \frac{1}{2E_1} \int_{-\pi/d}^{\pi/d} \frac{e^{ik\nu d}}{\cosh \gamma d + \cos kd} dk \qquad (4.37)$$

Only the even part of the exponential need be retained. The integral can easily be put into a standard form.[2] The result gives:

$$B(\nu) = B(0) e^{-\gamma|\nu|d} \qquad (4.38)$$

---

[2] Standard form:

$$\int_{-\pi}^{\pi} \frac{\cos nx}{1 + p\cos x} dx = \frac{2\pi}{\sqrt{1-p^2}} \left[\frac{\sqrt{1-p^2}-1}{p}\right]^n$$

Thus the component of the wave function in successive cells decreases exponentially with increasing distance from the impurity, as would be intuitively expected for a bound state.

An exact solution of the model problem has also been obtained by Koster and Slater (1954b) for a single impurity in a three-dimensional simple cubic crystal for which the energy may be expressed as

$$E = E_0 + 2E_1(\cos k_x a + \cos k_y a + \cos k_z a) \qquad (4.39)$$

(a is the lattice constant). In this case, we have in place of Eq. (4.30):

$$\frac{\Omega V_0}{(2\pi)^3} \int_{-\pi/a}^{\pi/a} dk_x \int_{-\pi/a}^{\pi/a} dk_y \int_{-\pi/a}^{\pi/a} dk_z \, [E_0 - E + 2E_1(\cos k_x a + \cos k_y a + \cos k_z a)]^{-1}$$

$$= \frac{V_0'}{\pi^3} \int_0^{\pi} dx \int_0^{\pi} dy \int_0^{\pi} dz \, [E' - (\cos x + \cos y + \cos z)]^{-1} = -1 \qquad (4.40)$$

In the second line of (4.40), we have put $x = k_x a$, etc.; $E' = (E_0 - E)/2|E_1|$, and $V_0' = V_0/2|E_1|$. This integral may be simplified by introducing an auxiliary variable t. Instead of the integral in (4.40), we write

$$\frac{V_0'}{\pi^3} \int_0^{\infty} dt \int_0^{\pi} dx \int_0^{\pi} dy \int_0^{\pi} dz \, \exp\{-[E' - (\cos x + \cos y + \cos z)]t\}$$

The integral may be simplified with the substitution of an integral expression for the Bessel function of imaginary argument $I_0(t)$ (Jeffreys and Jeffreys, 1950)

$$I_0(t) = \frac{1}{\pi} \int_0^{\pi} e^{t \cos x} dx \qquad (4.41)$$

Then we have

$$V_0' \int_0^{\infty} dt \, e^{-E't} I_0^3(t) \, dt = -1 \qquad (4.42)$$

This equation must be solved numerically. We can gain insight into the nature of the solution by some simple approximations.

If V_0' is large and negative, we expect the split-off state to lie well below the band, which means that E' must be large. Then the integral may be approximated by retaining only the lowest term in $I_0(t)$, which is unity. Then we get in this limit:

$$\frac{V_0'}{E'} = -1, \quad \text{or} \quad E = E_0 + V_0 \qquad (4.43)$$

The other limit is more difficult. In the limit of large t, $I_0(t) \propto e^t$. The integral in Eq. (4.42) must then diverge if $E' < 3$, so that no solution is obtained. The limiting condition $E' = 3$ corresponds to $E = E_0 + 6E_1$, which is the energy of the state at the bottom of the band ($k = 0$). The impurity state must fall below the band. For $E' = 3$, the integral in (4.42) has been evaluated by Koster and Slater, who find for this case that $V_0 \approx -4|E_1|$. We see that, in contradiction to the one-dimensional example, there is a minimum value of V_0 below which a split-off state will not be formed. It is possible to obtain an expansion of the integral in (4.42) in the neighborhood of $E' = 3$. When this is done, one finds that the energy of the bound state depends quadratically on the difference between the potential, and the value that first produces a bound state.

To show this, we define the quantity $J(E')$ as the integral in (4.40):

$$J(E') = \int_0^\pi dx \int_0^\pi dy \int_0^\pi dz [E' - (\cos x + \cos y + \cos z)]^{-1} \qquad (4.44)$$

We wish to find the value of this integral for E' slightly larger than 3. Thus we calculate

$$J(E') - J(3) = \int_0^\pi dx \int_0^\pi dy \int_0^\pi dz \times$$

$$\left[\frac{1}{E' - (\cos x + \cos y + \cos z)} - \frac{1}{3 - (\cos x + \cos y + \cos z)} \right]$$

If we expand the cosines in powers of x, then the integrals are dominated by the values at small x, y, z. This in turn suggests that we go to spherical coordinates, and make the upper limit infinite.

$$J(E') - J(3) = \frac{4\pi}{8} \int_0^\infty \left[\frac{1}{E' - 3 + r^2/2} - \frac{1}{r^2/2} \right] r^2 \, dr$$

This procedure is valid when $E' - 3$ is small. The factor of 8 in the denominator results since only one octant of a sphere is considered. The integral can now be reduced to an elementary one, and we find

$$J(E') - J(3) = \frac{-\pi^2}{\sqrt{2}} (E' - 3)^{1/2} \tag{4.45}$$

This result is substituted in (4.40), from which we find

$$E' - 3 = 2\pi^2 \left[\frac{J(3)}{\pi^3} + \frac{1}{V_0'} \right]^2 \tag{4.46}$$

We do not get a bound state unless $V_0' \leq -\pi^3/J(3)$. We are only interested in small values of $E' - 3$, in which case V_0' will not differ greatly from the initial value. The bracket may then be expanded to give

$$E' - 3 = 2\pi^2 \left(\frac{J(3)}{\pi^3} \right)^4 \left[V_0' + \frac{\pi^3}{J(3)} \right]^2 \tag{4.47}$$

This confirms the quadratic dependence of E' previously mentioned. The quantity $J(3)/\pi^3$ has the value 0.493.

An alternative attack on the point impurity problem in which one works directly with the difference equation for the Wannier functions is of some interest, particularly in view of recent applications to the problem of impurity states in metals (Wolff, 1961a; Clogston, 1962). Here we wish to describe not only the localized states outside the band but the virtual states or resonances in the band as well. The method of solution was given by Lax (1954). A Green's function is found for Eq. (4.24), which reads (for a stationary state)

$$\sum_{\nu, l} \{ [\mathscr{E}_n(\mathbf{R}_\mu - \mathbf{R}_\nu) - E\delta_{\mu\nu}]\delta_{nl} + U_n(\mathbf{R}_\mu, \mathbf{R}_\nu)\} B_l(\mathbf{R}_\nu) = 0 \tag{4.48}$$

Let us consider the function

$$\mathscr{G}_n(\mathbf{R}_\nu - \mathbf{R}_j) = \frac{\Omega}{(2\pi)^3} \int d^3k \, \frac{e^{i\mathbf{k} \cdot (\mathbf{R}_\nu - \mathbf{R}_j)}}{E - E_n(\mathbf{k})} \qquad (4.49)$$

Discussion of the treatment of the singularities of the denominator will be postponed temporarily. Let us calculate

$$\sum_\nu {}' [\mathscr{E}_n(\mathbf{R}_\mu - \mathbf{R}_\nu) - E\delta_{\mu\nu}] \mathscr{G}_n(\mathbf{R}_\nu - \mathbf{R}_j)$$

$$= \sum_\nu {}' \frac{\Omega}{(2\pi)^3} \int d^3k' \, e^{i\mathbf{k}' \cdot (\mathbf{R}_\mu - \mathbf{R}_\nu)} (E_n(\mathbf{k}') - E) \mathscr{G}_n(\mathbf{R}_\nu - \mathbf{R}_j)$$

$$= \frac{-\Omega^2}{(2\pi)^6} \int d^3k' \int d^3k \, e^{-i(\mathbf{k} \cdot \mathbf{R}_j - \mathbf{k}' \cdot \mathbf{R}_\mu)} \left(\sum_\nu {}' e^{-i(\mathbf{k}' - \mathbf{k}) \cdot \mathbf{R}_\nu} \right) \frac{E_n(\mathbf{k}') - E}{E_n(\mathbf{k}) - E}$$

$$= \frac{-\Omega}{(2\pi)^3} \int d^3k' \, e^{-i\mathbf{k} \cdot (\mathbf{R}_j - \mathbf{R}_\mu)} = -\delta_{\mu,j}$$

We have employed summation and integral relations from Appendix 2 [Eqs. (A2.9b) and (A2.10)]. Evidently $\mathscr{G}_n$ is a Green's function for the nth band.

It is easy to verify that

$$B_n(\mathbf{R}_j) = e^{i\mathbf{k} \cdot \mathbf{R}_j} \delta_{n,m} + \sum_{\nu,\mu} {}' \mathscr{G}_n(\mathbf{R}_j - \mathbf{R}_\nu) U_n(\mathbf{R}_\nu, \mathbf{R}_\mu) B_n(\mathbf{R}_\mu) \qquad (4.50)$$

is a formal equation, similar to an integral equation, which replaces (4.48). The first term, which represents an incoming wave in the mth band, is a solution of the homogeneous equation. Except for this term, which exists only when E lies inside the band, Eq. (4.50) is the same as (4.35a) for the Slater-Koster model.

The calculation of the Green's function may be difficult. First, it is necessary to decide how one intends to go around the singularities of the denominator. The Green's function is not well defined until this is specified. One can obtain standing wave solutions or propagating solutions. The propagating wave solutions are the more relevant for our purposes. These are obtained by inserting a small, positive imaginary part in the denominator, which is allowed to go to zero after evaluation

$$\mathscr{G}_n(\mathbf{R}_\nu - \mathbf{R}_j) = \lim_{\varepsilon \to 0^+} \frac{\Omega}{(2\pi)^3} \int d^3k \frac{e^{i\mathbf{k}\cdot(\mathbf{R}_\nu - \mathbf{R}_j)}}{E - E_n(\mathbf{k}) + i\varepsilon} \qquad (4.51)$$

The quantity $\mathscr{G}_n(0)$ is of considerable interest. This can be expressed formally in terms of the density of states, $G(E)$, for a volume Ω, and a single spin direction which is given by

$$G(E) = \frac{\Omega}{(2\pi)^3} \frac{d}{dE} \int d^3k$$

in which the integral includes all $\mathbf{k}$ space in which the energy is E or less.

$$\mathscr{G}_n(0) = \lim_{\varepsilon \to 0^+} \int \frac{G(E')\,dE'}{E - E' + i\varepsilon} = P \int \frac{G(E')}{E - E'}\,dE - i\pi G(E) \qquad (4.52)$$

in which the symbol P indicates that the principal value of the integral is to be obtained.

We also require the asymptotic form of the Green's function. In the limit of large $|R|$, Eq. (4.51) for $\mathscr{G}_n(\mathbf{R})$ may be evaluated by the method of stationary phase. A general expression has been given by Koster (1954) and Lifshitz (1956). We will confine our attention to the situation in which E is near the bottom of a spherical band. We approximate $E(\mathbf{k})$ as:

$$E(\mathbf{k}) = E_0 + E_2 \mathbf{k}^2$$

Then, for large R

$$\mathscr{G}_n(\mathbf{R}) = \lim_{\varepsilon \to 0^+} \frac{\Omega}{2\pi^2 R} \int_0^\infty \frac{k' \sin k' R}{E - E_0 - E_2 k'^2 + i\varepsilon}\,dk' = \frac{-\Omega\, e^{ikR}}{4\pi E_2 R} \qquad (4.53)$$

where $k = [(E - E_0)/E_2]^{1/2}$.

We now consider the Slater-Koster model in which the perturbation is localized in the cell at the origin

$$U_{nl}(\mathbf{R}_\nu, \mathbf{R}_\mu) = V_0\, \delta_{ln}\, \delta_{\mu 0}\, \delta_{\nu 0} \qquad (4.54)$$

It is possible to solve the equation (4.50) for $B_n(\mathbf{R}_j)$ in this model. The result is (taking the incident wave to be in the nth band).

$$B_n(\mathbf{R}_j) = e^{i\mathbf{k}\cdot\mathbf{R}_j} + \frac{V_0\,\mathscr{G}_n(\mathbf{R}_j)}{1 - V_0\,\mathscr{G}_n(0)} \tag{4.55}$$

We are interested in the limit of large $\mathbf{R}_j$, in which the asymptotic expression for $\mathscr{G}_n$ may be used

$$B_n(\mathbf{R}_j) = e^{i\mathbf{k}\cdot\mathbf{R}_j} - \frac{V_0\Omega}{4\pi E_2(1 - V_0\,\mathscr{G}_n(0))}\,\frac{e^{ikR_j}}{R_j} \tag{4.56}$$

The solution evidently contains an incident wave and a spherical scattered wave. The scattering is s-wave because the perturbing potential is localized in the cell at the origin. The conventional relation between the scattering amplitude and the phase shift may be applied to determine the phase shift. If $f(\theta)$ is the scattering amplitude, we have (L. I. Schiff, 1955)

$$f(\theta) = \frac{1}{2ik}\sum_{l=0}^{\infty}(2l+1)(e^{2i\delta_l}-1)P_l(\cos\theta) \tag{4.57}$$

In the present case, only the term with $l = 0$ is present, and the scattering amplitude is just the coefficient of the scattered wave. The equation for δ_0 is

$$\frac{e^{i\delta_0}\sin\delta_0}{k} = \frac{-V_0\Omega}{4\pi E_2(1 - V_0\,\mathscr{G}_n(0))} \tag{4.58}$$

If use is made of the expression for the density of states of a spherical band in the form,

$$G(E) = \frac{\Omega k}{4\pi^2 E_2}$$

Eq. (4.58) may be solved and a real δ_0 determined.

$$\delta_0 = \tan^{-1}\left[\frac{-\pi V_0 G(E)}{1 - I(E)V_0}\right] \tag{4.59}$$

in which $I(E)$ is the real part of $\mathscr{G}_n(0)$: namely

$$\mathscr{G}_n(0) = I(E) - i\pi G(E)$$

An important sum rule has been derived by Friedel (1958) concerning the phase shifts δ_l for more realistic impurity potentials. Suppose an impurity center is introduced into a metal. The center has a nuclear charge Z units greater than that of the host. We suppose the scattered wave, when analyzed into its component angular momenta shows a set of phase shifts δ_l (we do not restrict ourselves to the Slater-Koster model at this point): Suppose further that the metal is in the form of a sphere whose radius (a macroscopic quantity) is R. All the electron wave functions are quantized within this sphere. For the lth partial scattered wave, we have

$$kR + \delta_l(E) = N\pi$$

where N is an integer. For a wave of wave vector $\mathbf{k} + \Delta\mathbf{k}$, the smallest change on the right can be π. Hence, the interval in k between two successive waves is

$$\Delta k = \frac{\pi}{R + (d\delta_l/dk)}$$

The number of states per unit increment of k is just the reciprocal of this quantity. Evidently the perturbation changes the number of states by an amount $1/\pi (d\delta_l/dk)$. The total change in the number of states up to some particular value of k, considering all l values and the $(2l+1)$ substates for each l, is just $(1/\pi)\, \Sigma_l (2l+1)\delta_l(k)$. In order that the perturbation be screened at large distances, it is necessary to bring $Z/2$ states below the Fermi level, since each screening "bound" level can be occupied by one electron of each spin. Hence we have

$$\frac{2}{\pi} \sum (2l+1)\delta_l(E_\mathrm{F}) = Z \tag{4.60}$$

This is the Friedel sum rule.

We shall now discuss the behavior of $I(E)$ and the phase shift δ_0 in the Slater-Koster model. The determination of $I(E)$ is, in general, quite difficult since the density of states $G(E)$ may be quite complicated.[3] Some general features may be determined, however, without a detailed calculation. The density of states is finite, zero outside of a finite interval on the energy axis (bounded by E_0 and E_m, say), continuous and positive everywhere in that interval. Then for $E < E_0$, $I(E)$ is negative, but increasing (decreasing in magnitude). For $E > E_m$, $I(E)$ is positive

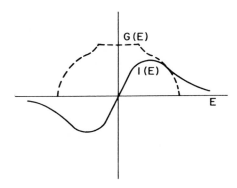

FIG. 19. $I(E)$, the Hilbert transform of the density of states is shown schematically for a simple $G(E)$.

and decreasing. Hence the qualitative behavior of the function must be as shown in Fig. 19. Additional maxima and minima might conceivably be present for functions $G(E)$ more complex than the one illustrated here.

---

[3] For the $E(k)$ given in (4.39), the expressions for $I(E)$ and $G(E)$ may be reduced to single integrals, which have been evaluated numerically by Koster and Slater (1954b)

$$I(E') = \int_0^\infty J_0^3(t) \sin E' t \, dt$$

$$G(E') = \frac{1}{\pi} \int_0^\infty J_0^3(t) \cos E' t \, dt$$

in which J_0 is an ordinary Bessel function and $E' = (E_0 - E)/2|E_1|$.

The behavior of the phase shift as a function of energy is governed by the intersections of $I(E)$ with $1/V_0$. There are evidently either zero or an even number of these. If an intersection occurs for $E < E_0$, a bound state exists, as was shown in the previous section (we always consider V_0 negative). For such a condition $\mathscr{G}_n(0)$ is real, and the k which occurs in the asymptotic expansion of $\mathscr{G}_n(k)$ is purely imaginary, so that the wave function coefficients $B(\mathbf{R})$ decrease exponentially. For a weaker potential the point of intersection moves to energies $E > E_0$. There is then no bound state, but rather a resonance. At the intersection the phase shift has the value $\pi/2$. This corresponds, when δ is increasing, to a maximum of the scattering amplitude and, hence, of the cross section. A virtual state is said to exist under these circumstances. If the potential is strong enough so that a bound state exists, $I(E_0)V_0 > 1$, so that the phase shift at the bottom of the band is π. This behavior is characteristic of phase shifts in the presence of bound states in general. The consistency of this result with the Friedel sum rule will be observed: Of course, if the impurity potential is sufficiently weak, neither bound states nor resonances will exist.

A resonant state is not arbitrarily sharp, but has a certain width. This may be defined by comparing the density of states near the resonance with that which would be expected for a hypothetical level of complex energy $E_0 - i\Gamma$. The level width, Γ, is found to be (for the Koster-Slater model)

$$\Gamma = \frac{Im\,\mathscr{G}_n(0)}{Re\left(\dfrac{d}{dE}\,\mathscr{G}_n(0)\right)_{E_0}} = -\pi G(E_0)\left(\frac{dI}{dE}\right)^{-1}_{E_0}$$

in which E_0 is the energy of the resonance. Since Γ must be positive, we see that a resonance can occur only when dI/dE is negative.

4.3 The Effective Mass Equation

The Koster-Slater model of an impurity center has yielded a soluble problem but not a practical procedure for determining the positions of actual impurity levels in real crystals. We wish to develop an equation in this section which will enable the approximate calculation of the influence of localized perturbations on real crystals. The starting point is Eq. (4.22), which is the one-electron Schrödinger equation in the representation of Luttinger and Kohn. We shall be concerned with stationary state solutions (energy E) of (4.22). A principal objective is to obtain an equation which pertains to a single band.

The potential and the momentum are the two sources of interband matrix elements in (4.22). The matrix elements of the potential can be written as:

$$(n\mathbf{k}|U|n'\,\mathbf{k}') = \int e^{i(\mathbf{k}'-\mathbf{k})\cdot\mathbf{r}}\,U(\mathbf{r})\,u_n^*(\mathbf{r})\,u_{n'}(\mathbf{r})\,d^3r \qquad (4.61)$$

$$= \sum_m B_m^{nn'} \int e^{i(\mathbf{k}'-\mathbf{k}-\mathbf{K}_m)\cdot\mathbf{r}}\,U(\mathbf{r})\,d^3r$$

$$= (2\pi)^3 \sum_m B_m^{nn'}\,U(\mathbf{k}-\mathbf{k}'+\mathbf{K}_m)$$

in which $U(\mathbf{k}-\mathbf{k}'+\mathbf{K}_m)$ is a Fourier coefficient of the potential U:

$$U(\mathbf{k}) = \frac{1}{(2\pi)^3}\int_{\substack{\text{entire}\\\text{crystal}}} e^{-i\mathbf{k}\cdot\mathbf{r}}\,U(\mathbf{r})\,d^3r \qquad (4.62)$$

and we have substituted (4.14). At this point we assume that the potential U is sensibly constant within a unit cell, so that the Fourier coefficients of the potential in (4.48) which involve nonzero reciprocal lattice vectors $\mathbf{K}_m$ may be discarded. This is a fundamental limiting approximation of the present method. To the extent this can be done, only the term $B_0^{nn'}$ survives in (4.61), and this quantity is given in (4.16). We have

$$(n\mathbf{k}|U|n'\,\mathbf{k}') = \delta_{nn'}\,U(\mathbf{k}-\mathbf{k}') \qquad (4.63)$$

Substitution of Eq. (4.63) converts Eq. (4.22) into the form

$$\left[E_n + \frac{\hbar^2 \mathbf{k}^2}{2m} - E\right] A_n(\mathbf{k}) + \frac{\hbar \mathbf{k}}{m}\cdot\sum_{n'}\mathbf{p}_{nn'}\,A_{n'}(\mathbf{k}) + \qquad (4.64)$$

$$\int d^3k'\,U(\mathbf{k}-\mathbf{k}')A_n(\mathbf{k}') = 0$$

The only remaining interband matrix elements are those associated with the momentum.

The procedure developed for removing these elements depends on the assumption that $\mathbf{k}$ in (4.64) is small: Terms higher than second order in $\mathbf{k}$ will be neglected. A unitary transformation can then be constructed which removes the offending element in (4.64). The technique could be extended to higher order terms.

In general, if we have an eigenvalue equation

$$H\psi = E\psi \qquad (4.65)$$

and make the unitary transformation

$$\psi = e^{iS}\varphi \qquad (4.66)$$

in which S is Hermitean; the transformed eigenvalue equation

$$\bar{H}\varphi = E\varphi \qquad (4.67)$$

is obtained in which the transformed Hamiltonian $\bar{H}$ is given by

$$\bar{H} = e^{-iS} H e^{iS} \qquad (4.68)$$

If S is in some sense "small," the exponential functions in (4.68) can be expanded. Terms of third and higher order in S are neglected.

$$\bar{H} = (1 - iS - \tfrac{1}{2}S^2)H(1 + iS - \tfrac{1}{2}S^2) = H + i[H, S] - \tfrac{1}{2}[[H, S], S] \qquad (4.69)$$

in which the square bracket denotes the commutator:

$$[H, S] = HS - SH$$

Let us introduce the abbreviations H_0, H_1 through:

$$(n\mathbf{k}|H_0|n'\,\mathbf{k}') = \left(E_n + \frac{\hbar^2 \mathbf{k}^2}{2m}\right)\delta_{nn'}\,\delta(\mathbf{k} - \mathbf{k}') \qquad (4.70a)$$

$$(n\mathbf{k}|H_1|n'\,\mathbf{k}') = \hbar\frac{\mathbf{k}\cdot\mathbf{p}_{nn'}}{m}\,\delta(\mathbf{k} - \mathbf{k}') \qquad (4.70b)$$

Equation (4.64) is of the form (4.65) with

$$H = H_0 + H_1 + U \qquad (4.71)$$

The off diagonal terms of (4.64) can be eliminated to first order by a unitary transformation provided S is given through

CHAPTER 4. POINT IMPURITIES AND EXTERNAL FIELDS

$$i[H_0, S] = -H_1 \qquad (4.72)$$

The matrix elements of S in the representation employed are determined to be

$$(n\mathbf{k}|S|n'\ \mathbf{k}') = \frac{i(n\mathbf{k}|H_1|n'\ \mathbf{k}')}{E_n - E_{n'}} = \frac{i\ \mathbf{k}\cdot \mathbf{p}_{nn'}}{m\omega_{nn'}}\delta(\mathbf{k}-\mathbf{k}') \qquad \text{for} \qquad n \neq n'$$

$$= 0 \qquad \text{if} \qquad n = n' \qquad (4.73)$$

where $\hbar\omega_{nn'} = E_n - E_{n'}$.

The transformed Hamiltonian, including terms of second order in S can now be found. Note that

$$i[H_1, S] - \tfrac{1}{2}[[H_0, S], S] = \tfrac{1}{2}i[H_1, S]$$

We obtain

$$\bar{H} = H_0 + U + i[U, S] - \tfrac{1}{2}[[U, S], S] + \tfrac{1}{2}i[H_1, S] \qquad (4.74)$$

The matrix elements of the last term in (4.74) can be evaluated easily, since both H_1 and S are diagonal with respect to $\mathbf{k}$ and $\mathbf{k}'$:

$$(n\mathbf{k}|[H_1, S]|n'\ \mathbf{k}') = \delta(\mathbf{k}-\mathbf{k}') \sum_{n''} [(n\mathbf{k}|H_1|n''\ \mathbf{k})(n''\ \mathbf{k}|S|n'\ \mathbf{k}) -$$

$$(n\mathbf{k}|S|n''\ \mathbf{k})(n''\ \mathbf{k}|H_1|n'\ \mathbf{k})]$$

$$= i\hbar\delta(\mathbf{k}-\mathbf{k}') \sum_{n''} \frac{(\mathbf{k}\cdot\mathbf{p}_{nn''})(\mathbf{k}\cdot\mathbf{p}_{n''n'})}{m^2}\left[\frac{1}{\omega_{n''n'}} + \frac{1}{\omega_{n''n}}\right] \qquad (4.75)$$

The matrix elements of $[U, S]$ are found by a similar procedure (note that U is diagonal in the band index, but not in $\mathbf{k}$).

$$(n\mathbf{k}|[U, S]|n'\ \mathbf{k}') = \int d^3k''(n\mathbf{k}|U|n\mathbf{k}'')(n\mathbf{k}''|S|n'\ \mathbf{k}') - \qquad (4.76)$$

$$\int d^3k''(n\mathbf{k}|S|n'\ \mathbf{k}'')(n'\ \mathbf{k}''|U|n'\ \mathbf{k}')$$

$$= i(\mathbf{k}'-\mathbf{k})\cdot\frac{\mathbf{p}_{nn'}}{m\omega_{nn'}}U(\mathbf{k}-\mathbf{k}') \qquad \text{for} \qquad n \neq n'$$

$$= 0 \qquad \text{if} \qquad n = n'$$

This term contains interband matrix elements and is of first order in k. However, the momentum matrix elements are reduced by a factor which is the ratio of a Fourier coefficient of potential to an interband energy difference. Further, the contribution from such terms can only come as a second order perturbation, since there are no diagonal matrix elements. In second order, this factor U/ω is effectively squared. We will suppose that such contributions are negligible. This implies we also will neglect the second order term $[[U, S], S]$. Further, to second order in k, we can neglect the off-diagonal (in the band index) elements in (4.75). These could be eliminated by a further unitary transformation which would introduce corrections to the Hamiltonian of higher than second order in k.

Consequently, we have in place of (4.64) (we write $C = e^{-iS} A$ for the wave function):

$$\left[E_n + \frac{\hbar^2 k^2}{2m} + \hbar \sum_{n''} \frac{(\mathbf{k} \cdot \mathbf{p}_{nn''})(\mathbf{k} \cdot \mathbf{p}_{n''n})}{m^2 \omega_{nn''}} - E \right] C_n(\mathbf{k}) + \quad (4.77)$$

$$\int d^3k\, U(\mathbf{k} - \mathbf{k}')\, C_n(\mathbf{k}') = 0$$

We now observe that the first three terms in (4.77) are equivalent to the expression given in (1.37) for the energy as a function of wave vector to second order in k, so we replace those terms by $E_n(\mathbf{k})$:

$$(E_n(\mathbf{k}) - E) C_n(\mathbf{k}) + \int d^3k'\, U(\mathbf{k} - \mathbf{k}') C_n(\mathbf{k}') = 0 \quad (4.78)$$

This equation resembles the ordinary Schrödinger equation in momentum space for one particle in the field of the potential U. There is one principle difference that the effective mass m^*, rather than the free electron mass m_0, is involved. All the effects of the crystal potential are incorporated in the effective mass.

The reduction of (4.78) to a differential equation in ordinary space is, however, only approximate. We define a function $F_n(\mathbf{r})$ by

$$F_n(\mathbf{r}) = \int e^{i\mathbf{k} \cdot \mathbf{r}} C_n(\mathbf{k})\, d^3k \quad (4.79)$$

The integration in (4.79) includes only the first Brillouin zone, rather than all of **k**-space as is customary in the Fourier transformation. The manipulation of Fourier transformations is usually aided by the delta function, which is ordinarily defined by

$$\delta(\mathbf{r} - \mathbf{r}') = \frac{1}{(2\pi)^3} \int_{\substack{\text{all}\\ \mathbf{k}-\text{space}}} e^{i\mathbf{k}\cdot(\mathbf{r}-\mathbf{r}')} d^3k$$

In the present case, it is desirable to define a δ-like function

$$\Delta(\mathbf{r} = \mathbf{r}') = \frac{1}{(2\pi)^3} \int_{BZ} e^{i\mathbf{k}\cdot(\mathbf{r}-\mathbf{r}')} d^3k \qquad (4.80)$$

We are concerned here with arbitrary values of **r** and **r**'. A similar object has been introduced in Appendix 2, Eq. (A2.10), for the case in which **r** and **r**' are direct lattice vectors. The function[4] Δ evidently has the property that

$$\int_{\text{all space}} \Delta(\mathbf{r}) d^3r = 1 \qquad (4.81)$$

For $|\mathbf{r}|$ large compared to a lattice spacing, $\Delta(\mathbf{r})$ oscillates and decreases as r^{-3}. For functions of position $f(\mathbf{r})$ which are slowly varying and hence extend over many cells, we may treat $\Delta(\mathbf{r})$ as a delta function

$$\int \Delta(\mathbf{r} - \mathbf{r}') f(\mathbf{r}') d^3r' \approx f(\mathbf{r}) \qquad (4.82)$$

One can qualitatively estimate that the error involved is of the order of $(a/a_f)^2$ where a is the lattice constant, and a_f is some measure of the "extent" of $f(\mathbf{r})$. This is comparable to the error introduced by neglect of the higher Fourier coefficients of the impurity potential and of the neglect of the commutator $[U, S]$. All these errors are negligible for slowly varying impurity potentials.

---

[4] For a simple cubic lattice (lattice parameter a) we find

$$\Delta(\mathbf{r}) = \frac{1}{\pi^3} \frac{\sin \pi x/a \, \sin \pi y/a \, \sin \pi z/a}{xyz}$$

4.3 THE EFFECTIVE MASS EQUATION

To accomplish the transformation of (4.78) to real space, we multiply by $e^{i\mathbf{k}\cdot\mathbf{r}}$ and integrate over the Brillouin zone. Let us consider the term

$$\int E_n(\mathbf{k})C_n(\mathbf{k})\,e^{i\mathbf{k}\cdot\mathbf{r}}d^3k \tag{4.83}$$

$E_n(\mathbf{k})$ can be expressed as

$$E_n(\mathbf{k}) = E_n + \sum_{i,j} \alpha_{ij} k_i k_j \tag{4.84}$$

where α_{ij} is the reciprocal effective mass tensor given implicitly in (4.77). Terms of higher than second order in $\mathbf{k}$ are neglected. On substitution into (4.83) we obtain

$$E_n F_n(\mathbf{r}) + \alpha_{ij}\int k_i k_j e^{i\mathbf{k}\cdot\mathbf{r}} C_n(\mathbf{k})\,d^3r \tag{4.85}$$

$$= E_n F_n(\mathbf{r}) + \sum_{i,j}\alpha_{ij}\left(\frac{1}{i}\frac{\partial}{\partial x_i}\right)\left(\frac{1}{i}\frac{\partial}{\partial x_j}\right)F_n(\mathbf{r}) = E_n\left(\frac{1}{i}\nabla\right)F_n(\mathbf{r})$$

The expression $E_n[(1/i)\nabla]$ means that we are to substitute $(1/i)\,\partial/\partial x_j$ for k_j wherever k_j appears in the expression for the energy as a function of $\mathbf{k}$.

The last term of (4.78) is transformed as follows:

$$\iint U(\mathbf{k}-\mathbf{k}')\,e^{i\mathbf{k}\cdot\mathbf{r}}C_n(\mathbf{k}')\,d^3k\,d^3k' \tag{4.86}$$

$$= \frac{1}{(2\pi)^3}\iiint U(\mathbf{r}')\,e^{-i(\mathbf{k}-\mathbf{k}')\cdot\mathbf{r}'}\,e^{i\mathbf{k}\cdot\mathbf{r}}C_n(\mathbf{k}')\,d^3r'\,d^3k\,d^3k'$$

$$= \int d^3r'\,U(\mathbf{r}')\,\Delta(\mathbf{r}-\mathbf{r}')F_n(\mathbf{r}') = U(\mathbf{r})F_n(\mathbf{r})$$

In the last step, we have used the approximation of Eq. (4.82). We finally obtain the transformed effective mass equation

$$\left(E_n\left(\frac{1}{i}\nabla\right) - E\right)F_n(\mathbf{r}) + U(\mathbf{r})F_n(\mathbf{r}) = 0 \tag{4.87}$$

To obtain the wave function, we return to Eq. (4.18). The transformation (4.66) implies that

$$A_n(\mathbf{k}) = \sum_{n'} \int (n\mathbf{k}|e^{iS}|n'\,\mathbf{k}')C_{n'}(\mathbf{k}')\,d^3k' \qquad (4.88)$$

$$= C_n(\mathbf{k}) - \frac{\mathbf{k}}{m} \cdot \sum_{n'} \frac{\mathbf{p}_{nn'}}{\omega_{nn'}} C_{n'}(\mathbf{k}) + \dots$$

The leading term in A_n is just C_n, and we neglect the first order correction. To this order

$$\Psi(\mathbf{r}) = \sum_n \int C_n(\mathbf{k})\,e^{i\mathbf{k}\cdot\mathbf{r}} u_n(\mathbf{r})\,d^3k = \sum_n u_n(\mathbf{r}) F_n(\mathbf{r}) = \sum \psi_n(\mathbf{r}) F_n(\mathbf{r})$$

$$(4.89)$$

In the last step of (4.89), we have used the fact that our band extremum occurs at $k = 0$. This equation does not connect the bands. If we are interested in the wave function associated with a particular impurity level under the conduction band, for instance ($n = c$), we have, finally,

$$\Psi = \psi_c(0, \mathbf{r}) F(\mathbf{r}) \qquad (4.90)$$

The impurity wave function is then an oscillatory band wave function modified by a slowly varying, but exponentially decreasing, envelop function F.

The effective mass equation (4.87) will not be valid in regions of space where the potential is rapidly varying — as for instance in the vicinity of a cell containing an impurity atom.

The extension of this procedure to the case in which the band extremum is not at the origin, but is nondegenerate is easy. We see from (4.8) that we now make expansions in the quantity $\delta\mathbf{k}$, where $\delta\mathbf{k} = \mathbf{k} - \mathbf{k}_0$. The derivation proceeds just as before. In place of (4.84), we have

$$E_n(\mathbf{k}) = E_n(\mathbf{k}_0 + \delta\mathbf{k}) = E_n(\mathbf{k}_0) + \sum \alpha_{ij}\,\delta k_i\,\delta k_j \qquad (4.91)$$

4.3 THE EFFECTIVE MASS EQUATION

This time we replace δk_i by $(1/i)\partial/\partial x_i$, so that the extension of (4.87) is

$$\left[E_n\left(\mathbf{k_0} + \frac{1}{i}\nabla\right) - E\right]F_n(\mathbf{r}) + U(\mathbf{r})F_n(\mathbf{r}) = 0 \qquad (4.92)$$

This equation is interpreted as requiring an expansion of the energy to second order in $(1/i)\nabla$.

When the band extremum is not located at $\mathbf{k} = 0$, there generally will be more than one extremum of the same energy. If the potential U has the point symmetry of the crystal, then impurity states associated with each extremum will have the same energy, and it is necessary to form linear combinations of these solutions. The appropriate linear combination which gives, for instance, the state of lowest energy is not determined within the effective mass theory as presented, but rather requires consideration of the first order corrections. On physical grounds, one generally expects a completely symmetric linear combination to have lowest energy.

We will briefly discuss the application of the effective mass formalism to real crystals. Consider, for example, the case of a donor impurity with a single excess electron in a semiconductor (for instance, phosphorus in silicon or arsenic in germanium). In this case, the potential $U(r)$ at large distances from the impurity will be that of a single point charge, screened by the dielectric constant, κ, of the host crystal. The use of the dielectric constant in this connection seems quite plausible on physical grounds when the electron is far from the impurity center, and has been rigorously justified by Kohn (1958) (see Appendix 3). Of course, the potential will be different near the impurity. First, we consider the case of a spherical band, effective mass m^*. Then, if we set the zero of energy to be the conduction band minimum, Eq. (4.87) becomes

$$-\frac{\hbar^2}{2m^*}\nabla^2 F - \frac{e^2}{\kappa r}F = EF \qquad (4.93)$$

This is, of course, just a simple hydrogenic problem, for which the energies are

$$E_n = -\frac{m^* e^4}{2\kappa^2 n^2 \hbar^2} \qquad (4.94)$$

If we take $\kappa = 16$, and an average effective mass $m^* = 0.12$, which are roughly appropriate for germanium, we find
$$E_n = -(0.0064/n^2) \text{ ev}$$
It is interesting to note that the effective "Bohr radius" for the lowest orbit is $\kappa \hbar^2/m^* e^2$, which is about 70 Å. The orbit includes thousands of cells. Experimental impurity ionization energies are slightly less than twice as large (in magnitude), about 0.010 ev. The worst approximation here is the use of the average effective mass, and this can be improved. We let the x-axis lie along the $(1, 1, 1)$ axis through the band minimum (effective mass m_l in this direction; m_t in the directions perpendicular to it). Then we get in place of (4.80)
$$\left[-\frac{\hbar^2}{2m_l}\frac{\partial^2}{\partial x^2} - \frac{\hbar^2}{2m_t}\left(\frac{\partial^2}{\partial y^2} + \frac{\partial^2}{\partial z^2}\right)\right]F - \frac{e^2}{\kappa r}F = EF \quad (4.95)$$
The exact solution to this equation is not known, but a variation calculation may be performed using a trial function for the lowest state
$$F = (a^2 b/\pi)^{1/2} \exp\{-[a(y^2 + z^2) + bx^2]^{1/2}\} \quad (4.96)$$
This form is exact in the limit of a spherical band. The parameters a and b are to be varied. The variational calculation with this function has been performed by several authors (Kittel and Mitchell, 1954; Lampert, 1955; Luttinger and Kohn, 1955). It is found that the ground state energy is -0.0090 ev in somewhat better agreement with experiment.

If one wishes to discuss acceptor states in semiconductors such as silicon and germanium, a further complication arises because of the degeneracy of the valence band at $\mathbf{k} = 0$. We turn to a discussion of the form of effective mass equation near such a degeneracy. Consider the expansion (4.18) for the perturbed wave function. We shall reserve the indices $_l$ and $_{l'}$ to indicate members of the degenerate set of functions, and denote the remaining functions with the index m. The analysis which leads to Eq. (4.64) is still valid, except that since there are, by hypothesis, no momentum matrix elements between members of the degenerate set, we have (on setting $n = l$)

$$\left(E_l + \frac{\hbar^2 \mathbf{k}^2}{2m} - E\right)A_l(\mathbf{k}) + \frac{\hbar \mathbf{k}}{m} \cdot \sum_m{}' \mathbf{p}_{lm} A_m(\mathbf{k}) + \int d^3k' \, U(\mathbf{k} - \mathbf{k}')A_l(\mathbf{k}') = 0$$
(4.97)

4.3 THE EFFECTIVE MASS EQUATION

The momentum matrix elements are eliminated to first order in k by a unitary transformation as in (4.66), (4.73), and (4.75). The elements of S are given in (4.60) as before. However, the A_l are all of the same order of magnitude, so it is necessary to retain the terms in (4.75) which connect them. We have instead of (4.77), a set of equations of the form:

$$\left[E_l + \frac{\hbar^2 k^2}{2m} - E\right] C_l + \hbar \sum_{l'} \left[\sum_m \frac{(\mathbf{k}\cdot\mathbf{p}_{lm})(\mathbf{k}\cdot\mathbf{p}_{ml'})}{m^2 \omega_{lm}}\right] C_{l'} +$$
$$\int d^3k\, U(\mathbf{k}-\mathbf{k}') C_l(\mathbf{k}') = 0 \qquad (4.98)$$

There is one such equation for each value of l. If the perturbing potential, U, were zero, the same determinantal equation would be obtained for the energies as in Section 1.7. If we introduce rectangular components of $\mathbf{k}$, k^α and k^β, and take the zero of energy at the unperturbed value E_l, the equations may be abbreviated as

$$\sum_{l',\alpha,\beta} [D_{ll',\alpha\beta} k^\alpha k^\beta - E\delta_{ll'}] C_{l'} + \int d^3k\, U(\mathbf{k}-\mathbf{k}') C_l(\mathbf{k}') = 0 \qquad (4.99)$$

In (4.99) the coefficient $D_{ll',\alpha\beta}$ stands for:

$$D_{ll',\alpha\beta} = \frac{\hbar^2}{2m}\delta_{ll'}\delta_{\alpha\beta} + \hbar \sum_m \frac{p_{lm}^\alpha p_{ml'}^\beta}{m^2 \omega_{lm}} \qquad (4.100)$$

and p_{lm}^α is the αth rectangular component of the vector matrix element $\mathbf{p}_{lm}$. The quantities $D_{ll'}$ must be determined from a band calculation or from experiments such as cyclotron resonance. The transformation back to ordinary space may now be applied as in Eq. (4.79). A set of coupled differential equations (as many as there are degenerate functions) results, which replace the single equation (4.85):

$$\sum_{l',\alpha,\beta}\left[D_{ll',\alpha\beta}\left(\frac{1}{i}\frac{\partial}{\partial x_\alpha}\right)\left(\frac{1}{i}\frac{\partial}{\partial x_\beta}\right) + U(\mathbf{r})\delta_{ll'}\right] F_{l'}(\mathbf{r}) = EF_l(\mathbf{r}) \qquad (4.101)$$

These equations generalize the effective mass theory to degenerate bands. The leading term in the wave function is, in analogy with (4.89)

$$\Psi = \sum_l \psi_l(\mathbf{r}) F_l(\mathbf{r}) \qquad (4.102)$$

The solution of Eq. (4.101), even by approximate techniques is much more difficult than in the case of (4.87). Kohn and Schechter (1955) have applied a variational technique to obtain the ionization energy and low-lying excited states of acceptors in germanium and silicon. They find, in the case of germanium, an ionization energy of 0.0089 ev, in fair agreement with the experimental values which range from 0.0102 to 0.0112 ev depending on the impurity.

For both acceptors and donors, the discrepancies between the effective mass theory and experiment are principally due to the departure of the potential near an impurity from the simple form $e^2/\kappa r$ previously used. Obviously, screening of the impurity potential by the dielectric constant of the host crystal cannot be expected to occur close to the impurity. The corrections of this sort are of course smaller for excited states of the excess electron or hole (for p states, the region where the perturbing potential is strong will be avoided) than for the ground state. Kohn and Luttinger have been able to determine the corrections to the effective mass theory for the wave function at a donor impurity, which has been studied experimentally through observations of hyperfine structure in spin resonance experiments.

4.4 The Steady Magnetic Field

The subject of the behavior of "Bloch electrons" in external fields, whether magnetic or electric, is subtle and full of difficulties. The problems arise from two sources: (1) the expressions for the interaction energy contain the electron coordinate, so that the periodicity of the Hamiltonian is lost, and (2) the interaction is not localized. The steady magnetic field will be discussed first because in this case (as opposed to that of the electric field), bound state solutions of the free-particle Schrödinger equation exist.

The Hamiltonian for a single electron in the presence of a magnetic field can be obtained by replacing the momentum operator $\mathbf{p}$ wherever it appears in the Hamiltonian for zero field by $\mathbf{P} = \mathbf{p} + e\mathbf{A}$, where $\mathbf{A}$ is the

4.4 THE STEADY MAGNETIC FIELD

vector potential, and by including a term $(e\hbar/2m)\boldsymbol{\sigma}\cdot\mathbf{B}$ to represent the energy of the spin. The charge e is taken to be positive in these expressions. We use mks units in which the vector potential $\mathbf{A}$ is related to the flux density $\mathbf{B}$ (in units of webers/m²) by

$$\mathbf{B} = \nabla \times \mathbf{A} \tag{4.103}$$

$$H = \frac{1}{2m}(\mathbf{p} + e\mathbf{A})^2 + V + \frac{e\hbar}{2m}\boldsymbol{\sigma}\cdot\mathbf{B} \tag{4.104}$$

The vector potential $\mathbf{A}$ is not uniquely determined by the field, since the gradient of any scalar function of position may be added to $\mathbf{A}$ without changing $\mathbf{B}$. Such a change in $\mathbf{A}$ is referred to as a gauge transformation. A change in phase of the wave function must accompany a gauge transformation of the vector potential. For a field which does not depend on position, a possible choice for $\mathbf{A}$ is:

$$\mathbf{A} = \tfrac{1}{2}\mathbf{B}\times\mathbf{r} \tag{4.105}$$

Since weak magnetic fields are of principal interest, one might attempt to treat the interaction in perturbation theory. This is not possible, however, for wave functions which are indefinitely extended in space, as plane waves and Bloch functions are, because the interaction term can become arbitrarily large. Neither are analogies with classical mechanics fruitful, since the magnetic susceptibility of a classical system of free charges enclosed in a container is zero. This follows because the magnetic field cannot change the energy of a classical particle so that the partition function (which is a function of the energy) does not depend on the field. Consequently, the problem must be attacked entirely from the point of view of quantum mechanics.

What one would like to have is a generalization of the effective mass equation (4.87) to include a magnetic field. This would enable a reduction of the actual Schrödinger equation with the Hamiltonian (4.104) to a free particle equation, which is much easier to solve. In the following, we will indicate the extent to which such a reduction is possible.

It is usually convenient to express the total magnetic susceptibility of a material as the sum of three terms:

$$\chi_{\text{total}} = \chi_{\text{spin}}^{(\text{band})} + \chi_{\text{diamagnetic}}^{(\text{band})} + \chi_{\text{diamagnetic}}^{(\text{core})} \tag{4.106}$$

in which $\chi_{\text{spin}}^{(\text{band})}$ represents the (paramagnetic) susceptibility of the spins of the electrons in unfilled bands, $\chi_{\text{diamagnetic}}^{(\text{band})}$ is the diamagnetic susceptibility associated with the "orbital" motion of the band electrons, and $\chi_{\text{diamagnetic}}^{(\text{core})}$ is the contribution from the core electrons in closed shells. The latter term may be estimated easily; we shall not discuss it here. For the Hamiltonian (4.104) in which the coefficient of $\boldsymbol{\sigma}$ does not depend on position, the wave function can be expressed as the product of a space function and a spin function, and the energy can be represented as the sum of a contribution from the orbital motion and from the spin, the latter being just $\pm e\hbar B/2m$. Thus, for the present we may disregard the interaction of the spin with the field and concentrate attention on the terms responsible for the orbital diamagnetism.

This treatment will follow the work of Luttinger and Kohn (1955; Kjeldaas and Kohn, 1957; Kohn, 1959). For an alternative derivation of a "one band" Hamiltonian, see Wannier and Fredkin (1962) and Roth (1962). We wish to look at the Hamiltonian (4.104), with the spin term omitted, in the representation of Luttinger and Kohn. As a preliminary to the body of the calculation, it is desirable to obtain the matrix representation of the (rectangular) coordinate vector $\mathbf{r}$ which enters in the Hamiltonian through the vector potential:

$$(n\mathbf{k}|\mathbf{r}|n'\,\mathbf{k}') = \int e^{i(\mathbf{k}'-\mathbf{k})\cdot\mathbf{r}} u_n^*(\mathbf{k}_0,\mathbf{r})\mathbf{r} u_{n'}(\mathbf{k}_0,\mathbf{r})\, d^3r \qquad (4.107)$$

$$= i\nabla_{\mathbf{k}} \int e^{i(\mathbf{k}'-\mathbf{k})\cdot\mathbf{r}} u_n^*(\mathbf{k}_0,\mathbf{r}) u_{n'}(\mathbf{k}_0,\mathbf{r})\, d^3r$$

$$= i\nabla_{\mathbf{k}}\, \delta(\mathbf{k}'-\mathbf{k})\delta_{nn'}$$

It is convenient to introduce the symbol $(\mathbf{k}|Q|\mathbf{k}')$ to designate the matrix elements of an operator Q on a basis of plane waves:

$$(\mathbf{k}|Q|\mathbf{k}') \equiv \frac{1}{(2\pi)^3}\int e^{-i\mathbf{k}\cdot\mathbf{r}} Q e^{i\mathbf{k}'\cdot\mathbf{r}}\, d^3r \qquad (4.108)$$

Then we have for $\mathbf{r}$

$$(n\mathbf{k}|\mathbf{r}|n'\,\mathbf{k}') = (\mathbf{k}|\mathbf{r}|\mathbf{k}')\delta_{nn'} \qquad (4.109)$$

Similarly, we obtain for the momentum $\mathbf{p}$:

$$(n\mathbf{k}|\mathbf{p}|n'\,\mathbf{k}') = \hbar\mathbf{k}\delta_{nn'}\,\delta(\mathbf{k}'-\mathbf{k}) + \mathbf{p}_{nn'}\,\delta(\mathbf{k}'-\mathbf{k}) \qquad (4.110)$$

$$= \delta_{nn'}(\mathbf{k}|\mathbf{p}|\mathbf{k}') + \mathbf{p}_{nn'}\,\delta(\mathbf{k}'-\mathbf{k})$$

in which $\mathbf{p}_{nn'}$ is defined by (4.21). We can now calculate the matrix element of P

$$(n\mathbf{k}|\mathbf{P}|n'\ \mathbf{k}') = (\mathbf{k}|\mathbf{P}|\mathbf{k}')\delta_{nn'} + \mathbf{p}_{nn'}\,\delta(\mathbf{k}' - \mathbf{k}) \quad (4.111)$$

since the vector potential depends on coordinates only, and hence is diagonal in bands in this representation. The singular character of $(\mathbf{k}|\mathbf{P}|\mathbf{k}')$ is given through (4.107). We can now determine the matrix elements of the kinetic energy

$$\left(n\mathbf{k}\left|\frac{\mathbf{P}^2}{2m}\right|n'\ \mathbf{k}'\right) = \frac{1}{2m}\sum_{n''}\int d^3k''(n\mathbf{k}|\mathbf{P}|n''\ \mathbf{k}'')(n''\ \mathbf{k}''|\mathbf{P}|n'\ \mathbf{k}') \quad (4.112)$$

$$= \left(\mathbf{k}\left|\frac{\mathbf{P}^2}{2m}\right|\mathbf{k}'\right)\delta_{nn'} + \frac{\mathbf{p}_{nn'}}{m}\cdot(\mathbf{k}|\mathbf{P}|\mathbf{k}') + \sum_{n''}\frac{\mathbf{p}_{nn''}\mathbf{p}_{n''n'}}{2m}\delta(\mathbf{k} - \mathbf{k}')$$

The potential energy V, which appears in the Hamiltonian (4.91) is the periodic potential of the crystal, which has no matrix elements between functions $\chi_n(\mathbf{k}, \mathbf{r})$ and $\chi_{n'}(\mathbf{k}', \mathbf{r})$ unless $\mathbf{k} - \mathbf{k}'$ is zero or a reciprocal lattice vector, which can occur only when $\mathbf{k}$ and $\mathbf{k}'$ are on the surface of the Brillouin zone. We will not consider this case.

$$(n\mathbf{k}|V|n'\ \mathbf{k}') = V_{nn'}\,\delta(\mathbf{k}' - \mathbf{k}) \quad (4.113)$$

Hence we find

$$(n\mathbf{k}|H|n'\ \mathbf{k}') = [(1/2m)(\mathbf{p}^2)_{nn'} + V_{nn'}]\delta(\mathbf{k}' - \mathbf{k}) + \frac{\mathbf{p}_{nn'}}{m}\cdot(\mathbf{k}|\mathbf{P}|\mathbf{k}') + \quad (4.114)$$

$$(\mathbf{k}|\mathbf{P}^2/2m|\mathbf{k}')\delta_{nn'}$$

The first two terms on the right-hand side of (4.114) evidently combine to give the energy in the absence of the magnetic field

$$\frac{1}{2m}(\mathbf{p}^2)_{nn'} + V_{nn'} = E_n\,\delta_{nn'} \quad (4.115)$$

so that (4.101) simplifies to:

$$(n\mathbf{k}|H|n'\ \mathbf{k}') = [E_n\,\delta(\mathbf{k}' - \mathbf{k}) + (\mathbf{k}|\mathbf{P}^2/2m|\mathbf{k}')]\delta_{nn'} + \mathbf{p}_{nn'}\cdot(\mathbf{k}|\mathbf{P}/m|\mathbf{k}')$$

$$(4.116)$$

We now wish to remove the off-diagonal (in the band index) elements by the unitary transformation procedure employed in Section 4.3. The Hamiltonian of 4.116 consists of three parts:

$$H = H_0 + H_1 + H_2 \tag{4.117a}$$

where

$$(n\mathbf{k}|H_0|n'\,\mathbf{k}') = E_n\,\delta_{nn'}\,\delta(\mathbf{k}-\mathbf{k}') \tag{4.117b}$$

$$(n\mathbf{k}|H_1|n'\,\mathbf{k}') = \mathbf{p}_{nn'} \cdot (\mathbf{k}|\mathbf{P}/m|\mathbf{k}') \tag{4.117c}$$

$$(n\mathbf{k}|H_2|n'\,\mathbf{k}') = (\mathbf{k}|\mathbf{P}^2/2m|\mathbf{k}')\delta_{nn'} \tag{4.117d}$$

The transformation for which we are looking must have the form (4.68)

$$\bar{H} = e^{-iS_1} H e^{iS_1}$$

in which S_1 is chosen so as to remove the off-diagonal elements to first order. S_1 must then be given through Eqs. (4.72) and (4.73)

$$(n\mathbf{k}|S_1|n'\,\mathbf{k}') = i\frac{(n\mathbf{k}|H_1|n'\,\mathbf{k}')}{\hbar\omega_{nn'}} = i\frac{\mathbf{p}_{nn'}}{\hbar\omega_{nn'}} \cdot (\mathbf{k}|\mathbf{P}/m|\mathbf{k}') \tag{4.118}$$

in which $\hbar\omega_{nn'} = E_n - E_{n'}$ as before. If we include terms of second order, the transformed Hamiltonian is given by an expression similar to (4.74):

$$\bar{H} = H_0 + H_2 + \frac{i}{2}[H_1, S_1] + \ldots \tag{4.119}$$

The matrix elements of the commutator in (4.119) are:

$$(n\mathbf{k}|[H_1, S_1]|n'\,k') = \frac{i}{m^2\hbar}\sum_{n''}[\mathbf{p}_{nn''} \cdot (\mathbf{k}|\mathbf{PP}|\mathbf{k}') \cdot \mathbf{p}_{n''n'}]\left(\frac{1}{\omega_{n''n'}} + \frac{1}{\omega_{n''n}}\right) \tag{4.120}$$

The off-diagonal elements are of second order in the cannonical momentum $\mathbf{P}$. These may be removed by a further unitary transformation with an Hermitean matrix S_2, in which the commutator $\frac{1}{2}i[H_1, S_1]$ is treated in the same manner previously applied to H_1, namely,

$$i[H_0, S_2] = -\tfrac{1}{2}i[H_1, S_1] \tag{4.121}$$

4.4 THE STEADY MAGNETIC FIELD

This leads to

$$(n\mathbf{k}|S_2|n'\,\mathbf{k}') = \frac{-1}{2m^2\hbar^2\omega_{nn'}} \sum_{n''} \mathbf{p}_{nn''} \cdot (\mathbf{k}|\mathbf{PP}|\mathbf{k}') \cdot \mathbf{p}_{n''n'} \left(\frac{1}{\omega_{n''n}} + \frac{1}{\omega_{n''n'}}\right)$$

$$= 0 \quad \text{if} \quad n = n' \tag{4.122}$$

A series of successive unitary transformations can be carried out to eliminate the off-diagonal elements to arbitrary order in **P**. (Note that if we wish to go beyond the second order, it is necessary to include higher order terms from S_1, such as $[H_2, S_1]$, etc.) To obtain the portion of the effective Hamiltonian matrix which is diagonal in the band index, including all second order terms, it is necessary to retain only the diagonal part of (4.120). If we substitute into (4.119), the result is:

$$(n\mathbf{k}|H|n'\,\mathbf{k}') = \delta_{nn'}(\mathbf{k}|H_n|\mathbf{k}')$$

where

$$(\mathbf{k}|H_n|\mathbf{k}') = E_n\,\delta(\mathbf{k}-\mathbf{k}') + \sum_{\alpha\beta} E_n{}^{\alpha\beta}(\mathbf{k}|P^\alpha P^\beta|\mathbf{k}') \tag{4.123}$$

in which $E_n^{\alpha\beta}$ is the expansion coefficient for the energy if the energy is expanded to second order in $\mathbf{k}$ (in the absence of a magnetic field), and P^α, etc. is a rectangular component of **P**

$$E_n{}^{\alpha\beta} = \frac{1}{2m}\delta_{\alpha\beta} + \sum_s \frac{p_{ns}^\alpha p_{sn}^\beta}{\hbar m^2 \omega_{ns}} \tag{4.124}$$

To second order in **P**, this is the desired result: The Hamiltonian in the presence of the magnetic field is obtained by replacing $\mathbf{k}$ in the expansion of $E(\mathbf{k})$ by $\mathbf{P}/\hbar$.

Kjeldaas and Kohn have evaluated the fourth order term

$$\sum E_n{}^{\alpha\beta\gamma\delta}(\mathbf{k}|P^\alpha P^\beta P^\gamma P^\delta|\mathbf{k}') \tag{4.125}$$

which appears in that case. More complicated quantities are involved in (4.125) than the coefficients of the fourth order terms in the expansion of $E(\mathbf{k})$. (The latter only determines certain sums of the $E_n^{\alpha\beta\gamma\delta}$.)

Before the unitary transformation e^{iS} was applied, the equation satisfied by the expansion coefficients $A_n(\mathbf{k})$ of the wave function (see Eq. (4.18)) in the representation of Luttinger and Kohn, was

$$\sum_n \int [(n\mathbf{k}|H|n'\,\mathbf{k}') - E\delta_{nn'}\,\delta(\mathbf{k}-\mathbf{k}')]A_{n'}(\mathbf{k}')\,d^3k' = 0$$

After transformation, the new expansion coefficients $C_n(\mathbf{k})$, which are related to the A by $C = e^{-iS} A$, satisfy

$$(E_n - E)C_n(\mathbf{k}) + \sum_{\alpha,\beta} E_n{}^{\alpha\beta} \int (\mathbf{k}|P^\alpha P^\beta|\mathbf{k}')C_n(\mathbf{k}')\,d^3k' = 0 \qquad (4.126)$$

The transformation back to ordinary space is accomplished through the function $F_n(\mathbf{r})$ defined in (4.79). We multiply (4.126) by $e^{i\mathbf{k}\cdot\mathbf{r}}$ and integrate over $\mathbf{k}$. The crucial term is the following:

$$\iint e^{i\mathbf{k}\cdot\mathbf{r}}(\mathbf{k}|P^\alpha P^\beta|\mathbf{k}')C_n(\mathbf{k}')\,d^3k\,d^3k' \qquad (4.127)$$

$$= \frac{1}{(2\pi)^3} \iiint e^{i\mathbf{k}\cdot(\mathbf{r}-\mathbf{r}')}[P^\alpha P^\beta\, e^{i\mathbf{k}'\cdot\mathbf{r}'}]C_n(\mathbf{k}')\,d^3k\,d^3k'\,d^3r'$$

$$= \iint \Delta(\mathbf{r}-\mathbf{r}')[P^\alpha P^\beta e^{i\mathbf{k}'\cdot\mathbf{r}'}]C_n(\mathbf{k}')\,d^3k'\,d^3r'$$

$$= P^\alpha P^\beta \int e^{i\mathbf{k}'\cdot\mathbf{r}} C_n(\mathbf{k}')\,d^3k' = P^\alpha P^\beta F_n(\mathbf{r})$$

In working through (4.127), we have used: (1) the definition (4.108) of the matrix element $(\mathbf{k}|P^\alpha P^\beta|\mathbf{k}')$, (2) the definition (4.80) of the function Δ, (3) the function Δ has been treated as a δ-function, and finally (4) the definition (4.79) of $F_n(\mathbf{r})$. The quantity P^α is a differential operator:

$$P^\alpha = \frac{\hbar}{i}\frac{\partial}{\partial x^\alpha} + eA^\alpha$$

Hence we have in place of (4.113)

$$(E_n - E)F_n(\mathbf{r}) + \sum_{\alpha\beta} E_n{}^{\alpha\beta} P^\alpha P^\beta F_n(\mathbf{r}) = 0 \qquad (4.128)$$

4.4 THE STEADY MAGNETIC FIELD

This is the effective mass equation in ordinary space in the presence of a magnetic field. If spin is included according to (4.104), the function $F_n(\mathbf{r})$ must be interpreted as a two-component spinor, and we have

$$(E_n - E)F_n(\mathbf{r}) + \sum_{\alpha\beta} E_n{}^{\alpha\beta} P^\alpha P^\beta F_n(\mathbf{r}) + \frac{e\hbar}{2m_0} \boldsymbol{\sigma} \cdot \mathbf{B} F_n(\mathbf{r}) = 0 \quad (4.129)$$

These equations are more complicated than one might at first realize. The components of the vector $\mathbf{P}$ do not commute with each other, so that one must be very careful to preserve the order. The commutator is

$$P^\alpha P^\beta - P^\beta P^\alpha = \frac{e\hbar}{i}\left(\frac{\partial A^\beta}{\partial x_\alpha} - \frac{\partial A^\alpha}{\partial x_\beta}\right) = \frac{e\hbar}{i}\sum_\gamma B^\gamma \varepsilon_{\gamma\alpha\beta} \quad (4.130)$$

in which B^γ is the γ (rectangular) component of the magnetic field and $\varepsilon_{\gamma\alpha\beta}$ is the antisymmetric Levi-Civita symbol (which has the value $+1$ or -1 according as the arrangement of numbers $\gamma\alpha\beta$ is an even or odd permutation of 1, 2, 3). It is convenient to express the second term in (4.129) in terms of its symmetric and antisymmetric parts

$$\sum_{\alpha\beta} E_n{}^{\alpha\beta} P^\alpha P^\beta = \frac{1}{4} \sum_{\alpha\beta} \bigg[(E_n{}^{\alpha\beta} + E_n{}^{\beta\alpha})(P^\alpha P^\beta + P^\beta P^\alpha) + \quad (4.131)$$

$$\frac{e\hbar}{i}(E_n{}^{\alpha\beta} - E_n{}^{\beta\alpha})\sum_\gamma \varepsilon_{\gamma\alpha\beta} B^\gamma \bigg]$$

The second term on the right-hand side of (4.131) can be expressed as

$$\frac{e\hbar}{2m_0} \mathbf{M} \cdot \mathbf{B} \quad (4.132)$$

where we have defined the vector $\mathbf{M}$, dual to the antisymmetric tensor whose components are $(E_n^{\alpha\beta} - E_n^{\beta\alpha})$, by:

$$M^\gamma = \frac{m_0}{2i} \sum_{\alpha\beta} (E_n{}^{\alpha\beta} - E_n{}^{\beta\alpha})\varepsilon_{\gamma\alpha\beta} \quad (4.133)$$

If we substitute (4.131) and (4.132) into (4.129), we get

$$(E_n - E)F_n(\mathbf{r}) + \frac{1}{4}\sum_{\alpha\beta}(E_n{}^{\alpha\beta} + E_n{}^{\beta\alpha})(P^\alpha P^\beta + P^\beta P^\alpha)F_n(\mathbf{r}) + \qquad (4.134)$$

$$\frac{e\hbar}{2m_0}(\boldsymbol{\sigma} + \mathbf{M})\cdot \mathbf{B}F_n(\mathbf{r}) = 0$$

We can show that **M** has the effect of changing the effective g-factor of the electron. This discussion is, however, rather complex, and we shall postpone it until we have analyzed a simple problem.

It will be shown later that **M** is zero if spin orbit coupling is neglected. We shall make this approximation and shall further consider the hypothetical case in which the band is spherical, that is,

$$E_n{}^{\alpha\beta} = \frac{1}{2m^*}\delta_{\alpha\beta} \qquad (4.135)$$

We will put $E_n = 0$, and drop the index n. It is convenient to take the magnetic field, B, along the z-direction, and to choose a gauge slightly different from that of (4.105):

$$A_x = -By, \qquad A_y = A_z = 0 \qquad (4.136)$$

Then

$$P_1 = \frac{\hbar}{i}\frac{\partial}{\partial x} - eBy; \qquad P_2 = \frac{\hbar}{i}\frac{\partial}{\partial y}; \qquad P_3 = \frac{\hbar}{i}\frac{\partial}{\partial z} \qquad (4.137)$$

With these simplifications, Eq. (4.121) takes the form

$$-\frac{\hbar^2}{2m^*}\left[\left(\frac{\partial}{\partial x} - \frac{ieBy}{\hbar}\right)^2 + \frac{\partial^2}{\partial y^2} + \frac{\partial^2}{\partial z^2}\right]F(\mathbf{r}) = EF(\mathbf{r}) \qquad (4.138)$$

This equation is separable in rectangular coordinates. Let us put

$$F(\mathbf{r}) = e^{i(k_x x + k_z z)}g(y) \qquad (4.139)$$

This leads to

$$\frac{d^2 g}{dy^2} + \frac{2m^*}{\hbar^2}\left[E - \frac{\hbar^2 k_z{}^2}{2m^*} - \frac{1}{2m^*}(\hbar k_x - eBy)^2\right]g = 0 \qquad (4.140)$$

4.4 THE STEADY MAGNETIC FIELD

It is convenient to define

$$y_0 = \frac{\hbar k_x}{eB} \qquad (4.141)$$

Then we can put (4.140) into the form

$$\frac{d^2 g}{dy^2} + \frac{2m^*}{\hbar^2}[\varepsilon - \tfrac{1}{2} m^* \omega_c^2 (y - y_0)^2] g(y) = 0 \qquad (4.142)$$

with

$$\omega_c = \frac{eB}{m^*}, \qquad \varepsilon = E - \frac{\hbar^2 k_z^2}{2m^*} \qquad (4.143)$$

Equation (4.142) is the equation for a simple harmonic oscillator of frequency ω_c, with the equilibrium point located at y_0. The eigenvalue, ε_l, is given by

$$\varepsilon_l = (l + \tfrac{1}{2})\hbar \omega_c \qquad (4.144)$$

where l is any positive integer, including zero. The quantity ω_c is usually called the cyclotron frequency. The energy in the field, neglecting spin, is given by

$$E = \frac{\hbar^2 k_z^2}{2m^*} + (l + \tfrac{1}{2})\hbar \omega_c \qquad (4.145)$$

The result (4.145) is quite remarkable. The continuous, three-dimensional parabolic band structure from which we started has been split up into a series of lines (the oscillator levels), which we can associate with the classical circular motion of the electron in a plane perpendicular to the magnetic field, plus a one-dimensional parabolic term coming from the free electron behavior in a direction parallel to the field. The discrete levels are known as Landau levels (Landau, 1930). The energy of the lowest state is no longer zero, but has been raised to $\tfrac{1}{2}\hbar \omega_c$.

The Landau levels are highly degenerate. To estimate the degeneracy, let the system be contained in a large rectangular box with sides of length L_x, L_y, and L_z. The number of possible values of k_i (where $i = x, y,$ or z) in a small interval, Δk_i, is $L_i \Delta k_i / 2\pi$. All values for k_x are permissible provided that the "orbit center" y_0 lies within the box: $-L_y/2 \leqslant y_0 \leqslant L_y$. Here we neglect the extent of the "orbit" relative to the size of the container. From this we can determine the range of allowed values for k_x

$$\frac{-eBL_y}{2\hbar} \leqslant k_x \leqslant \frac{eBL_y}{2\hbar} \qquad (4.146)$$

The number of levels in a single Landau level is then

$$L_x \Delta k_x / 2\pi = eBL_xL_y/2\pi\hbar \qquad (4.147)$$

If we next consider an interval Δk_z in k_z, the number of levels in a Landau level and Δk_z is

$$\frac{eBV}{4\pi^2 \hbar} \Delta k_z \qquad (4.148)$$

in which V is the volume of the box. We note that the degeneracy is proportional to B.

The function $g(y)$, which is a solution of (4.142), is an harmonic oscillator function:

$$g_l(y) = \left(\frac{2\pi^{1/2}\alpha}{\Omega\, 2^l\, l!}\right)^{1/2} H_l[\alpha(y - y_0)] \exp\left[-\tfrac{1}{2}\alpha^2(y - y_0)^2\right] \qquad (4.149)$$

in which $\alpha^2 = m^* \omega_c/\hbar = eB/\hbar$, H_l is a Hermite polynomial, and l is the oscillator quantum number which appeared in (4.145). We can obtain a measure of the radius of the orbit by computing the root mean square value of y. After a simple calculation, we get

$$y_{\text{rms}} = [\langle(y - y_0)^2\rangle]^{1/2} = (l + \tfrac{1}{2})^{1/2}/\alpha = (\hbar/eB)^{1/2} (l + \tfrac{1}{2})^{1/2} \qquad (4.150)$$

In a magnetic field of 1 weber/m^2 (10^4 gauss), we find y_{rms} is about 180 Å for the lowest state. This is certainly small enough to justify our neglect of the "orbit radius" compared to L_y in calculating the density of states; it is also large enough to justify replacement of $\Delta(\mathbf{r} - \mathbf{r}')$ by $\delta(\mathbf{r} - \mathbf{r}')$ in (4.127).

The Landau levels are separated by an energy

$$\frac{\hbar eB}{m^*} = 1.1577 \times 10^{-4} \text{ ev}\, \frac{(B)}{(m^*/m_0)} \qquad (4.151)$$

4.4 THE STEADY MAGNETIC FIELD

where B is in webers/m^2, and m_0 is the free electron mass. For low fields, and materials with effective mass ratios of the order of unity, this energy is small compared to thermal energies except at the very lowest temperatures: the quantization of levels can usually be ignored. However, if we consider materials with effective mass ratios of the order of 10^{-2}, in strong fields, the splitting of the Landau levels can become quite significant.

We now return to discussion of the vector **M** whose rectangular components were defined in (4.133). From (4.124), we see that

$$E_n{}^{\alpha\beta} - E_n{}^{\beta\alpha} = \sum_s \frac{(p_{ns}^\alpha p_{sn}^\beta - p_{ns}^\beta p_{sn}^\alpha)}{\hbar m^2 \omega_{ns}} \tag{4.152}$$

Since **p** is an Hermitean operator, $p_{ns}^\alpha = (p_{sn}^\alpha)^*$. Since the time reversal operator, which changes **k** into $-$ **k** is equivalent to complex conjugation for wave functions which do not include spin, we may always choose the wave function $u_n(0, \mathbf{r})$ to be real. Then the matrix element $\mathbf{p}_{ns}$ will be purely imaginary, and (4.139), and consequently **M**, must vanish. This argument holds, however, only if spin orbit coupling is neglected.

Let the portion of the Hamiltonian due to spin orbit coupling be

$$H_{so} = \frac{\hbar}{4m^2 c^2} \boldsymbol{\sigma} \cdot (\nabla V \times \mathbf{p}) \equiv \boldsymbol{\sigma} \cdot \mathbf{h}$$

Relativistic effects other than spin orbit coupling are neglected. In the presence of a magnetic field, we replace **p** by $\mathbf{p} + e\mathbf{A}$ as before

$$H_{so} = \frac{\hbar}{4m^2 c^2} \{\boldsymbol{\sigma} \cdot \nabla V \times (\mathbf{p} + e\mathbf{A})\} \tag{4.153}$$

If this term is added to the Hamiltonian of (4.104), one can easily show that the formal theory remains unchanged except that the matrix elements p_{ns} must be replaced by

$$\Pi_{ns} = \int u_n^* \left(\mathbf{p} + \frac{\hbar}{4m^2 c^2} \boldsymbol{\sigma} \times \nabla V\right) u_s d^3r \tag{4.154}$$

The contribution of the $\boldsymbol{\sigma} \times \nabla V$ term to Π_{ns} is usually negligible. Of more significance for the present problem is the modification of the wave function at $k = 0$, u_n, by the spin orbit coupling. We will take this into account through first order perturbation theory. Let $u_n^{(0)}$ be the unperturbed wave function. Then we have

$$u_n = u_n^{(0)} + \boldsymbol{\sigma} \cdot \sum_l \frac{\mathbf{h}_{nl}}{\hbar \omega_{nl}} u_l^0 \tag{4.155}$$

in which

$$\mathbf{h}_{nl} = \frac{(2\pi)^3}{\Omega_0} \int u_n^{*(0)} \mathbf{h} u_l^{(0)} d^3r \tag{4.156}$$

If we substitute (4.155) into the definition of the matrix element

$$\mathbf{p}_{ns} = \frac{(2\pi)^3}{\Omega_0} \int u_n^* \mathbf{p} u_s \, d^3r$$

the vector **M** can be calculated to first order in the spin orbit coupling. **M** is found to be proportional to the spin operator σ.

$$\mathbf{M}_n = \boldsymbol{\sigma} \cdot \mathbf{G}_n \tag{4.157}$$

in which $\mathbf{G}_n$ is a second rank tensor. An expression for $\mathbf{G}_n$ may be deduced from the work of Roth (1960). We will not discuss the details here. We put

$$\frac{\mathbf{g}_n}{2} = (\mathbf{1} + \mathbf{G}_n) \tag{4.158}$$

in which **1** is the unit dyadic. Then we may write the basic equation (4.134) as

$$(E_n - E)F_n + \tfrac{1}{4} \sum_{\alpha\beta} (E_n{}^{\alpha\beta} + E_n{}^{\beta\alpha})(P^\alpha P^\beta + P^\beta P^\alpha)F_n + \tag{4.159}$$

$$\frac{e\hbar}{4m_0} \boldsymbol{\sigma} \cdot \mathbf{g}_n \cdot \mathbf{B} F_n = 0$$

For the case of the parabolic band previously considered, the energy levels for a magnetic field in the z-direction are

$$E = \frac{\hbar^2 k_z^2}{2m} + (l + \tfrac{1}{2})\hbar\omega_c + \frac{e\hbar}{2m_0} gBm_s \tag{4.160}$$

where m_s is the spin quantum number: $m_s = \pm \tfrac{1}{2}$.

4.5 THE MAGNETIC SUSCEPTIBILITY OF FREE ELECTRONS

The components of **g** are sensitive to the details of the band structure due to the presence of energy denominators in (4.152) and (4.155). In materials such as InSb where there are large spin orbit splittings and small band gaps, the effective g factor may be very large (-50 has been reported in this case).

The theory of the g-factor of electrons in the alkali metals has been discussed by Yafet (1952, 1957). Studies concerning semiconductors have been reported by Luttinger (1956), Roth, Lax, and Zwerdling (1959), Roth (1960), and Liu (1961, 1962). Cohen and Blount (1960) have discussed the g-factor of electrons in bismuth.

Up to the present, we have tacitly assumed that we are concerned with a band whose maximum or minimum is located at $\mathbf{k} = 0$. In the case that the band extremum is not located at $\mathbf{k} = 0$, but at some point $\mathbf{k}_0$, the prescription of the effective mass theory is simple: We merely expand the energy in powers of $\delta \mathbf{k} = \mathbf{k} - \mathbf{k}_0$, and then replace δk^α by $(1/i)(\partial/\partial x_\alpha) + (e/\hbar)A^\alpha$. In the case of a band degeneracy, the theory proceeds in analogy to the case of the impurity level. In place of (4.128), we get instead

$$\sum_{l',\alpha,\beta} D_{ll'\alpha\beta}\left(\frac{1}{i}\frac{\partial}{\partial x_\alpha} + \frac{eA^\alpha}{\hbar}\right)\left(\frac{1}{i}\frac{\partial}{\partial x_\beta} + \frac{eA^\beta}{\hbar}\right)F_{l'}(\mathbf{r}) = (E - E_l)F_l(\mathbf{r})$$

(4.161)

Computations with this equation for practical cases of degenerate valence bands, as in germanium and silicon, are quite difficult (see Goodman, 1961; Evtuhov, 1962) and will not be considered here.

4.5 The Magnetic Susceptibility of Free Electrons

In this section, we shall calculate the magnetic susceptibility of a system of free electrons (with effective mass m^*) at low temperatures. This is particularly important in the discussion of the de Haas-van Alphen effect, which furnishes one of the most important techniques for the experimental investigation of Fermi surfaces. We will follow, to a large extent, the treatment of Sondheimer and Wilson (1951; Wilson, 1953).

In classical mechanics, it can be shown quite generally that the magnetic susceptibility of a system of charges is zero (the Bohr-van

Leeuwen theorem). It can be easily seen that this must be correct, since the energy of a particle is, in classical theory, unaffected by a magnetic field. Hence the free energy is also independent of the field, and the susceptibility is zero. In quantum theory the energies are, however, changed by application of the field, and it is possible to calculate a nonzero susceptibility.

The calculation begins with determination of the free energy, which is given for the case of Fermi statistics by the expression (Wilson, 1953)

$$F = U - ST = N\mu - kT \sum_l \ln(1 + e^{(\mu - E_l)/kT}) \qquad (4.162)$$

The dependence on the magnetic field is contained implicitly through E_l (which is given in 4.160). The quantity μ is the chemical potential (or Fermi energy); it is determined from the requirement that N, the number of electrons, be given by

$$N = \sum n_s \qquad (4.163)$$

This condition is evidently equivalent to:

$$\frac{\partial F}{\partial \mu} = 0 \qquad (4.164)$$

The magnetization, M, is given by

$$M = -\frac{1}{\mu_0}\frac{\partial F}{\partial H} = -\frac{\partial F}{\partial B} \qquad (4.165a)$$

in which μ_0 is the permeability of free space; and the magnetic susceptibility, χ, is defined by

$$\chi = M/H \qquad (4.165b)$$

The calculation proceeds by relating the free energy to the "classical" partition function

$$Z(\beta) = \sum_l e^{-\beta E_l} \qquad (4.166)$$

4.5 THE MAGNETIC SUSCEPTIBILITY OF FREE ELECTRONS

(where $\beta = 1/kT$). We define an auxiliary function $g(E)$ by

$$g(E) = \ln(1 + e^{(\mu - E)/kT}) \tag{4.167}$$

It is convenient to introduce functions $z(E)$ and $\varphi(\beta)$ through

$$\frac{Z(\beta)}{\beta^2} = \int_0^\infty z(E)\, e^{-\beta E}\, dE \tag{4.168a}$$

$$\varphi(\beta) = \int_0^\infty g(E)\, e^{-\beta E}\, dE \tag{4.168b}$$

According to the theory of the Laplace transform, we have

$$z(E) = \frac{1}{2\pi i} \int_{c-i\infty}^{c+i\infty} e^{Es}\, \frac{Z(s)}{s^2}\, ds \tag{4.169a}$$

$$g(E) = \frac{1}{2\pi i} \int_{c-i\infty}^{c+i\infty} \varphi(s)\, e^{Es}\, ds \tag{4.169b}$$

The contour of integration is parallel to the imaginary axis. The constant c must be chosen so that all of the singularities of the integrand are on the left, but is otherwise arbitrary.

$$F = N\mu - kT \sum_l g(E_l) \tag{4.170}$$

$$= N\mu - \frac{kT}{2\pi i} \int_{c-i\infty}^{c+i\infty} \sum_l e^{E_l s}\, \varphi(s)\, ds$$

$$= N\mu - \frac{kT}{2\pi i} \int_{c-i\infty}^{c+i\infty} \frac{Z(-s)}{s^2}\, s^2\, \varphi(s)\, ds$$

260 CHAPTER 4. POINT IMPURITIES AND EXTERNAL FIELDS

Since $s^2 \varphi(s)$ is the Laplace transform of $\partial^2 g/\partial E^2$, the final integral can be expressed as:

$$F = N\mu - kT \int_0^\infty z(E) \frac{\partial^2 g}{\partial E^2} dE$$

Now

$$\frac{\partial^2 g}{\partial E^2} = -\frac{1}{kT} \frac{\partial f}{\partial E}$$

where $f(E)$ is the Fermi function,

$$f(E) = [e^{(E-\mu)/kT} + 1]^{-1}$$

so

$$F = N\mu + \int_0^\infty z(E) \frac{df}{dE} dE \qquad (4.171)$$

At low temperatures df/dE is almost a delta function, so that the essential part of the problem is just the calculation of the inverse Laplace transform of $Z(\beta)/\beta^2$.

Now we determine $Z(\beta)$ from (4.166). The energies E_l are given by

$$E_l = \frac{\hbar^2 k^2}{2m} + (l + \tfrac{1}{2})\hbar\omega_c \pm \frac{g\hbar\omega_0}{4} \qquad (4.172)$$

in which $\omega_0 = eB/m_0$ is the cyclotron resonance frequency for free electrons, whereas ω_c contains the effective mass m^*. The $(\pm)$ sign results from the differing directions of spin. We use the result of (4.148) to give the number of states in an interval Δk_z. Then

$$Z(\beta) = \frac{eBV}{4\pi^2 \hbar} \sum_{\text{spin}} \sum_{l=0}^\infty e^{-\beta(l+1/2)\hbar\omega_c} e^{\pm g\hbar\omega_0/4} \int_{-\infty}^\infty dk_z \exp\left(-\frac{\beta\hbar^2 k_z^2}{2m^*}\right) \qquad (4.173)$$

$$= \frac{eBV}{2\pi\hbar^2} \left(\frac{m^*}{2\beta\pi}\right)^{1/2} \frac{\cosh g\hbar\omega_0/4}{\sinh \beta\hbar\omega_c/2}$$

4.5 THE MAGNETIC SUSCEPTIBILITY OF FREE ELECTRONS

This result is to be substituted into (4.169)

$$z(E) = \frac{1}{2\pi i} \int e^{E\beta} \frac{Z(\beta)}{\beta^2} d\beta = \hbar\omega_c V \left(\frac{m^*}{2\pi\hbar^2}\right)^{3/2} \frac{1}{2\pi i} \int_{c-i\infty}^{c+i\infty} \frac{e^{E\beta} \cosh g\hbar\omega_0/4}{\beta^{5/2} \sinh \beta\hbar\omega_c/2} d\beta$$

(4.174)

The integrand has poles on the imaginary axis at points $\beta\hbar\omega_c/2 = n\pi i$ (n is any integer, positive or negative, but not zero), and a branch point at the origin. It is convenient to change the path of integration into the

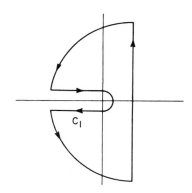

FIG. 20. Contour of integration for Eq. (4.174).

contour shown in Fig. 20. Since the contribution from the large arcs tends to zero as they are indefinitely enlarged, we need to consider only the contributions from the residues and from the paths parallel to the negative real axis. The contribution from a single residue is

$$(-1)^{n+1} 2\pi i \left(\frac{\hbar\omega_c}{2}\right)^{3/2} \frac{e^{i(2n\pi E/\hbar\omega_c - \pi/4)}}{(n\pi)^{5/2}} \cos\left(\frac{n\pi g m^*}{2m_0}\right)$$

The sum of the residues is evidently

$$-2\pi i \left(\frac{\hbar\omega_c}{2}\right)^{3/2} 2 \cdot \sum_{n=1}^{\infty} \frac{(-1)^n \cos(2n\pi E/\hbar\omega_c - \pi/4)}{(n\pi)^{5/2}} \cos\left(\frac{n\pi g m^*}{2m_0}\right) \quad (4.175)$$

It is this oscillatory part of $z(E)$, coming from the poles of the integrand, which produces the oscillatory behavior of the magnetic susceptibility observed in the de Haas-van Alphen effect.

To obtain the integral along that part of the contour parallel to the negative real axis in the limit that the radius of the small circle tends to zero (contour C_1), we observe first that, for small x,

$$\operatorname{csch} x = \frac{1}{x}\left(1 - \frac{x^2}{6} + \frac{7x^4}{360} + \cdots\right)$$

so that we get a contribution (with $z = \beta\hbar\omega_c/2$)

$$\frac{1}{2\pi i}\left(\frac{\hbar\omega_c}{2}\right)^{3/2} \int_{C_1} e^{2zE/\hbar\omega_c}\left[\frac{1}{z^{7/2}} + \frac{1}{z^{3/2}}\left(\frac{m^{*2}g^2}{8m_0^2} - \frac{1}{6}\right) + O(z^{1(2)})\right]dz \quad (4.176)$$

The integrals in (4.176) may be evaluated through the use of the identity (Jeffreys and Jeffreys, 1950, p. 401) that

$$\frac{1}{2\pi i}\int_{C_1} z^{-m} e^{zt}\, dz = \frac{t^{m-1}}{\Gamma(m)}$$

Then (4.176) becomes

$$\frac{2E^{5/2}}{\hbar\omega_c}\cdot\frac{8}{15\sqrt{\pi}} + \frac{\hbar\omega_c}{2}\frac{2E^{1/2}}{\sqrt{\pi}}\left(\frac{m^{*2}g^2}{8m_0^2} - \frac{1}{6}\right) + \cdots \quad (4.177)$$

Higher terms in the series can be neglected, provided (as we shall be able to infer subsequently) $\mu \gg \hbar\omega_c$. (These terms give rise to a field dependence of the normal diamagnetism and paramagnetism.) If we now combine the terms (4.175) and (4.177), we find for $z(E)$

$$z(E) = V\left(\frac{m^*}{2\pi\hbar^2}\right)^{3/2}\left[\frac{16E^{5/2}}{15\sqrt{\pi}} - \frac{(\hbar\omega_c)^2 E^{1/2}}{\sqrt{\pi}}\left(\frac{m^{*2}g^2}{8m_0^2} - \frac{1}{6}\right) + \cdots \right. \quad (4.178)$$

$$\left. - \frac{(\hbar\omega_c)^{5/2}}{2^{1/2}}\sum_{n=1}^{\infty}\frac{(-1)^n \cos(2\pi nE/\hbar\omega_c - \pi/4)}{(n\pi)^{5/2}}\cos\left(\frac{n\pi g m^*}{2m_0}\right)\right]$$

The free energy is now to be determined from (4.171). At low temperatures ($kT \ll \mu$) we may, with respect to the first two terms,

4.5 THE MAGNETIC SUSCEPTIBILITY OF FREE ELECTRONS

replace df/dE by $-\delta(E-\mu)$ (we thereby ignore a weak quadratic temperature dependence). The oscillatory term must be handled with greater care. We can write

$$\frac{df}{dE} = -\frac{1}{kT}\frac{1}{4\cosh^2[(E-\mu)/2kT]}$$

Then we consider the integral

$$-\frac{1}{4kT}\int_0^\infty \frac{\cos(2n\pi E/\hbar\omega_c - \pi/4)\,dE}{\cosh^2[(E-\mu)/2kT]} = -\tfrac{1}{4}\,\mathrm{Re}\, e^{i(2\pi n\mu/\hbar\omega_c - \pi/4)} \int_{-\infty}^\infty \frac{e^{2\pi i nkTy/\hbar\omega_c}}{\cosh^2(y/2)}\,dy$$

(4.179)

where we have defined $y = (E-\mu)/kT$, and extended the lower limit of the integration on y from $-\mu/kT$ to $-\infty$. The error in this replacement is of order $e^{-\mu/kT}$, which is negligible. The integral is a tabulated one (see Erdelyi, 1954, p. 31), and gives

$$-\frac{2\pi^2 nkT}{\hbar\omega_c}\frac{\cos(2\pi n\mu/\hbar\omega_c - \pi/4)}{\sinh(2\pi^2 nkT/\hbar\omega_c)}$$

Then the free energy per unit volume becomes [from (4.171)]

$$F = N\mu - \frac{16\mu^{5/2}}{15\sqrt{\pi}}\left(\frac{m^*}{2\pi\hbar^2}\right)^{3/2}\left[1 - \frac{15}{16}\left(\frac{\hbar\omega_c}{\mu}\right)^2\left(\frac{m^{*2}g^2}{8m_0^2} - \frac{1}{6}\right)\right. \quad (4.180)$$

$$\left. - \frac{15}{8(2^{1/2})}\left(\frac{kT}{\hbar\omega_c}\right)\left(\frac{\hbar\omega_c}{\mu}\right)^{5/2}\sum_{n=1}^\infty \frac{(-1)^n}{n^{3/2}}\frac{\cos(2\pi n\mu/\hbar\omega_c - \pi/4)\cos(n\pi gm^*/2m_0)}{\sinh(2\pi^2 nkT/\hbar\omega_c)}\right]$$

The magnetization is found by differentiating (4.180) as required by (4.165). Observe that

$$\frac{\partial F}{\partial B} = \frac{e}{m^*}\frac{\partial F}{\partial \omega_c}$$

In differentiating the oscillatory part of the free energy we neglect the contributions from all except the cosine term: this will be dominant under the conditions of interest. Then

$$M = M_0 \left[\left(\frac{3m^{*2} g^2}{4m_0^2} - 1 \right) - \frac{6\pi k T}{\hbar \omega_c} \left(\frac{2\mu}{\hbar \omega_c} \right)^{1/2} \times \right. \quad (4.181)$$

$$\left. \sum_n \frac{(-1)^n}{n^{1/2}} \frac{\sin (2\pi n \mu / \hbar \omega_c - \pi/4) \cos (n\pi g m^* / 2m_0)}{\sinh (2\pi^2 nkT / \hbar \omega_c)} \right]$$

in which

$$M_0 = \frac{e}{6\pi^2 \hbar} \left(\frac{m^* \mu}{2} \right)^{1/2} \quad \omega_c = \frac{e^2}{12\pi^2 \hbar} \left(\frac{2\mu}{m^*} \right)^{1/2} B \quad (4.182)$$

The susceptibility is immediately found from (4.165b). The leading term is the paramagnetic susceptibility due to the electron spins; the second term is the steady diamagnetic term which is one third as large when the effective mass ratio is unity and $g = 2$: these terms are independent of temperature. The oscillatory term is significant at low temperatures and high fields: it produces the de Haas-van Alphen effect.

The chemical potential, μ, in the presence of the field must now be determined from (4.164) and (4.180). We find

$$N = \frac{1}{3\pi^2} \left(\frac{2m^* \mu}{\hbar^2} \right)^{3/2} \left[1 - \frac{3}{16} \left(\frac{\hbar \omega_c}{\mu} \right)^2 \left(\frac{m^{*2} g^2}{8m_0^2} - \frac{1}{6} \right) + \right. \quad (4.183)$$

$$\left. \frac{3\pi k T}{\hbar \omega_c} \left(\frac{\hbar \omega_c}{2\mu} \right)^{3/2} \sum_{n=1}^{\infty} \frac{(-1)^n}{n^{1/2}} \frac{\sin (2\pi n \mu / \hbar \omega_c - \pi/4) \cos (n\pi g m^* / 2m_0)}{\sinh (2\pi^2 nkT / \hbar \omega_c)} \right]$$

It is necessary to solve (4.183) to determine μ in terms of N. It is usually sufficient, however, to include only the first term; the relation is then evidently the same as for the free electron gas. In this case, we may set $\mu = \hbar^2 k_F^2 / 2m^*$, in which k_F is the wave vector on the Fermi surface. Then we obtain for M_0 in (4.182)

$$M_0 = \frac{e^2 k_F}{12\pi^2 m^*} B \quad (4.184)$$

It is interesting to consider the physical origin of the diamagnetic effects. The application of the external field causes the band structure to break up (for directions of **k** perpendicular to the field) into a set of Landau levels. For fields which are not too strong, the Fermi level can be regarded as constant. The lowest level now has energy $\hbar\omega_c/2$, instead of zero, and the average energy is increased by the field. Hence there is a steady diamagnetic contribution, which will dominate the spin paramagnetism if the effective mass is small. As the field increases, the spacing between the Landau levels increases so that the levels are gradually "forced through" the Fermi level and depopulated. (At the same time the degeneracy of the Landau levels increases.) The average energy of the system fluctuates as this "forcing through" occurs. The amplitude of the oscillations increases with increasing field, and the period is proportional to B^{-1}. The amplitude is a rapidly decreasing function of $kT/\hbar\omega_c$, which implies that observation of the oscillations requires low temperatures.

The extension of these results to the case of an anisotropic effective mass is straightforward (Blackman, 1938); but for arbitrary nonparabolic band structures, the problem is much more difficult (Lifshitz and Kosevich, 1956). The period of oscillations will be determined by the semiclassical arguments of Onsager (1952) in the next section.

4.6 The Cyclotron Frequency for an Arbitrary Fermi Surface

A rigorous determination of the energy levels of an electron in a magnetic field for a situation in which the band structure in the absence of a field is some arbitrary $E(\mathbf{k})$ can be based, in principle, on the one-band Hamiltonians of Kohn and Wannier (see Section 4.4). Such a treatment is generally impractical and it is customary to refer to a semiclassical approach (Onsager, 1952; Lifshitz and Kosevich, 1956) For the gauge of Eq. (4.136), the commutation rule for the components of the quantity $P = (\hbar/i)\nabla + eA$, which is given in (4.130), is

$$[P_x, P_y] = \frac{e\hbar}{i} B \qquad (4.185)$$

$$[P_y, P_z] = [P_x, P_z] = 0$$

We wish to draw an analogy between the first of equations (4.185), and the ordinary cannonical commutation rule between coordinates and momenta

$$[p_k, q_j] = \frac{\hbar}{i} \delta_{kj} \qquad (4.186)$$

by making the identification

$$p_x = P_x; \qquad q_x = P_y/eB \qquad (4.187)$$

We apply the semiclassical quantum condition

$$\frac{1}{2\pi} \oint P_x \, dq_x = (n + \gamma)\hbar \qquad (4.188)$$

in which n is a positive integer and γ is some constant "phase factor" ($\gamma = \frac{1}{2}$ for free electrons). We substitute (4.187), and find

$$\oint P_x \, dP_y = 2\pi(n + \gamma)\hbar eB = A \qquad (4.189)$$

The integral in (4.189) runs along the curve bounding a cross section of a surface of constant energy perpendicular to the field, and has a value equal to the area, A, of this cross section.[5] The area of the cross section depends on the energy of the state whose quantum number is n. We will suppose that in the limit of large energies, the separation between states is $\hbar\omega_c$ (this is the definition of the cyclotron frequency)

$$\frac{dE}{dn} = \hbar\omega_c \qquad (4.190)$$

Differentiation of Eq. (4.189) with respect to energy yields:

$$\omega_c = 2\pi eB \left(\frac{dA}{dE}\right)^{-1} \qquad (4.191)$$

We can easily see that ω_c is the circular frequency of rotation of a classical orbit on a cross section, perpendicular to the magnetic field, of a surface of constant energy (Pippard, 1960). The Lorentz force on a particle in a field is evB so that the time required for the momentum to change by dp is dp/evB. The time required to complete a revolution is then

---

[5] The area mentioned above is in "momentum" space. To go to the ordinary reciprocal lattice, a scale factor of $\hbar^2$ is required: A' (area in k-space) $= A/\hbar^2 = 2\pi(n + \gamma)eB/\hbar$.

4.6 CYCLOTRON FREQUENCY FOR ANY FERMI SURFACE

$$T = \frac{1}{eB} \oint \frac{dp}{v} \quad (4.192)$$

Now consider similar cross sections of surfaces of constant energy E and $E + dE$. The separation between the surfaces is dE/V_p, $E = dE/v$. The area of the annular ring between the surfaces is

$$dA = dE \oint \frac{dp}{v} \quad (4.193)$$

where dp is a momentum increment along the ring. Evidently

$$T = \frac{1}{eB} \frac{dA}{dE}, \quad \text{and} \quad \omega_c = \frac{2\pi}{T} = 2\pi e B \left(\frac{dA}{dE}\right)^{-1} \quad (4.194)$$

as required.

Further, we can infer very simply the existence of oscillations in the magnetic susceptibility. Since $\hbar\omega_c/\mu$ is quite small in usual circumstances, the Fermi energy will be essentially independent of the field. Consider a particular level of oscillator quantum number n. For small fields, the orbits of this n are small, lying well inside the Fermi surface. We consider a cross section of a surface of constant energy E, corresponding to n, of thickness dk_z, and area A. As the field increases, the area of the cross section grows according to (4.191). When it equals the area of the corresponding cross section of the Fermi surface A_0, the oscillator level will be half depopulated. This occurs when

$$2\pi(n + \gamma) = \frac{A_0(p_z)}{\hbar eB} \quad (4.195)$$

The magnetization will vary with the state of depletion of the level with energy nearest the Fermi energy. A periodic dependence of magnetization on field is to be expected as successive levels come up to the Fermi surface and depopulate. Thus we expect that the susceptibility contains (at least, as lowest harmonic) a term proportional to

$$\sin \frac{A_0(p_z)}{\hbar eB} \quad (4.196)$$

So far, we have considered a particular cross section corresponding to a definite value of p_z. The contribution of all cross sections must be summed. Since the sine is a rapidly oscillating function (usually, $\hbar eB \ll A_0$), the contributions from different cross sections will tend to cancel. The resultant will be governed by the region of Fermi surface for which the area is an extremum with respect to p_z: in examining experimental susceptibility measurements, we find a periodic behavior as given by (4.196), with A_0 an extremal cross section.

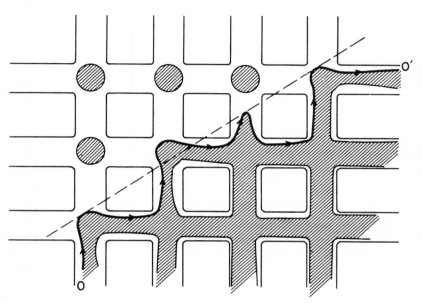

FIG. 21. Section of Fermi surface in the extended zone scheme. The magnetic field is tilted slightly with respect to the crystal axes. Regions of electron orbits (*upper left*) are separated from regions of hole orbit (*bottom right*) by an open orbit O, O' (Chambers, 1960).

Note that these results give the correct behavior for the free electron gas. In that case, $A_0 = \pi p_f^2 = 2\pi m^* \mu$, so we get

$$\frac{A_0}{\hbar eB} = \frac{2\pi\mu}{\hbar\omega_c} \qquad (4.197)$$

as required.

In summary, the fundamental period of de Haas-van Alphen oscillations of the magnetic susceptibility is $A_0/\hbar eB$, where A_0 is the area of an extremal cross section of the Fermi surface perpendicular to the field.

It is possible, however, that the intersection of the Fermi surface with a plane may not be a closed curve, if the Fermi surface touches the Brillouin zone. The situation may be most easily visualized in the extended zone scheme (see Fig. 21). Such intersections are referred to as "open orbits," and are particularly important in the determination of the magnetoresistance.

Finally, we note that in sufficiently high magnetic fields, the field itself is predominant in determining the motion of an electron. The lattice potential then is only a small perturbation, which tends to remove the degeneracy of the Landau levels and therefore broadens them. When this situation occurs, electron orbits may pass through Brillouin zone boundaries where energy gaps would normally be present. This effect was named magnetic breakdown by Cohen and Falicov (1961), and has been studied by Blount (1962b) and Pippard (1962) as well. It can be shown that magnetic breakdown is possible if $\hbar\omega_c E_F > E_g^2$ where E_g is the gap across a zone face.

4.7 The Steady Diamagnetic Susceptibility: Arbitrary Band Structure

Section 4.5 contained a "quasi-exact" derivation of the magnetic susceptibility of free electrons, which was seen to contain a steady susceptibility, and a term involving a superposition of oscillatory components. The diamagnetic part, $\chi_{d,s}$ of the steady susceptibility is

$$\chi_{d,s} = -\frac{M_0}{H} = -\frac{e^2 \mu_0 k_F}{12\pi^2 m^*} \qquad (4.198)$$

in which μ_0 is here the permeability of free space ($4\pi \times 10^{-7}$ in mks units). The generalization of the oscillatory term to arbitrary band structures was considered in Section 4.6. We now wish to outline the corresponding generalization of (4.198).

The calculation of the diamagnetic susceptibility for an arbitrary band structure was first carried through to a complete, although formal, expression by Hebborn and Sondheimer (1960). Earlier studies were made by Peierls (1933a,b), Wilson (1953), Adams (1953), and Kjeldaas and Kohn (1956). Alternative approaches have been given by Enz (1960), Roth (1962), and Blount (1962b). The calculation is quite laborious, and we shall necessarily omit many of the details.

As before, we begin by considering the classical partition function, which is given by (4.166). Unfortunately, we do not have an explicit formula for the energy levels in the presence of a field, and so cannot carry out an explicit summation to produce an expression analogous to (4.173). Instead, we expand the partition function in powers of B, and retain terms of the order of B^2. The radius of convergence of such an expansion will be governed by the location of the first pole. Thus the expansion technique can yield only the steady diamagnetism.

The partition function may be generally defined through

$$Z = \text{Tr}\{e^{-\beta H}\} \qquad (4.199)$$

in which H is the complete Hamiltonian for the system. This will easily be seen to reduce to (4.166) when the trace is evaluated in a coordinate system in which the basic vectors are eigenvectors of H. In the present case, it is convenient to express the Hamiltonian, using the gauge of (4.136) as

$$H = H_0 + H_1 + H_2 \qquad (4.200)$$

in which

$$H_0 = \frac{p^2}{2m} + V(r) \qquad (4.201\text{a})$$

$$H_1 = \frac{i\hbar eBy}{m}\frac{\partial}{\partial x} = i\hbar\omega_0 y \frac{\partial}{\partial x} \qquad (4.201\text{b})$$

$$H_2 = \frac{e^2 B^2}{2m} y^2 \qquad (4.201\text{c})$$

We will evaluate the trace in a representation in which H_0 is diagonal. The basic functions are the Bloch functions of (4.2) and (4.3).

It is necessary to be very careful in expanding the operator $e^{-\beta H}$, since H_0 does not commute with H_1 and H_2. To understand the technique

4.7 THE STEADY DIAMAGNETIC SUSCEPTIBILITY

let us consider, following Goldberger and Adams (1952), an exponential of the form $e^{-(a+b)}$, where a and b are any two noncommuting operators: Define

$$u(s) = e^{-(a+b)s} \tag{4.202}$$

Then $u(s)$ satisfies the differential equation

$$\frac{\partial u(s)}{\partial s} = -(a+b)u(s); \qquad u(0) = 1 \tag{4.203}$$

Now define an operator $v(s)$ through

$$u(s) = e^{-as} v(s) \tag{4.204}$$

We find that

$$\frac{dv(s)}{ds} = -e^{as} b e^{-as} v(s); \qquad v(0) = 1 \tag{4.205}$$

It is desirable to replace the differential equation and boundary condition by an integral equation: This is seen to be:

$$v(s) = 1 - \int_0^s ds' \, e^{as'} b e^{-as'} v(s') \tag{4.206}$$

This equation may be solved by iteration. Successive terms involve increasing powers of b. The iterative procedure may thus be seen to be useful in the case in which b is in some sense a "small" operator compared to a. The iterative solution of (4.206) which includes terms of second order in b is

$$v(s) = 1 - \int_0^s ds' \, e^{as'} b \, e^{-as'} + \int_0^s ds' \int_0^{s'} ds'' \, e^{as'} b \, e^{-a(s'-s'')} b \, e^{-as''} \tag{4.207}$$

We can now obtain the desired expansion for $u(s)$. The expression is rendered a bit more convenient by the introduction of variables s_1, s_2 through $s' = s s_1$, $s'' = s' s_2$. We wish to obtain $u(1)$

$$u(1) = e^{-(a+b)} = e^{-a} - \int_0^1 ds_1 \, e^{-(1-s_1)a} \, b \, e^{-s_1 a} + \tag{4.208}$$

$$\int_0^1 s_1 \, ds_1 \int_0^1 ds_2 \, e^{-(1-s_1)a} \, b \, e^{-s_1(1-s_2)a} \, b \, e^{-s_1 s_2 a} + \ldots$$

The general term in the expansion (4.208) may be written down and expressed in a reasonably simple form using the so-called "ordered product." We shall not, however, discuss the general properties of the expansion. The terms presented explicitly in (4.208) suffice for the present problem.

It can be seen on general grounds that the term in the trace which is linear in B, resulting from the application of the linear term of (4.202) to H_1, must vanish. Such a term would yield a contribution to the magnetization independent of the applied field. The second order terms are of two kinds: (1) a term resulting from use of H_2 in the linear term of (4.208), and (2) the quadratic term in (4.208) in which $b = \beta H_1$. Hence, we write

$$Z = Z_0 + Z_2 + Z_{11} \tag{4.209}$$

in which Z_0 is the partition function in the absence of the field. Our definition of the trace is (for an arbitrary operator A)

$$\text{Tr}(A) = \sum_n \int d^3k \int d^3r \, \psi_n^*(\mathbf{k}, \mathbf{r}) A \psi_n(\mathbf{k}, \mathbf{r})$$

Thus

$$Z_2 = -\beta \sum_n \int d^3k \int d^3r \int_0^1 ds \, \psi_n^*(\mathbf{k}, \mathbf{r}) \, e^{-(1-s)\beta H_0} H_2 \, e^{-s\beta H_0} \psi_n(\mathbf{k}, \mathbf{r}) \tag{4.210}$$

Z_{11} results from the linear term H_1 in second order:

$$Z_{11} = \beta^2 \sum_n \int d^3k \int d^3r \int_0^1 s_1 \, ds_1 \int_0^1 ds_2 \, \psi_n^*(\mathbf{k}, \mathbf{r}) \, e^{-(1-s_1)\beta H_0} H_1 \times \tag{4.211}$$

$$e^{-s_1(1-s_2)\beta H_0} H_1 \, e^{-s_1 s_2 \beta H_0} \psi_n(\mathbf{k}, \mathbf{r})$$

4.7 THE STEADY DIAMAGNETIC SUSCEPTIBILITY

These expressions may be reduced to the following:

$$Z_2 = \beta \sum_n \int d^3k \int_0^1 ds\, e^{-(1-s)\beta E_n(\mathbf{k})} \langle n\mathbf{k}|H_2|n\mathbf{k}\rangle e^{-s\beta E_n(\mathbf{k})} \qquad (4.212a)$$

$$Z_{11} = \beta^2 \sum_{n,n'} \int d^3k \int d^3k' \int_0^1 s_1\, ds \int_0^1 ds_2\, e^{-(1-s_1)\beta E_n} \langle n\mathbf{k}|H_1|n'\,\mathbf{k}'\rangle \times \qquad (4.212b)$$

$$e^{-s_1(1-s_2)\beta E_{n'}} \langle n'\,\mathbf{k}'|H_1|n\mathbf{k}\rangle e^{-s_1 s_2 \beta E_n(\mathbf{k})}$$

In the expression for Z_{11}, we have used the completeness relation

$$\sum_n \int d^3k \psi_n^*(\mathbf{k}, \mathbf{r}')\psi_n(\mathbf{k}, \mathbf{r}) = \delta(\mathbf{r} - \mathbf{r}') \qquad (4.213)$$

The matrix elements of H_1 and H_2 are singular functions of **k** (see 4.107). For this reason the exponentials of the energy are left on the right of the matrix elements.

The explicit evaluation of the formulas for Z_{11} and Z_2 is quite complicated and will not be discussed here (see Hebborn and Sondheimer, 1960, for details). The free energy is computed from the partition function with the use of (4.169) and (4.171). This proceeds rather simply: Evaluation of Z_1 and Z_2 leads to expressions which depend on β through terms of the form $\beta e^{-\beta E_n}$ and $\beta^2 e^{-\beta E_n}$. We may write schematically

$$Z = Z_0 + B^2 \sum_n \int d^3k \,[a_n(\mathbf{k})\beta\, e^{-\beta E_n} + b_n(\mathbf{k})\beta^2\, e^{-\beta E_n}] \qquad (4.214)$$

in which $a_n(\mathbf{k})$ and $b_n(\mathbf{k})$ are complicated functions involving derivatives of the energy and the wave functions with respect to **k**. Let us consider the integrals over β and E involved in the computation of the free energy. We find that

$$\frac{1}{2\pi i}\int_0^\infty dE \int_{c-i\infty}^{c+i\infty} \frac{d\beta}{\beta^2}\,[a_n(\mathbf{k})\beta e^{-\beta E_n} + b_n(\mathbf{k})\beta^2\, e^{-\beta E_n}]\, e^{\beta E}\, \frac{df}{dE}\, dE \qquad (4.215)$$

$$= \frac{1}{2\pi i}\int_0^\infty dE \int_{c-i\infty}^{c+i\infty} d\beta \left[-a_n(\mathbf{k})f(E) + b_n(\mathbf{k})\frac{df}{dE}\right] e^{-\beta(E_n - E)}$$

$$= \int_0^\infty dE\, \delta(E_n - E) \left[-a_n(\mathbf{k})f(E) + b_n(\mathbf{k}) \frac{df}{dE} \right]$$

$$= -a_n(\mathbf{k})f(E_n) + b_n(\mathbf{k}) \frac{df}{dE}(E_n)$$

In the first step in (4.208), we have integrated by parts with respect to E in the first term; and in the second step have used:

$$\frac{1}{2\pi i} \int_{c-i\infty}^{c+i\infty} d\beta\, e^{\beta(E_n - E)} = \frac{e^{c(E_n - E)}}{2\pi} \int_{-\infty}^{\infty} e^{ix(E_n - E)} dx = \delta(E_n - E)$$

Thus, the contribution to the susceptibility from these terms is just

$$\chi = 2\mu_0 \sum_n \int d^3k \left[-a_n(\mathbf{k}) f(E_n) + b_n(\mathbf{k}) \frac{df}{dE_n} \right] \tag{4.216}$$

The calculation of Hebborn and Sondheimer yields the following very complex expression:

$$\chi = \chi_1 + \chi_2 + \chi_3 + \chi_4$$

$$\chi_1 = \frac{e^2 \mu_0}{48\pi^3 \hbar^2} \sum_n \int \left\{ \left[\frac{\partial^2 E_n}{\partial k_1^2} \frac{\partial^2 E_n}{\partial k_2^2} - \left(\frac{\partial^2 E_n}{\partial k_1 \partial k_2} \right)^2 + \right. \right. \tag{4.217}$$

$$\left. \left. \frac{3}{2} \left(\frac{\partial E_n}{\partial k_1} \frac{\partial^3 E_n}{\partial k_1 \partial k_2^2} + \frac{\partial E_n}{\partial k_2} \frac{\partial^3 E_n}{\partial k_1^2 \partial k_2} \right) \right] \frac{\partial f_0(E_n)}{\partial E_n} d^3k \right.$$

$$\chi_2 = \frac{e^2 \mu_0}{4\pi^3 m} \sum_n \int \left[\frac{m_0}{\hbar^2} \frac{\partial E_n}{\partial k^2} \left\{ \int \left| \frac{\partial u_n}{\partial k_1} \right|^2 d\tau_0 - |X_{nn}|^2 \right\} - \right. \tag{4.218}$$

$$k_1 \int \left(\frac{\partial u_n^*}{\partial k_1} \frac{\partial u_n}{\partial k_2} + \frac{\partial u_n}{\partial k_1} \frac{\partial u_n^*}{\partial k_2} \right) d\tau_0 + 2X_{nn} W_n +$$

$$i \int \left(\frac{\partial u_n^*}{\partial k_1} \frac{\partial^2 u_n}{\partial x \partial k_2} - \frac{\partial u_n}{\partial k_1} \frac{\partial^2 u_n^*}{\partial x \partial k_2} \right) d\tau_0 \left] \frac{\partial E_n}{\partial k_2} \frac{\partial f_0(E_n)}{\partial E_n} d^3k \right.$$

4.7 THE STEADY DIAMAGNETIC SUSCEPTIBILITY

$$\chi_3 = -\frac{e^2 \mu_0}{4\pi^3 m} \sum_n \int \left[\left[\frac{im}{\hbar^2} X_{nn} \int \left(\frac{\partial u_n^*}{\partial k_1} \frac{\partial u_n}{\partial k_2} - \frac{\partial u_n}{\partial k_1} \frac{\partial u_n^*}{\partial k_2} \right) d\tau_0 + \right. \right. \tag{4.219}$$

$$2 \sum_{l \neq n} \frac{X_{ln}}{E_l - E_n} \left\{ \int \left(\frac{\partial u_l}{\partial y} \frac{\partial u_n^*}{\partial k_1} - \frac{\partial u_l}{\partial x} \frac{\partial u_n}{\partial k_2} \right) d\tau_0 + \right.$$

$$\left. \left. k_1 Y_{nl} - k_2 X_{nl} + \frac{m}{\hbar^2} \frac{\partial E_n}{\partial k_1} Y_{nl} \right\} \right] \frac{\partial E_n}{\partial k_2} f_0(E_n) d^3k$$

$$\chi_4 = -\frac{e^2 \mu_0}{4\pi^3 m} \sum_n \int \left[\left\{ \int \left| \frac{\partial u_n}{\partial k_2} \right|^2 d\tau_0 + 2 W_n \frac{\partial X_{nn}}{\partial k_2} - \right. \right. \tag{4.220}$$

$$\frac{2\hbar^2}{m} \sum_{l \neq m} \frac{1}{E_l - E_n} \left| k_1 Y_{ln} - \int \frac{\partial u_l^*}{\partial x} \frac{\partial u_n}{\partial k_2} d\tau_0 \right|^2 \right\} f_0(E_n) +$$

$$\left. \frac{\hbar^2}{m} (W_n)^2 \frac{\partial f_0(E_n)}{\partial E_n} \right] d^3k$$

In these equations, the quantities X_{nl}, Y_{nl}, W_n are defined as follows[6]:

$$X_{nl} = i \int u_n^* \frac{\partial u_l}{\partial k_1} d\tau_0 \qquad Y_{nl} = i \int u_n^* \frac{\partial u_l}{\partial k_2} d\tau_0$$

$$W_n = k_1 Y_{nn} - \frac{1}{2} \int \left(\frac{\partial u_n^*}{\partial x} \frac{\partial u_n}{\partial k_2} + \frac{\partial u_n}{\partial x} \frac{\partial u_n^*}{\partial k_2} \right) d\tau_0$$

The integrals with respect to τ_0 extend over the unit (Wigner-Seitz) cell. The magnetic field is, of course, along the z-direction.

In the free electron limit, u_n is a constant so that only the first term survives. If one makes use of the delta function character of $\partial f_0/\partial E$, it is found that Eq. (4.198) is obtained, as required. The tight binding limit of very narrow energy bands is also of interest: then the derivatives of the energy with respect to the components of k vanish exponentially. Only χ_4 survives in this limit, and Hebborn and Sondheimer show that

---

[6] The normalization of the Bloch functions employed by Hebborn and Sondheimer in Eqs. (4.217–4.220) is different from the convention of Eq. (4.4). They choose

$$\frac{\Omega}{(2\pi)^3} \int \psi_n^{*\prime}(\mathbf{k}', \mathbf{r}) \psi_n(\mathbf{k}, \mathbf{r}) d^3r = \delta_{nn'} \delta(\mathbf{k} - \mathbf{k}')$$

the standard result is obtained for the atomic diamagnetism. Finally, it has been shown that the expressions possess the necessary invariance and symmetry properties. These complex formulas have not yet been evaluated for a real material.

4.8 The Steady Electric Field

The study of the motion of electrons in a steady electric field has turned out, perhaps rather surprisingly, to be quite complicated. The field contributes to the Hamiltonian a term

$$U = e\mathscr{E} \cdot \mathbf{r} \qquad (4.221)$$

where $\mathscr{E}$ is the electric field strength. As in the previous sections, the charge e is taken as a positive number for electrons. It is evident in

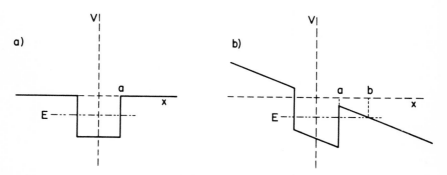

FIG. 22. Effect on an electric field on a square well in one dimension.

this case, as in the discussion of the steady magnetic field, that there will be difficulties associated with the application of perturbation theory because, for sufficiently large distances, the perturbation becomes arbitrarily large, no matter how weak is the field. Straightforward application of perturbation theory is not possible, and important physical quantities may not possess a power series in the field strength. There is, in addition, another complication not present in the discussion of the magnetic field. Strictly speaking, there are no bound states when the Hamiltonian contains a term of the form of (4.221). To see this

4.8 THE STEADY ELECTRIC FIELD

consider the potential energy diagrams of Fig. 22, which show the effect of an electric field on a square well potential in one dimension. It is seen that for any energy E which, in the absence of the field, produces a bound state in the well, there will be no bound state when the field is present because the electron has a finite probability of tunneling out to the right. On the right in Fig. 22b, the wave function will be (in the WKB approximation)

$$\frac{A}{(E - e\mathscr{E} x)^{1/4}} \exp\left[i(2m)^{1/2}\left\{\frac{2}{3e\mathscr{E}\hbar}(E - e\mathscr{E} x)^{3/2}\right\}\right] \quad (4.222)$$

This is not normalizable. The transmission coefficient through the barrier, which is the ratio of the square magnitude of the function outside the well to that in the well, is, in the WKB approximation, approximately (Bohm, 1951)

$$\exp\left(-\frac{4}{3}(2m)^{1/2}\frac{E^{3/2}}{\hbar e\mathscr{E}}\right) \quad (4.223)$$

One notices that this expression, although very small for small $\mathscr{E}$, does not possess a power series expansion in $\mathscr{E}$. Of course in very many practical situations, the applied field is sufficiently weak so that the life time of a bound state is very long indeed. Tunneling is, however, essential in explaining the phenomena of cold field emission from solids.

We will base our study of a Bloch electron in an electric field on the time dependent Schrödinger equation. Because an electron may move through a large region of **k**-space under the influence of the field, it is desirable to employ the crystal momentum representation (4.6). We suppose that the electric field is in the x-direction, and that this is also the direction of a reciprocal lattice vector.

It is necessary to determine the matrix elements of the coordinate x in the crystal momentum representation

$$\langle n'\,\mathbf{k}'|x|n\mathbf{k}\rangle = \int e^{i(\mathbf{k}-\mathbf{k}')\cdot\mathbf{r}} u_{n'}^*(\mathbf{k}',\mathbf{r}) x u_n(\mathbf{k},\mathbf{r})\,d^3r$$

$$= \frac{1}{i}\left[\frac{\partial}{\partial k_x}\int e^{i(\mathbf{k}-\mathbf{k}')\cdot\mathbf{r}} u_{n'}^*(\mathbf{k}',\mathbf{r})u_n(\mathbf{k},\mathbf{r})\,d^3r - \int e^{i(\mathbf{k}-\mathbf{k}')\cdot\mathbf{r}} u_{n'}^*(\mathbf{k}',\mathbf{r})\frac{\partial}{\partial k_x}u_n(\mathbf{k},\mathbf{r})\,d^3r\right]$$

The quantity $u_{n'}(\partial/\partial k_x)u_n$ is a periodic function, so that we obtain

$$\langle n\mathbf{k}|x|n'\,\mathbf{k}'\rangle = i\,\frac{\partial}{\partial k_x}\,[\delta(\mathbf{k}-\mathbf{k}')\delta_{nn'}] + X_{nn'}\,\delta(\mathbf{k}-\mathbf{k}') \qquad (4.224)$$

where

$$X_{nn'} = \frac{(2\pi)^3}{\Omega_0}\,i\int u_n{}^*(\mathbf{k},\mathbf{r})\,\frac{\partial}{\partial k_x}\,u_{n'}(\mathbf{k},\mathbf{r})\,d^3r \qquad (4.225)$$

and the integral in (4.225) includes a single cell. The calculation of the matrix element has been handled in a fashion which apparently ignores vital questions of convergence: this problem has been discussed carefully by Blount (1962a).

The Schrödinger equation now has the form

$$\left[E_n(\mathbf{k}) + ie\mathscr{E}\,\frac{\partial}{\partial k_x} - i\hbar\,\frac{\partial}{\partial t}\right]\varphi_n(\mathbf{k}) + e\mathscr{E}\sum_{n'} X_{nn'}(\mathbf{k})\varphi_{n'}(\mathbf{k}) = 0 \qquad (4.226)$$

An important general result can be obtained immediately. If we multiply (4.226) by $\varphi_n{}^*(\mathbf{k})$, subtract from the resulting equation its complex conjugate, and then sum on the band index n, the term involving the $X_{nn'}$ disappears on summation. This follows because $X_{n'n}^* = X_{nn'}$. We then obtain

$$\left[-e\mathscr{E}\,\frac{\partial}{\partial k_x} + \hbar\,\frac{\partial}{\partial t}\right]\sum_n |\varphi_n(\mathbf{k})|^2 = 0 \qquad (4.227)$$

The general solution of this equation is

$$\sum_n |\varphi_n(\mathbf{k})|^2 = G\left(k_x + \frac{e\mathscr{E}t}{\hbar},\,k_y,\,k_z\right) \qquad (4.228)$$

where G is an arbitrary function of its argument. This result implies that the centroid of the electron wave packet moves through $\mathbf{k}$-space in accord with the relation

$$\frac{dk_x}{dt} = -\frac{e\,\mathscr{E}}{\hbar} \qquad (4.229)$$

This is just what would be expected in classical mechanics for a particle of charge e and momentum $\hbar \mathbf{k}$.

We now examine Eq. (4.229) in more detail. The diagonal matrix element X_{nn} will not vanish at a general point of the Brillouin zone, nor need it vanish at symmetry points unless the group of the wave vector contains the inversion. This matrix element gives rise to a displacement of the band structure which is linear in the applied field. We can incorporate this shift into the first term of (4.226) by defining:

$$E_n^{(1)}(\mathbf{k}) = E_n(\mathbf{k}) + e\mathscr{E} X_{nn} \tag{4.230}$$

Then (4.226) becomes

$$\left[E_n^{(1)}(\mathbf{k}) + ie\mathscr{E} \frac{\partial}{\partial k_x} - i\hbar \frac{\partial}{\partial t} \right] \varphi_n(\mathbf{k}) + e\mathscr{E} \sum_{\substack{n' \\ n \neq n'}} X_{nn'} \varphi_{n'}(\mathbf{k}) = 0 \tag{4.231}$$

The motion of an electron in a given band is described by the terms in the square bracket; the off-diagonal elements produce transitions between bands. These transitions comprise the phenomena of tunneling. We will neglect the interband terms for the present, and consider the resulting, reduced equation for functions we now call $\varphi_n^{(1)}$.

$$\left[E_n^{(1)}(\mathbf{k}) + ie\mathscr{E} \frac{\partial}{\partial k_x} - i\hbar \frac{\partial}{\partial t} \right] \varphi_n^{(1)}(\mathbf{k}) = 0 \tag{4.232}$$

An acceleration theorem may be deduced from this equation. The group velocity associated with a wave packet composed of states belonging to a single band centered around some particular $\mathbf{k}$ is

$$\mathbf{v}_n = \frac{1}{\hbar} \nabla_\mathbf{k} E_n(\mathbf{k}) \tag{4.233}$$

When an electric field is applied, states of different $\mathbf{k}$ are mixed according to (4.5), and the velocity of the wave packet is altered. We define an "acceleration" by

$$\frac{d}{dt} \langle \mathbf{v}_n \rangle = \frac{1}{\hbar} \frac{d}{dt} \int \varphi_n^*(\mathbf{k}) \nabla_\mathbf{k} E_n(\mathbf{k}) \varphi_n(\mathbf{k}) \, d^3k \tag{4.234}$$

The expression for the "acceleration" may be reduced to an interesting form through the use of (4.232)

$$\frac{d}{dt} \langle v_n \rangle = \frac{1}{\hbar} \int \left[\frac{\partial \varphi_n^*}{\partial t} V_k E_n \varphi_n + \varphi_n^* V_k E_n \frac{\partial \varphi_n}{\partial t} \right] d^3k$$

$$= \frac{1}{\hbar} \left(\frac{e\mathscr{E}}{\hbar} \right) \int \left[\frac{\partial \varphi_n^*}{\partial k_x} V_k E_n \varphi_n + \varphi_n^* V_k E_n \frac{\partial \varphi_n}{\partial k_x} \right] d^3k$$

$$= - \frac{e\mathscr{E}}{\hbar^2} \int \varphi_n^* \frac{\partial}{\partial k_x} (V_k E_n) \varphi_n \, d^3k \qquad (4.235)$$

The third line of (4.235) is obtained from the second through an integration by parts. We may interpret this result by defining the acceleration associated with a state of wave vector $\mathbf{k}$, $\boldsymbol{\alpha}_n(\mathbf{k})$ by

$$\frac{d}{dt} \langle v_n \rangle = \int \varphi_n^* \boldsymbol{\alpha}_n(\mathbf{k}) \varphi_n \, d^3k \qquad (4.236)$$

This acceleration is:

$$\boldsymbol{\alpha}_n(\mathbf{k}) = - \frac{e\mathscr{E}}{\hbar^2} \frac{\partial}{\partial k_x} V_k E_n \qquad (4.237a)$$

or, in the case of an arbitrarily directed field,

$$\boldsymbol{\alpha}_n(\mathbf{k}) = - \frac{e}{\hbar^2} (\mathscr{E} \cdot V_k) V_k E_n \qquad (4.237b)$$

If we make use of the definition (1.40) of the reciprocal effective mass tensor, we have for the jth rectangular component of the acceleration, $\alpha_{nj}(\mathbf{k})$

$$\alpha_{nj}(\mathbf{k}) = - e \sum_i \mathscr{E}_i \left(\frac{1}{m^*} \right)_{ij} \qquad (4.238a)$$

or

$$\sum_j m_{lj}^* \alpha_{nj}(\mathbf{k}) = - e \mathscr{E}_l \qquad (4.238b)$$

4.8 THE STEADY ELECTRIC FIELD

where $\mathscr{E}_i$, etc., is the ith rectangular component of the field strength. These results are just what would be expected in classical mechanics for a particle characterized by a tensor effective mass.

Following Kane (1959b) and Argyres (1962), we may obtain a stationary solution of (4.232) which may be used in the discussion of tunneling. The functions $\varphi_n^{(1)}$ are assumed to have the time dependence $\exp[(-i/\hbar)Et]$. Then we obtain

$$\left[E_n^{(1)}(\mathbf{k}) + ie\mathscr{E}\frac{\partial}{\partial k_x}\right]\varphi_n^{(1)}(\mathbf{k}) = E\varphi_n^{(1)}(\mathbf{k}) \tag{4.239}$$

The solution of this equation may be obtained immediately:

$$\varphi_n^{(1)}(\mathbf{k}) = \kappa^{-1/2}\exp\left\{-\frac{i}{e\mathscr{E}}\int_0^{k_x}[E - E_n^{(1)}(\mathbf{k}')]\,dk_x'\right\}\delta(k_y - k_{y0})\delta(k_z - k_{z0}) \tag{4.240}$$

The normalization constant, κ, is the length of the line (k_x, k_{y0}, k_{z0}) lying within the Brillouin zone. We assume for simplicity that this line is a reciprocal lattice vector. The energy eigenvalue, E, is then determined by the condition that the wave function must be the same at the end points of this line, since these are equivalent points. Then the change in the phase angle in a distance κ must be an integral multiple of 2π.

$$\frac{1}{e\mathscr{E}}\int_{-\kappa/2}^{\kappa/2}[E - E_n^{(1)}(\mathbf{k}')]\,dk_x' = 2\nu\pi$$

where ν is a positive or negative integer. We will henceforth designate the eigenvalues in the presence of the field by the index ν:

$$E_\nu = \nu\frac{2\pi e\mathscr{E}}{\kappa} + \frac{1}{\kappa}\int_{-\kappa/2}^{\kappa/2}E_n^{(1)}(\mathbf{k}')\,dk_x' \tag{4.241}$$

The energy spectrum in the presence of the electric field contains a series of discrete (Stark) levels, separated by $2\pi e\mathscr{E}/\kappa$. If the lattice parameter is d, this separation is of the order $e\mathscr{E}d$. This is quite small compared to typical band energies for fields less than 10^5 volts/cm. The existence of these levels was pointed out by Wannier (1959, 1960).

We now wish to consider the theory of tunneling in time-dependent perturbation theory, following the approach of Kane (1959b). Note that we do not treat the entire electric field as a perturbation, but only that part connecting different bands. The one band problem has already been solved. Other studies of tunneling are those of Zener (1934), Houston (1940), McAfee *et al.* (1951), Franz (1958a), Keldysh (1958, 1958a), Price and Radcliffe (1959), Argyres (1962), and Fredkin and Wannier (1962). There are still some controversial problems in the theory, however, and further work will be required before the problem can be regarded as solved. A more complete discussion of the calculation of the tunneling current will be found in Appendix IV.

The probability per unit time, w, of making a transition from band v to band c is given by the usual formula:

$$w = \frac{2\pi}{\hbar} |M_{cv}|^2 \rho(E) \qquad (4.242)$$

where M_{cv} is the matrix element for the transition

$$M_{cv} = \int \varphi_c^{*(1)}(\mathbf{k}) [e\mathscr{E} X_{cv}(\mathbf{k})] \varphi_v^{(1)}(\mathbf{k}) d^3k \qquad (4.243)$$

$$= \frac{e\mathscr{E}}{\kappa} \int X_{cv}(\mathbf{k}) \exp\left\{\frac{i}{eE} \int_0^{k_x} [E_c^{(1)}(\mathbf{k}') - E_v^{(1)}(\mathbf{k}')] dk_x'\right\} dk_x$$

The tunneling connects states of equal energy. The quantity $\rho(E)$ is the density of states for the transition. Considerable controversy has surrounded the use of this formula since, as we have seen above, the levels are discrete, and (4.242) is derived on the assumption of a continuum of levels. Kane choses the density of states to be

$$\rho(E) = \frac{1}{\Delta E} = \frac{\kappa}{2\pi e\mathscr{E}} \qquad (4.244)$$

where ΔE is the interval in energy between two Stark levels whose quantum numbers ν and ν' differ by unity. This use of a density of states was criticized by Price and Radcliffe, but Argyres has shown that a correct answer results provided that small oscillatory terms in tunneling current due to the discrete nature of the Stark levels can be disregarded.

The rate of tunneling is computed in the following way. The transition probability, w, is multiplied by the density of states in **k**-space which is, in this case, $2\kappa/(2\pi)^3$ (the factor of 2 takes account of spins), and is integrated over the transverse momentum components. If n is the number of electrons which transfer from the valence band to the conduction band per second per unit volume, we have

$$n = \frac{2\kappa}{(2\pi)^3} \int w \, dk_y \, dk_z \tag{4.245}$$

One should note that the density of states for the normal band does not enter the calculation, as one might expect in a naive theory.

It is necessary to determine the matrix element M_{cv} to obtain an explicit expression for the tunneling current. The integrals can be evaluated approximately by the method of stationary phase, making use of the analytic properties of the function $E_c - E_v$. We note that it is possible to determine the quantity X_{cv} in the vicinity of a band edge from effective mass theory (Section 1.7). Details of the evaluation are given in Appendix IV. The final result for the tunneling current in the simple two-band model considered by Kane turns out to have a form quite similar to (4.223): It contains an exponential whose argument depends inversely on the electric field strength and is directly proportional to the 3/2 power of the band gap.

We note in conclusion that Adams (1957) and Wannier (1960) have proposed to eliminate the interband matrix elements $X_{nn'}$ from (4.226) by a cannonical transformation. Kane has shown, however, that the tunneling effect is substantially unaltered by this transformation.

4.9 Optical Properties of Semiconductors

In this and the following two sections, we will consider the interaction of a system of Bloch electrons with an alternating electric field. The first topic we will consider is the theory of interband transitions, which makes possible an explanation of the optical characteristics of semiconductors in terms of band theory. Ordinary semiconductors exhibit, in the visible or near infrared spectral regions, a rapid increase in the absorption of light (as the photon energy is increased) which is associated with the onset of band to band transitions. For frequencies below that of the fundamental absorption edge, the absorption is small close to the edge, but increases with increasing wavelength roughly as $1/\omega^2$ for long wavelengths owing to free carrier absorption.

We will ignore the possibility of exciton formation (bound states of an electron and a hole) and will discuss the calculation of the absorption constant. The Hamiltonian for an electron in an external field has been given previously:

$$H = \frac{(\mathbf{p} + e\mathbf{A})^2}{2m} + V(\mathbf{r})$$

In the standard semiclassical radiation theory, it is customary to neglect the portion of the Hamiltonian quadratic in $\mathbf{A}$, and to treat as a perturbation (in a gage in which $\nabla \cdot \mathbf{A} = 0$)

$$H' = \frac{e}{m}\mathbf{A} \cdot \mathbf{p} = -\frac{ie\hbar}{m}\mathbf{A} \cdot \nabla \tag{4.246}$$

Let us consider the matrix element of the perturbation between a final state $\psi_{n'}(\mathbf{k}', \mathbf{r})$ and an initial state $\psi_n(\mathbf{k}, \mathbf{r})$. We choose for the vector potential a monochromatic plane wave $A_0\,\boldsymbol{\epsilon}_s\,e^{i(\mathbf{s}\cdot\mathbf{r}-\omega t)}$, in which $\boldsymbol{\epsilon}_s$ is a polarization vector. Then

$$H'_{fi} = \frac{e}{m}\langle n'\,\mathbf{k}'|\mathbf{A}\cdot\mathbf{p}|n\mathbf{k}\rangle \tag{4.247}$$

$$= -\frac{ie\hbar}{m}A_0\,\boldsymbol{\epsilon}_s \cdot \left[\int e^{-i\mathbf{k}'\cdot\mathbf{r}}u_{n'}^*(\mathbf{k}',\mathbf{r})\,e^{i\mathbf{s}\cdot\mathbf{r}}\,\nabla\{e^{i\mathbf{k}\cdot\mathbf{r}}u_n(\mathbf{k},\mathbf{r})\}\,d^3r\right]$$

The quantity in brackets is of principal concern to us.

$$[\ldots] = \int e^{i(\mathbf{k}+\mathbf{s}-\mathbf{k}')\cdot\mathbf{r}}u_{n'}^*(\mathbf{k}',\mathbf{r})\nabla u_n(\mathbf{k},\mathbf{r})\,d^3r + \tag{4.248}$$

$$i\mathbf{k}\int e^{i(\mathbf{k}+\mathbf{s}-\mathbf{k}')\cdot\mathbf{r}}u_{n'}^*(\mathbf{k}',\mathbf{r})u_n(\mathbf{k},\mathbf{r})\,d^3r$$

The functions $u_{n'}^*\,\nabla u_n$ and $u_{n'}^*\,u_n$ are periodic in the crystal and so may be expanded in plane waves (whose wave vectors are reciprocal lattice vectors) as was done in Section 4.1 in discussing the orthogonality of the basis functions for the Kohn-Luttinger representation. Provided

4.9 OPTICAL PROPERTIES OF SEMICONDUCTORS

that the wavelength of the electromagnetic wave is sufficiently long so that if $\mathbf{k}'$ and $\mathbf{k}$ are within the zone $\mathbf{k} + \mathbf{s} - \mathbf{k}'$ is not a reciprocal lattice vector, we obtain

$$[\ldots] = \frac{(2\pi)^3}{\Omega} \delta(\mathbf{k} + \mathbf{s} - \mathbf{k}') \int_{\text{cell}} u_{n'}^*(\mathbf{k}, \mathbf{r}) \nabla u_n(\mathbf{k}, \mathbf{r}) \, d^3r \quad (4.249)$$

The integral includes a single unit cell whose volume is Ω. The delta function expresses the conservation of momentum in the absorption. For light in the infrared and visible regions of the spectrum, s is small compared to the dimensions of a Brillouin zone: in a simple cubic lattice of lattice parameter a, $s/K_1 = a/\lambda$, where λ is the wavelength of the light, and K_1 is the length of the first reciprocal lattice vector. Under normal circumstances, this ratio is of the order of 10^{-3}. It is then a rather good approximation to neglect $\mathbf{s}$. With reference to an energy band diagram, first order optical transitions are referred to as vertical.

It is necessary to compute the transition probability. One might expect some difficulty from the square of a delta function, which arises when one considers the square of the matrix element. However, one can show that the transition probability per unit volume per unit time is well defined and contains only one delta function of momentum conservation (Bethe et al., 1955). Further, since one must consider in a solid not transitions between discrete states but rather transitions between groups of states, it is necessary to integrate over ranges of states in $\mathbf{k}'$ and $\mathbf{k}$. When this is done, the delta function of momentum conservation renders one integration trivial, and the following expression for the transition probability is left.

$$w = \frac{2\pi}{\hbar} \left(\frac{2}{8\pi^3} \right) \int |\bar{H}_{fi}|^2 \, \delta(E_{n'}(\mathbf{k}) - \hbar\omega - E_n(\mathbf{k})) \, d^3k \quad (4.250)$$

(we count both directions of the spin) in which H_{fi} is a reduced matrix element:

$$\bar{H}_{fi} = -\frac{ie\hbar}{m} \frac{(2\pi)^3}{\Omega} A_0 \, \boldsymbol{\epsilon} \cdot \int_{\text{cell}} u_{n'}^*(\mathbf{k}, \mathbf{r}) \nabla u_n(\mathbf{k}, \mathbf{r}) \, d^3r \quad (4.251)$$

It is necessary to specify the band structure for further progress. A particularly interesting case is one in which the bands n and n' both have extrema at $\mathbf{k} = 0$. We put

286   CHAPTER 4. POINT IMPURITIES AND EXTERNAL FIELDS

$$E_n(\mathbf{k}) = -\frac{\hbar^2 \mathbf{k}^2}{2m_v^*} \ ; \qquad E_n'(\mathbf{k}) = E_g + \frac{\hbar^2 \mathbf{k}^2}{2m_c^*}$$

The bands, henceforth referred to as valence and conduction bands, are separated by a gap E_g at $\mathbf{k} = 0$, and have effective masses m_{v*} and m_{c*}, respectively. Then

$$E_n' - \hbar\omega - E_n = E_g + \frac{\hbar^2 k^2}{2\mu} - \hbar\omega \qquad (4.252)$$

in which $\mu = m_c^* m_v^*/(m_c^* + m_v^*)$. We shall further suppose that the matrix element is independent of energy. This will be a good approximation if the transition is allowed at $\mathbf{k} = 0$, and the gap is not too small. To perform the integral, we note that

$$\int_a^b g(x)\delta(f(x))\, dx = \sum_{x_0} g(x_0) \left|\frac{dx}{df}\right|_{x_0} \qquad (4.253)$$

in which x_0 is a root of $f(x)$, and the sum includes all the roots in the interval from a to b. In the present case, $g = 4\pi k^2$, f is given by (4.252), and we find:

$$w = \frac{(2\mu)^{3/2}}{\pi \hbar^4} |\bar{H}_{fi}|^2 (\hbar\omega - E_g)^{1/2} \qquad (4.254)$$

The important result here is the dependence of the transition probability on $(\hbar\omega - E_g)^{1/2}$. This energy dependence can be observed experimentally in absorption measurements. The absorption constant is defined as the ratio of the energy removed from the incident beam per unit time and per unit volume to the incident flux:

$$\alpha = (\hbar\omega) \times \frac{\text{number of transitions per unit volume and time}}{\text{incident flux}} \qquad (4.255)$$

The energy flux is interpreted as the product of the energy density and the velocity of flow. The energy density in the medium is $\varepsilon \mathscr{E}^2$, which when averaged over a cycle gives $\varepsilon \omega^2 A_0^2/2$, in mks units, with ε the permittivity, A_0 the amplitude of the vector potential, and ω the circular frequency of the wave. The propogation velocity is c/n, where n is the index of refraction, and dispersion is neglected. Note that $\varepsilon/n = n\varepsilon_0$, where ε_0 is the permittivity of free space. Hence

$$\alpha = \frac{2\hbar w}{\omega n \varepsilon_0 c A_0^2} \tag{4.256}$$

where w is the transition probability. It is convenient to introduce an effective momentum matrix element for the transition by

$$\hbar^2 \left| \frac{(2\pi)^3}{\Omega} \int_{\text{cell}} u_{n'}^* \, \boldsymbol{\epsilon} \cdot \nabla u_n \, d^3r \right|^2 = |\mathbf{p} \cdot \boldsymbol{\epsilon}|^2 \tag{4.257}$$

We can now combine (4.257), (4.256), (4.254), and (4.247) to obtain a final expression for the absorption coefficient for direct transitions

$$\alpha = K \frac{(2\mu)^{3/2}}{(\hbar\omega)^{1/2}} \left(1 - \frac{E_g}{\hbar\omega}\right)^{1/2} \quad \text{with} \quad K = \frac{2e^2}{\pi m^2 \hbar^2 n \varepsilon_0 c} |\mathbf{p} \cdot \boldsymbol{\epsilon}|^2 \tag{4.258}$$

If the transition between valence and conduction bands is forbidden by a selection rule at the band extrema, it will occur in the vicinity of the extrema as components of different symmetry are included in the wave function. In this case, the matrix element H_{fi} will generally be proportional to k. If this used in (4.250), a transition probability proportional to $(\hbar\omega - E_g)^{3/2}$ is found. The character of the transition, and information concerning the symmetry of valence and conduction band wave functions can thus be determined by observing the energy dependence of the absorption coefficient near the threshold.

It frequently happens, however, that the valence and conduction band extrema are located at different points in **k**-space. An optical transition between them usually requires, in this case, the assistance of a phonon to supply the additional momentum. Such processes, which are called indirect transitions (Bardeen *et al.*, 1957), may occur in two ways: (1) An electron in the valence band may absorb a photon and make a transition to an intermediate state in the conduction band of essentially the same wave vector, and then a phonon may be emitted or absorbed to yield the final state: (2) Alternatively, the photon may excite an electron from a valence band state directly below the conduction band minimum, with the hole being transferred to the valence band maximum by phonon emission or absorption. The final state is the same in both cases.

Indirect transitions are studied in second order perturbation theory. It is shown in elementary quantum mechanics (L. I. Schiff, 1955), that in second order, time-dependent perturbation theory, one must replace the matrix element H'_{fi} which appears in the usual expression for the transition probability by

$$\sum_m \frac{H'_{fm} H'_{mi}}{E_i - E_m} \qquad (4.259)$$

where the index m refers to the intermediate states. In an indirect transition one of the matrix elements has the form (4.247), in which the electron interacts with the electromagnetic field; and the other involves the electron-phonon coupling. The matrix element for the absorption of a phonon of wave vector $\mathbf{k}$ is proportional to $N_\mathbf{k}^{1/2}$, where $N_\mathbf{k}$ is the number of phonons already present with wave vector $\mathbf{k}$, while the matrix element for emission is proportional to $(N_\mathbf{k} + 1)^{1/2}$. Since, in thermal equilibrium at temperature T, $N_\mathbf{k}$ is proportional to $(e^{\theta/T} - 1)^{-1}$ (where the energy of the phonon has been written as $\kappa\theta$ and κ is Boltzmann's constant), there is a characteristic temperature dependence of the absorption constant for these transitions.

We usually cannot evaluate the sum in (4.259) owing to ignorance of the spectrum of intermediate states and of the electron-phonon interaction. Thus, it is not possible to give an expression for the transition probability comparable in detail to (4.258). We can, however, determine the dependence of the absorption constant on photon energy. To do this we integrate the delta function of energy conservation over a range of initial and final states. In the present case, the requirement of momentum conservation does not make one of the integrations trivial, because the phonons can take up what momentum is required. Consider a transition in which a phonon is absorbed. Let us choose the zero of energy at the conduction band minimum, and let E_f, E_i, and E_g represent the energy of states in the conduction and valence bands, and the energy gap, respectively. Energy conservation demands that

$$E_f = E_i + \hbar\omega + \kappa\theta \qquad (4.260)$$

Since the densities of states in the conduction and valence band are proportional to $\sqrt{E_f}$ and $\sqrt{-E_g - E_i}$, respectively [$(-)(E_g + E_i)$ is a positive number], if one assumes the sum in (4.259) to be independent of energy (the optical transition is allowed), one has to evaluate

$$\iint \sqrt{E_f}\sqrt{-E_g - E_i}\,\delta(E_i + \hbar\omega + \kappa\theta - E_f)\,dE_i\,dE_f \qquad (4.261)$$

The integral over E_i may be done immediately. The remaining integral has limits 0 and $\hbar\omega + \kappa\theta - E_g$ and is

$$\int_0^{\hbar\omega + \kappa\theta - E_g} [E_f(\hbar\omega + \kappa\theta - E_g - E_f)]^{1/2}\,dE_f \qquad (4.262)$$

We then obtain

$$\frac{1}{8}\pi(\hbar\omega + \kappa\theta - E_g)^2 \qquad (4.263)$$

It suffices to change the sign of $\kappa\theta$ to obtain the result for phonon emission. This expression may be combined with the factors involving the number of phonons to put the absorption coefficient in the form

$$\alpha = C\left[\frac{1}{1 - e^{-\theta/T}}(\hbar\omega - E_g - \kappa\theta)^2\,\eta(\hbar\omega - E_g - \kappa\theta) \right.$$
$$\left. + \frac{1}{e^{\theta/T} - 1}(\hbar\omega - E_g + \kappa\theta)^2\,\eta(\hbar\omega - E_g + \kappa\theta)\right] \qquad (4.264)$$

in which C contains the unknown factors and η is a unit step function. This formula was first given by Macfarlane and Roberts (1955). When $\alpha^{1/2}$ is plotted against $\hbar\omega$, the curve lies close to one straight line for $E_g - \kappa\theta < \hbar\omega < E_g + \kappa\theta$, and close to a steeper line for $\hbar\omega > E_g + \kappa\theta$. From analysis of the absorption in indirect transitions, it is possible to deduce the dependence of the energy gap on temperature, and by comparison with the vibrational spectrum (which is assumed to be known) to determine the separation in **k**-space of the valence and conduction band extrema.

4.10 Optical Absorption by Free Carriers

The optical properties at long wavelengths of semiconductors which contain appreciable numbers of free carriers are dominated by the contribution from the carriers. The importance of free electron absorption in metals is obvious.

We will generalize the previous discussion of the optical absorption in the following way. The current produced by an alternating electromagnetic field, described classically, in an electron system will be calculated and used to obtain a complex conductivity from which the transmission and reflection coefficients may be computed according to the standard procedures of classical electromagnetic theory. Only frequencies sufficiently high so that relaxation effects may be neglected will be considered. Our discussion follows that of Wilson (1936).

We begin with the time-dependent Schrödinger equation for a single particle:

$$i\hbar \frac{\partial \Psi}{\partial t} = H\Psi$$

The electric current density, **j**, is determined by the requirement that an equation of continuity be satisfied:

$$\nabla \cdot \mathbf{j} + \frac{\partial \rho}{\partial t} = 0 \tag{4.265}$$

in which ρ, the charge density, is $-e\psi^*\psi$. The Hamiltonian of the system is:

$$H = \frac{(\mathbf{p} + e\mathbf{A})^2}{2m} - e\Phi + V(\mathbf{r})$$

in which Φ is the scalar potential, and **A** the vector potential of the electromagnetic field. A standard calculation yields an expression for **j**:

$$\mathbf{j} = -\frac{e\hbar}{2im}(\psi^* \nabla\psi - \psi\nabla\psi^*) - \frac{e^2}{m}\mathbf{A}\psi^*\psi \tag{4.266}$$

If the electrons are distributed in states characterized by wave vectors **k**, such that there are $N(\mathbf{k})$ electrons in **k**, Eq. (4.266) applies to each state separately, and the total current density J is

$$J = \sum_{\mathbf{k}} N(\mathbf{k}) j(\mathbf{k}) \tag{4.267}$$

in which $j(\mathbf{k})$ is given by (4.266) with ψ replaced by $\psi(\mathbf{k}, \mathbf{r})$. Only occupied states are included.

We consider initially a monochromatic plane wave described by a vector potential

$$\mathbf{A} = \mathbf{A_0}\, e^{-i\omega t} \tag{4.268}$$

4.10 OPTICAL ABSORPTION BY FREE CARRIERS

The current is calculated according to time-dependent perturbation theory in which the perturbing Hamiltonian is the same as in (4.246):

$$H' = \frac{e}{m} \mathbf{A} \cdot \mathbf{p}$$

and we consider only terms of first order in A. We shall denote the eigenfunctions of the unperturbed Hamiltonian as $\psi_n(\mathbf{k}, \mathbf{r})$, or mores imply as ψ_n. The complete wave functions have the time dependence $e^{-i\omega_n t}$ with $\omega_n = E_n(\mathbf{k})/\hbar$. On account of the $\mathbf{k}$ selection rule discussed previously [Eq. (4.249)], it suffices to expand the perturbed wave function Ψ in terms of function $\psi_n(\mathbf{k}, \mathbf{r})$ for fixed $\mathbf{k}$, and thus, for the present, the $\mathbf{k}$ designation may be dropped. We write

$$\Psi = \sum c_n(t) \psi_n e^{-i\omega_n t} \quad (4.269)$$

To determine the c_n, we substitute this expression into the Schrödinger equation, multiply by $\psi_s^* e^{i\omega_s t}$, and integrate over all space. When use is made of the orthonormality of the ψ_n we obtain an equation for the c_n, which is:

$$i\hbar \frac{dc_s}{dt} = \sum_n c_n H'_{sn} e^{i\omega_{sn} t} \quad (4.270)$$

in which

$$\omega_{sn} = \omega_s - \omega_n \quad (4.271)$$

and

$$H'_{sn} = \int \psi_s^* H' \psi_n d^3r = H^0_{sn} e^{-i\omega t} \quad (4.272)$$

so that H^0_{sn} is independent of time.

Equations (4.270) may be solved in the approximation, valid through first order in A, that all the c_n on the right side may be neglected except the one (say c_0), which pertains to the initial state. Then c_0 may be set equal to unity, and we have

$$i\hbar \frac{dc_s}{dt} = H^0_{s0} e^{i(\omega_{s0} - \omega)t} \quad (s \neq 0) \quad (4.273)$$

CHAPTER 4. POINT IMPURITIES AND EXTERNAL FIELDS

This differential equation may be solved very easily. For the initial condition, we will assume the perturbation was turned on at $t = 0$, so that $c_s(0) = 0$. The solution is

$$c_s = \frac{H_{s0}^0}{\hbar(\omega_{s0} - \omega)} [1 - e^{i(\omega_{s0} - \omega)t}] \qquad (s \neq 0) \qquad (4.274)$$

The wave function Ψ which is to be used in (4.266) is normalized correctly up to second order in A. One finds

$$\mathbf{j} = -\frac{e\hbar}{2im} \Big\{ [\psi_0^* \nabla \psi_0 - \psi_0 \nabla \psi_0^*] + \sum [c_n(\psi_0^* \nabla \psi_n - \psi_n \nabla \psi_0^*) e^{-i\omega_n 0 t} +$$

$$c_n^*(\psi_n^* \nabla \psi_0 - \psi_0 \nabla \psi_n^*)] e^{i\omega_n 0 t} \Big\} - \frac{e^2}{m} A |\psi_0|^2 \qquad (4.275)$$

Terms of order A^2 and higher have been neglected. The expression can be simplified when the effect of the sum over $\mathbf{k}$, which is required by (4.267), is considered. Since $E(\mathbf{k}) = E(-\mathbf{k})$, $N(\mathbf{k}) = N(-\mathbf{k})$, and currents from states of $\mathbf{k}$ and $-\mathbf{k}$ will contribute equally. On the other hand, time reversal symmetry requires that, for spinless particles

$$\psi_n^*(\mathbf{k}) = \psi_n(-\mathbf{k}) \qquad (4.276)$$

Hence, if we sum over currents from states of wave vectors $\mathbf{k}$ and $-\mathbf{k}$, the first term disappears. (This is a special case of a general result: there is no net current in the ground state of any quantum system.) We will ignore the first term from now on. Next, we observe that the terms involved in the summation of (4.275) contain the difference of a quantity and its complex conjugate. Hence we may write

$$\mathbf{j} = -\frac{e\hbar}{m} \text{Im} \sum_n [c_n(\psi_0^* \nabla \psi_n - \psi_n \nabla \psi_0^*) e^{-i\omega_n 0 t}] - \frac{e^2}{m} A |\psi_0|^2 \qquad (4.277)$$

in which Im stands for "imaginary part of." We must now substitute the expression for c_n. Note that

$$c_{n0} e^{-i\omega_{n0} t} = \frac{H_{n0}^0 e^{-i\omega t}}{\hbar(\omega_{n0} - \omega)} [\cos(\omega - \omega_{n0})t - 1 + i \sin(\omega - \omega_{n0})t]$$

4.10 OPTICAL ABSORPTION BY FREE CARRIERS

In addition, it is legitimate to choose the wave functions ψ_0, ψ_n to be real. Then the matrix element H_{s0}^0 is imaginary, and we get[7] (set $H_{n0}'' = H_{n0}^0/i$)

---

[7] The argument in this respect is rather subtle. Suppose the contrary and let

$$\frac{e}{m}\frac{H_{n0}^0}{(\omega_{n0} - \omega)}(\psi_0^* \nabla\psi_n - \psi_n \nabla\psi_0^*) = a + ib \tag{1}$$

$$\cos(\omega - \omega_{n0})t - 1 = r; \qquad \sin(\omega - \omega_{n0})t = s.$$

Then

$$\mathbf{j}(\mathbf{k}) = -\operatorname{Im}\sum_n (a + ib)e^{-i\omega t}(r + is)$$

$$= \sum_n [(ar - bs)\sin\omega t - (br + as)\cos\omega t] \tag{2}$$

Now consider the current from the corresponding state of wave vector $-\mathbf{k}$. In this case we have, instead of the term $(a + ib)$

$$\frac{e}{m(\omega_{n0} - \omega)}\left[\int \psi_n^*(-\mathbf{k})H'\psi_0(-\mathbf{k})\,d^3r\right][\psi_0^*(-\mathbf{k})\nabla\psi_n(-\mathbf{k}) - \psi_n(-\mathbf{k})\nabla\psi_0^*(-\mathbf{k})] \tag{3}$$

With the aid of (4.276) this becomes

$$\frac{e}{m(\omega_{n0} - \omega)}\left[\int \psi_n(\mathbf{k})H'\psi_0^*(\mathbf{k})\,d^3r\right][\psi_0(\mathbf{k})\nabla\psi_n^*(\mathbf{k}) - \psi_n^*(\mathbf{k})\nabla\psi_0(\mathbf{k})] \tag{4}$$

Since $H' \propto i\nabla$ we have that

$$\int \psi_n(\mathbf{k})H'\psi_0^*(\mathbf{k})\,d^3r = -\left[\int \psi_n^*(\mathbf{k})H'\psi_0(\mathbf{k})\,d^3r\right]^*$$

so that (4) becomes $-(a + ib)^* = -a + ib$. Hence

$$\mathbf{j}(-\mathbf{k}) = \sum_n [(-ar - bs)\sin\omega t - (br - as)\cos\omega t] \tag{5}$$

and

$$\mathbf{j}(\mathbf{k}) + \mathbf{j}(-\mathbf{k}) = -2\sum_n b(r\cos\omega t + s\sin\omega t) \tag{6}$$

which is just the result which is obtained by making the wave function real.

CHAPTER 4. POINT IMPURITIES AND EXTERNAL FIELDS

$$\mathbf{j} = -\frac{e}{m} \sum_n \frac{H''_{n0}}{(\omega_{n0} - \omega)} [\sin(\omega - \omega_{n0})t \sin \omega t + (\cos(\omega - \omega_{n0})t - 1)\cos \omega t] \times$$

$$(\psi_0 \nabla \psi_n - \psi_n \nabla \psi_0) - \frac{e^2}{m}|\psi_0|^2 \mathbf{A} \quad (4.278)$$

It is necessary in any physical problem that the perturbing electromagnetic field be real. The time dependence of the vector potential must be $\cos \omega t$ (or $\sin \omega t$) rather than $e^{-i\omega t}$. The current for real A is one-half the sum of $\mathbf{j}$ as given above and a corresponding expression with ω replaced by $(-\omega)$. Hence we replace (4.278) by

$$\mathbf{j} = -\frac{e}{2m} \sum_n H''_{n0} \times \quad (4.279)$$

$$\left\{ \left[\frac{\cos(\omega - \omega_{n0})t}{\omega_{n0} - \omega} + \frac{\cos(\omega + \omega_{n0})t}{\omega + \omega_{n0}} - \frac{1}{\omega_{n0} - \omega} - \frac{1}{\omega_{n0} + \omega} \right] \cos \omega t - \left[\frac{\sin(\omega_{n0} - \omega)t}{\omega_{n0} - \omega} - \frac{\sin(\omega + \omega_{n0})t}{\omega_{n0} + \omega} \right] \sin \omega t \right\} (\psi_0 \nabla \psi_n - \psi_n \nabla \psi_0) -$$

$$\frac{e^2}{m}|\psi_0|^2 \mathbf{A}$$

We have now assumed that the vector potential is proportional to $\cos \omega t$. The electric field, $\mathscr{E}$, is

$$\mathscr{E} = -\frac{\partial \mathbf{A}}{\partial t} = \mathbf{A}_0 \omega \sin \omega t; \qquad \frac{\partial \mathscr{E}}{\partial t} = \mathbf{A}_0 \omega^2 \cos \omega t \quad (4.280)$$

The current consists of two parts: one in phase with the applied field (proportional to $\sin \omega t$), and the other out of phase with the field. Let these be designated as $\mathbf{j}_1$ and $\mathbf{j}_2$, respectively. The first of these is:

$$\mathbf{j}_1 = \frac{e^2}{2m^2 \omega i} \sum_n \left(\int \psi_n^* \mathscr{E} \cdot \mathbf{p} \psi_0 \, d^3 r \right) \left[\frac{\sin(\omega_{n0} - \omega)t}{(\omega_{n0} - \omega)} - \frac{\sin(\omega + \omega_{n0})t}{(\omega_{n0} + \omega)} \right] \times$$

$$[\psi_0 \nabla \psi_n - \psi_n \nabla \psi_0] \quad (4.281)$$

4.10 OPTICAL ABSORPTION BY FREE CARRIERS

In the limit of large times,

$$\lim_{t \to \infty} \frac{\sin(\omega_{n0} \pm \omega)t}{\omega_{n0} \pm \omega} = \pi\delta(\omega_{n0} \pm \omega) \qquad (4.282)$$

The portion of the total current which is in phase with the field is:

$$\mathbf{J}_1 = \frac{\pi e^2}{2m^2 \omega i} \sum_{\mathbf{k},n} \left(\int \psi_n^* \, \mathscr{E} \cdot \mathbf{p} \psi_0 \, d^3r \right) [\delta(\omega_{n0} - \omega) - \delta(\omega_{n0} + \omega)] \times$$

$$[\psi_0 \nabla \psi_n - \psi_n \nabla \psi_0] \qquad (4.283)$$

A conductivity is defined through the Fourier transform of the current and the field. Define

$$\mathscr{I}_1(\mathbf{q}) = \int \mathbf{J}_1 \, e^{i\mathbf{q}\cdot\mathbf{r}} \, d^3r \qquad (4.284)$$

We are concerned only with the long wavelength limit ($\mathbf{q} \to 0$) of the conductivity, since for optical frequencies the external field will not vary appreciably in distances of the order of a lattice spacing. Hence we compute $\mathscr{I}_1(0)$. The basic integral is

$$\frac{1}{i} \int (\psi_0 \nabla \psi_n - \psi_n \nabla \psi_0) \, d^3r = \frac{2}{i} \int \psi_0 \nabla \psi_n \, d^3r = \frac{2}{\hbar} \mathbf{p}_{0n} \qquad (4.285)$$

We obtain for $\mathscr{I}_1(0)$:

$$\mathscr{I}_1(0) = \frac{\pi e^2}{m^2 \omega \hbar} \sum_k N(\mathbf{k}) \sum_n \mathbf{p}_{0n}(\mathscr{E} \cdot \mathbf{p}_{n0}) [\delta(\omega_{n0} - \omega) - \delta(\omega_{n0} + \omega)] \qquad (4.286)$$

In a cubic crystal, the conductivity is isotropic, and the current is related to the field by Ohm's law

$$\mathscr{I}_1 = \sigma \mathscr{E} \qquad (4.287)$$

in which the (real) conductivity σ is given by

$$\sigma = \frac{\pi e^2}{m^2 \omega \hbar} \sum_k N(\mathbf{k}) \sum_n |\mathbf{p}_{0n}|^2 [\delta(\omega_{n0} - \omega) - \delta(\omega_{n0} + \omega)] \qquad (4.288)$$

It follows from the analysis of the previous section that the conductivity, regarded as a function of frequency, exhibits an edge followed by a rapid rise at the onset of an interband transition.

The portion of the current out of phase with the electric field is

$$\mathbf{j}_2 = -\frac{e^2}{m\omega^2}\left[|\psi_0|^2\frac{\partial \mathscr{E}}{\partial t} - \frac{1}{2im}\sum_n\left(\int \psi_n^*\frac{\partial \mathscr{E}}{\partial t}\cdot \mathbf{p}\psi_0\, d^3r\right)\times \right. \quad (4.289)$$

$$\left. \{\psi_0 \nabla\psi_n - \psi_n \nabla\psi_0\}\left\{\frac{1-\cos(\omega_{n0}-\omega)t}{\omega_{n0}-\omega} + \frac{1-\cos(\omega_{n0}+\omega)t}{\omega_{n0}+\omega}\right\}\right]$$

We must now sum over **k**. In the limit of large times, we can make use of the relation

$$\lim_{t\to\infty}\int f(x)\frac{(1-\cos xt)}{x}dx = P\int \frac{f(x)}{x}dx \quad (4.290)$$

in which P signifies the Cauchy principal value of the integral [provided $f(x)$ possesses a derivative at $x=0$], since the sum over occupied states is, in fact, an integral. Then we have

$$\mathbf{J}_2 = -\frac{e^2}{m\omega^2}\sum_{\mathbf{k}} N(\mathbf{k})\left\{|\psi_0|^2\frac{\partial\mathscr{E}}{\partial t} - \frac{1}{im}\sum_n\frac{\omega_{n0}}{\omega_{n0}^2-\omega^2}\left(\int \psi_n^*\frac{\partial\mathscr{E}}{\partial t}\cdot\mathbf{p}\psi_0\, d^3r\right)\times\right.$$

$$\left. [\psi_0\nabla\psi_n - \psi_n\nabla\psi_0]\right\} \quad (4.291)$$

In order to determine the dielectric constant, it is necessary to consider the Fourier transform of J_2 in the limit of long wavelengths. Since ψ_0 is normalized, we obtain for $\mathscr{J}_2(0)$ in a cubic crystal:

$$\mathscr{J}_2(0) = \frac{\partial\mathscr{E}}{\partial t}\left(-\frac{e^2}{m\omega^2}\right)\sum N(\mathbf{k})\left[1 - \frac{2}{\hbar m}\sum_n \omega_{n0}\frac{|\mathbf{p}_{0n}|^2}{\omega_{n0}^2-\omega^2}\right] \quad (4.292)$$

We can simplify this equation through the use of the sum rule (1.41b) for the effective mass (which we are here considering to be a scalar)

$$\frac{m}{m^*} = 1 + \frac{2}{\hbar m}\sum_n \frac{|\mathbf{p}_{0n}|^2}{\omega_{0n}}$$

4.10 OPTICAL ABSORPTION BY FREE CARRIERS

Let the coefficient of $\partial \mathscr{E}/\partial t$ in (4.292) be denoted as χ_e, the electric susceptibility of the electron system. If κ_e is the (dimensionless) dielectric constant of the material, and ε_0 is the permittivity of free space, then

$$\kappa_e = 1 + \frac{\chi_e}{\varepsilon_0} = 1 - \frac{N_c e^2}{m_{op}^* \varepsilon_0 \omega^2} + \frac{2e^2}{m^2 \hbar \varepsilon_0} \sum_k N(\mathbf{k}) \sum_n \frac{|\mathbf{p}_{0n}|^2}{\omega_{0n}(\omega_{0n}^2 - \omega^2)} \quad (4.293)$$

in which

$$\frac{1}{m_{op}^*} = \frac{1}{N_c} \sum_k \frac{N(\mathbf{k})}{m^*(\mathbf{k})} \quad (4.294)$$

and N_c is the number of conduction electrons per unit volume.

The quantity m_{op}^* can be interpreted as an optical effective mass (Cohen and Heine, 1958). The scalar effective mass $m^*(\mathbf{k})$ is related to the band structure by

$$\frac{1}{m^*(\mathbf{k})} = \frac{1}{3\hbar^2} \nabla_\mathbf{k}^2 E(\mathbf{k})$$

The sum over $\mathbf{k}$ in (4.294) can be replaced by an integration in the usual way:

$$\sum = \frac{1}{4\pi^3} \int d^3k$$

We obtain

$$\frac{1}{m_{op}^*} = \frac{1}{12\pi^3 \hbar^2 N_c} \int d^3k \, \nabla_\mathbf{k}^2 E(\mathbf{k}) \quad (4.295)$$

This expression may be transformed by Gauss's theorem into a surface integral of the velocity $v_\mathbf{k} = \nabla_\mathbf{k} E/\hbar$ over the Fermi surface. For a monovalent metal, we obtain finally (Cohen and Heine, 1958)

$$m_{op}^* = \hbar k_F^0 S_F^0 \left[\int dS_F \cdot v_\mathbf{k} \right]^{-1} \quad (4.296)$$

in which S_F^0 is the area, and k_F^0 the radius of a spherical free electron Fermi surface which contains N_c electrons. The optical effective mass determined here should be contrasted with the expression for the "thermal effective mass" which was obtained previously.

The optical properties of the material considered are determined from the solutions of Maxwell's equations for a material whose (real) dielectric constant and conductivity are given by (4.293) and (4.288) respectively. The absorption constant may be calculated from these quantities by standard techniques, which we will not discuss here. The behavior of the dielectric constant is, however, of considerable interest. The second term of (4.293) is analogous to the contribution to the dielectric constant from the free electrons of a classical plasma. We may define an effective plasma frequency, ω_p, for the material by

$$\omega_p^2 = \frac{N_c e^2}{m_{op}^* \varepsilon_0}. \tag{4.297}$$

The third term is the atomic polarizability, as modified in the metal, if the sum over $\mathbf{k}$ includes all fully occupied (as well as partially occupied) bands. One sees from (4.296) that a fully occupied band, which does not have a Fermi surface, makes no contribution to m_{op}^*. For frequencies ω much smaller than the threshold for interband transitions, Eq. (4.293) simplifies to

$$\kappa_e = 1 + \kappa_a - \frac{\omega_p^2}{\omega^2} \tag{4.298}$$

in which κ_a is the contribution from the atomic polarizability

$$\kappa_a = \frac{2e^2}{m^2 \hbar \varepsilon_0} \sum_{n,l} \frac{1}{4\pi^3} \int d^3k \, \frac{|\mathbf{p}_{ln}|^2}{\omega_{ln}^3} \tag{4.299}$$

The sum on l in (4.299) includes all occupied bands; the sum on n includes all bands, whether occupied or unoccupied. In this limit, the dielectric constant has the simple form $A - B/\omega^2$. At an absorption edge, however, a negative contribution to κ_e appears as a result of the change in sign of one of the denominators in (4.293). Therefore the (real) dielectric constant will usually have a maximum at a threshold for absorption.

The dependence of the dielectric constant on frequency specified by (4.298) has been observed in some cases. It is then possible to determine the optical effective mass, and the atomic polarizability. Cohen

(1958) has observed that it is possible to make a very good least squares fit to the experimental data of Ives and Briggs (1936, 1937a, b) concerning the refractive index of the alkali metals (except for lithium, which was not measured), with a function of the required form. The agreement is very much improved with the inclusion of κ_{a1} which had been neglected by Ives and Briggs. The atomic polarizability so obtained is in quite reasonable agreement (except for rubidium) with values of polarizabilities of alkali ions in alkali halides as previously determined by Tessman, Kahn, and Shockley (1953). The optical effective masses given by Cohen (1958) and Ehrenreich and Philipp (1962) for the alkali and noble metals are quoted in Table XXIII.

TABLE XXIII

EXPERIMENTAL OPTICAL EFFECTIVE MASSES

Metal	m_{op}^*	Metal	m_{op}^*
Na	1.01 ± 0.02	Cu	1.42 ± 0.05
K	1.08 ± 0.02	Ag	1.03 ± 0.06
Rb	1.08 ± 0.03	Au	0.98 ± 0.04
Cs	1.02 ± 0.02		

One noteworthy feature of Equation (4.298) is that the dielectric constant, κ_e, is negative for frequencies ω, below a critical value ω_c

$$\omega_c^2 = \frac{\omega_p^2}{1 + \kappa_a} \tag{4.300}$$

Under these circumstances, the metal is totally reflecting. It becomes transparent, however, when ω exceeds ω_c. This transition has been observed in the alkali metals. There is, of course, considerable absorption for $\omega > \omega_c$; this has been attributed by Butcher (1951) and by Cohen to interband transitions.

It has also been possible to determine effective masses of electrons in semiconductors from the dielectric constant (for example, see Spitzer and Fan, 1957).

There is an interesting relation between the optical effective mass, given by (4.296), and the thermal effective mass, m_t^*, defined in Section 3.3 in connection with the electron specific heat. Recall that

$$m_t^* = \frac{\hbar k_F}{S_F^0} \int \frac{dS_F}{v_k}$$

Hence

$$\frac{m_t^*}{m_{op}^*} = \frac{1}{S_F^{02}} \left[\int \frac{dS_F}{v_k} \right] \left[\int dS_F\, V_k \right] \qquad (4.301)$$

Averages of v_k and $1/v_k$ over the Fermi surface are involved. The Schwarz inequality, which applies to any two arbitrary functions f and g, states that

$$\left| \int dS_F\, fg \right|^2 \leqslant \left| \int dS_F\, f^2 \right| \left| \int dS_F\, g^2 \right| \qquad (4.302)$$

Hence

$$\left[\int dS_F\, v_k \right] \left[\int dS_F/v_k \right] \geqslant S_F^2 \qquad (4.303)$$

in which S_F is the area of the actual Fermi surface. Thus we have the inequality:

$$\frac{m_t^*}{m_{op}^*} \geqslant \left(\frac{S_F}{S_F^0} \right)^2 \qquad (4.304)$$

Equality holds only if the Fermi surface is spherical, which implies that v_k is constant over the surface.

In a monovalent metal, we must have $S_F \geqslant S_F^0$, and hence $m_t^* \geqslant m_{op}^*$ if the Fermi surface does not touch the Brillouin zone. However, if contact occurs, it is possible that $S_F < S_F^0$ since the area of contact is not actually a part of the Fermi surface. Interpretation of experimental results on the basis of these inequalities is not always certain on account of the imprecision of measurements. It seems, however, that in all the alkali metals, with the possible exception of lithium for which the data is really insufficient, one does have $m_t^* > m_0$, and therefore contact is unlikely.

4.10 OPTICAL ABSORPTION BY FREE CARRIERS

We now consider briefly, in a very phenomenologic fashion, the effect of including relaxation processes. In the previous discussion, the in-phase portion of the current results only from interband transitions because it is only in these that conservation of wave vector is possible. If phonon interactions are included, intraband transitions in which a photon is absorbed and a phonon is absorbed or emitted are possible. These processes are quite analogous to the interband indirect transitions previously discussed. A phenomenological theory of the resulting optical absorption, given originally by Drude (1902) and Zener (1933), can be worked out easily on the basis of the semiclassical equation

$$\frac{d\mathbf{v}}{dt} + \frac{\mathbf{v}}{\tau} = \frac{e\mathscr{E}}{m^*} = \frac{e\mathscr{E}_0}{m^*} \sin \omega t \quad (4.305)$$

in which $\mathscr{E}$ is the external electric field, which we have assumed to have the time dependence $\sin \omega t$, and τ is the relaxation time for electron-phonon scattering, which is assumed to be constant. This equation can easily be solved for the steady state velocity $\mathbf{v}$, from which the current density $J = Ne\mathbf{v}$ can be found:

$$\mathbf{J} = \frac{Ne^2 \tau \mathscr{E}_0}{m^*(1 + \omega^2 \tau^2)} (\sin \omega t - \omega \tau \cos \omega t) \quad (4.306)$$

As before, the current has an in-phase and an out-of-phase component. The (real) conductivity, σ, is

$$\sigma = \frac{Ne^2 \tau}{m^*(1 + \omega^2 \tau^2)} \quad (4.307)$$

The conductivity vanishes in the limit of no electron-phonon scattering ($\tau \to \infty$), or in the limit of very high frequency. It gives rise to the previously mentioned absorption. The dielectric constant is now

$$\kappa_e = \kappa_a + \frac{x_e}{\varepsilon_0} = \kappa_a - \frac{Ne^2}{m^* \varepsilon_0 \omega^2(1 + 1/\omega^2 \tau^2)} \quad (4.308)$$

in which we have, phenomenologically, included the constant polarizability of the lattice atoms in κ_a. In the limit of high frequencies, the result agrees with the quantum calculation for $m^* = m^*_{\text{op}}$ if interband transitions are not present.

The utility of this simple approach is that it can be readily generalized to include effects of external magnetic fields, which cause great difficulty in the quantum theory. The effects of relaxation are also immediately apparent.

4.11 Optical Properties in a Magnetic Field

We have previously considered the profound effect of a magnetic field upon an energy band system. These effects are manifest in the optical absorption, and in the present section we consider the modifications required in the discussion of Section 4.9 required by the presence of a magnetic field. The necessary extension of the theory has been given by Roth et al. (1959), and by Burstein et al. (1959).

We consider here nondegenerate bands of index c and v which are characterized by effective masses m_c^* and m_v^* conduction and valence bands, respectively. Since the valence band has negative curvature, in the presence of an external field B in the z-direction the energy levels in the vicinity of the extremal point, which we take to be $\mathbf{k} = 0$, are

$$E_n = E_g + \hbar\omega_c(l + \tfrac{1}{2}) + \frac{\hbar^2 k_z^2}{2m_c^*} + \frac{e\hbar}{2m_0} g_c B m_s, \qquad (4.309)$$

$$E_{n'} = -\hbar\omega_v(l + \tfrac{1}{2}) - \frac{\hbar^2 k_z^2}{2m_v^*} + \frac{e\hbar}{2m_0} g_v B m_s$$

The quantities ω_c and ω_v are the cyclotron frequencies for the conduction and valence bands respectively: $\omega_c = eB/m_c^*$, etc. m_s is the spin quantum number which may have the values $\pm \tfrac{1}{2}$. E_g is the energy gap in the absence of the field, and we have chosen the zero of energy to be the valence band maximum, also in the absence of the field. When the field is present, the minimum separation between the bands is

$$\Delta E = E_g + \tfrac{1}{2}\hbar(\omega_c + \omega_v) - \frac{\hbar eB}{4m_0}(g_c + g_v) \qquad (4.310)$$

The term involving the cyclotron frequencies is generally dominant so that the effect of the magnetic field is to increase the energy gap and hence to displace the absorption edge.

4.11 OPTICAL PROPERTIES IN A MAGNETIC FIELD

We wish to obtain the selection rules and absorption constant for the allowed transitions. Spin will be neglected in this discussion. The wave functions are, in lowest approximation, products of Bloch wave functions for a given band index n and oscillator wave functions for the Landau levels (index l)

$$\psi_n = u_n(\mathbf{r}) F_l^{(n)}(\mathbf{r}) \tag{4.311}$$

Let A_{ext} be the vector potential of the (steady) external field which is in the z-direction, and A_R be the same function for the radiation field. The complete Hamiltonian is

$$H = \frac{(\mathbf{p} + e\mathbf{A}_{\text{ext}} + e\mathbf{A}_R)^2}{2m} + V(\mathbf{r}) \tag{4.312}$$

Our wave functions are approximate solutions of the eigenvalue problem

$$\frac{(\mathbf{p} + e\mathbf{A}_{\text{ext}})^2}{2m} \psi_n + V\psi_n = E_n \psi_n$$

so that if the radiation field is weak, the perturbing Hamiltonian, H', is now

$$H' = \frac{e(\mathbf{p} + e\mathbf{A}_{\text{ext}})}{m} \cdot \mathbf{A}_R \tag{4.313}$$

The matrix element H'_{fi} which appears in the calculation of the absorption coefficient is given by

$$H'_{fi} = \frac{e}{m} \int_\infty u_c^*(\mathbf{r}) F_l^{(c)*}(\mathbf{r}) [(\mathbf{p} + e\mathbf{A}_{\text{ext}}) \cdot \mathbf{A}_R] u_v(\mathbf{r}) F_{l'}^{(v)}(\mathbf{r}) \, d^3r \tag{4.314}$$

in which c and v refer to the conduction and valence bands, respectively. The integral may be split into two parts.

$$H'_{fi} = \frac{e}{m} \left[\int_{\text{cell}} u_c^*(\mathbf{r}) \mathbf{p} \cdot \mathbf{A}_R \, u_v(\mathbf{r}) \, d^3r \int_\infty F_l^{(c)*}(\mathbf{r}) F_{l'}^{(v)}(\mathbf{r}) \, d^3r + \right. \tag{4.315}$$

$$\left. \int_{\text{cell}} u_c^*(\mathbf{r}) u_v(\mathbf{r}) \, d^3r \int_\infty F_l^{(c)*}(\mathbf{r}) (\mathbf{p} + e\mathbf{A}_{\text{ext}}) \cdot \mathbf{A}_R \, F_{l'}^{(v)}(\mathbf{r}) \, d^3r \right]$$

In order to obtain this equation, we assume that both the external field and the functions $F_l(\mathbf{r})$ are slowly varying functions of position compared to the Bloch functions $u(\mathbf{r})$. The first term is responsible for the ordinary optical transitions. It follows from (4.149) that the functions $F^{(c)}$ and $F^{(v)}$ are independent of the effective mass m^*, so that these functions are the same for the two bands. Since the functions F_l, $F_{l'}$ are orthonormal, we get only transitions between bands in which there is no change in the quantum number of the Landau levels. These are the transitions of interest in the optical problem. The $\Delta l = 0$ selection rule replaces the requirement that the y component of k be conserved. No change of k_x or k_z is permitted.

Since the Bloch functions u_n are also orthonormal, the second term induces transitions between Landau levels belonging to the same band. The wave functions F_l which are given in (4.139) and (4.149) involve exponentials and simple harmonic oscillator functions. From the momentum matrix elements for these wave functions, it follows that the selection rules for transitions induced by the term, which are observed in cyclotron resonance, are

$$\Delta k_x = 0, \quad \Delta k_z = 0, \quad \Delta l = \pm 1 \quad (4.316)$$

The condition $\Delta l = \pm 1$ indicates that the energy change in the transition will be $\pm \hbar \omega_{c,v}$, as would be expected.[8] If we make the dipole approximation for the radiation field in the first term, we obtain

$$H'_{fi} = \delta(k_x - k_x')\delta(k_z - k_z')[\bar{H}_{fi}\,\delta_{ll'} + I_{ll'}\,\delta_{nn'}] \quad (4.317)$$

where $\bar{H}_{fi}$ is given by (4.251) and

$$I_{ll'}\delta(k_x - k_x')\delta(k_z - k_z') = \frac{\Omega}{(2\pi)^3}\int F_l(\mathbf{r})(\mathbf{p} + e\mathbf{A}_{\text{ext}})\cdot \mathbf{A}_R F_{l'}(\mathbf{r})\,d^3r \quad (4.318)$$

For the present, we will consider only the optical transitions. We would like to apply Eq. (4.250) which determines the transition probability to calculate the absorption coefficient. We must, however, use the density

---

[8] This result is valid only when the radiation field, $\mathbf{A}_R$, is uniform over the "orbit." If the field varies appreciably in a distance comparable to the extent of the wave function, the $\Delta l = \pm 1$ rule does not apply, and transitions will occur for any integer value of Δl. Cyclotron resonance absorption will then occur for subharmonics of the fundamental frequency.

of states appropriate to a Landau level, which may be obtained from (4.148), except for an extra factor of 2 required by spin degeneracy. We then have in place of (4.250), for a fixed l (and unit volume),

$$w_l = \frac{eB}{2\pi^2 \hbar} \frac{2\pi}{\hbar} \int dk_z |\bar{H}_{fi}|^2 \delta\left(E_g + \hbar\omega_l + \frac{\hbar^2 k_z^2}{2\mu} - \hbar\omega\right) dk_z \quad (4.319)$$

$$= \frac{2eB}{\pi \hbar^3} \left(\frac{\mu}{2}\right)^{1/2} \frac{|\bar{H}_{fi}|^2}{\sqrt{\hbar\omega - (E_g + \hbar\omega_l)}}$$

in which

$$\omega_l = (l + \tfrac{1}{2})(\omega_c + \omega_v) \quad (4.320)$$

and $\mu^{-1} = m_c^{*-1} + m_v^{*-1}$.

The absorption coefficient is obtained from (4.256) after summing over all pairs of Landau level; for which the denominator is positive.

$$\alpha = \frac{2e^2(\mathbf{p} \cdot \boldsymbol{\epsilon})^2}{\pi m^2 \hbar^2 n\varepsilon_0 c} \frac{(2\mu)^{1/2} eB}{\omega} \sum [\hbar\omega - (E_g + \hbar\omega_l)]^{-1/2} \quad (4.321)$$

$$= K(2\mu)^{1/2} \left(\frac{eB}{\omega}\right) \sum_l [\hbar\omega - (E_g + \hbar\omega_l)]^{-1/2}$$

in which K is given in (4.258), and the matrix element $\mathbf{p}$ is defined in (4.257).

The onset of strong absorption is shifted from the energy $E_0 = E_g$, for which it occurs in the absence of a field, to

$$E_0 = E_g + \frac{\hbar}{2}(\omega_c + \omega_v) = E_g + \frac{\hbar eB}{2\mu} \quad (4.322)$$

This result is in agreement with (4.311) since we have neglected the spin interactions here. It is possible to determine μ experimentally by observing the displacement of the absorption treshold in a magnetic field.

One notes the presence of inverse square root singularities in the expression for the absorption coefficient. These result from the fact that the density of final states in the expression for the transition probability is, in this case, characteristic of a one-dimensional band. In reality, however, the predicted infinite peaks are not observed, not only because of natural line width considerations and instrumental resolution, but also because the Landau levels themselves are not arbitrarily sharp (Blount, 1962b). The effects of line broadening have been discussed in the papers cited: Burstein *et al.* (1959) and Roth *et al.* (1959). The result is that one observes a series of absorption peaks centered approximately around the frequencies for which the denominators vanish.

The effect of an external magnetic field on the absorption coefficient for indirect transitions has also been determined. In this case (Roth *et al.*, 1959) the absorption does not exhibit oscillations, but rather exhibits a series of steps. The theory has also been worked out for degenerate bands.

Extensive studies of this magneto-optical effect have been made for some semiconductors (Burstein *et al.*, 1959; Zwerdling *et al.*, 1959). The experimental data can be successfully analyzed and interpreted in terms of the theory; however, the theoretical analysis is somewhat more complicated than the preceeding discussion would suggest. This is primarily due, in the case of germanium, to the complicated structure of the valence band. The valence band energy levels in the presence of a magnetic field must be determined from a set of coupled differential equations, (4.161). In addition, it is necessary to take into account the deviations of the conduction band shape from a simple parabolic form, and the departure of the electron g-factor from the usual value of 2 caused by spin orbit coupling. In spite of these complications, it has been possible to determine the particular transitions between Landau levels which are responsible for the observed oscillations of the absorption coefficient. In germanium, the principal magneto-optic effects are associated with the direct transition between the valence and conduction bands centered at $\mathbf{k} = 0$ (rather than the indirect transition to the lowest conduction band at L). One therefore has a means of probing the band structure in a region not accessible to ordinary cyclotron resonance measurements because of lack of carriers. It was possible to make a very precise determination of the band gap at the zone center, and a reasonably accurate determination of the effective mass in the $\mathbf{k} = 0$ conduction band minimum.

The theory of optical absorption by localized impurity states (Bowlden, 1957) and in the presence of an external electric field (Franz, 1958b; Keldysh, 1958b; Callaway, 1963) has also been developed. In the latter case, one finds that the absorption edge is not sharp, but has an exponential tail in the gap, and that oscillations due to the discrete "Stark" levels are present.

Appendix 1

Some Symmetrized Linear Combinations of Plane Waves

The construction of symmetrized linear combinations of plane waves which transform according to a particular row of a specified irreducible representation of the group of a wave vector was discussed in Section 2.2. In this appendix the coefficients of the waves in such a combination are presented for some of the most interesting representations — those which contain s, p, or d spherical harmonics — for a point of full cubic symmetry.

The wave vector of a plane wave in such an expansion can be written as $(2\pi/A)(a, b, c)$ where A is the lattice parameter and $(2\pi/A)a$ is the x component, etc. All the waves of a given type which can participate in any combination can be found from any one of them by applying all the operators of the full cubic group and thus are vectors whose components can be specified by giving a permutation and/or a change in sign of the original a, b, c.

As an example of the use of the tables, suppose we require the combination of plane waves of the 200 type which transforms according to row "x" in representation Γ_{15}. Inspection of the tables shows that the coefficients for waves of the $(a, 0, 0)$ type are $(a, 0, 0), +$; $(\bar{a}, 0, 0), -$. The combination is

$$e^{4\pi i x/A} - e^{-4\pi i x/A}$$

Note that $\bar{a}$ means $-a$, etc.

It will be seen that in some cases, more than one combination which transforms according to a particular row can be formed from waves of a given type. In such cases, all the combinations have been given.

TABLE XXIV
Coefficients of the Plane Wave Combinations

Wave vectors	Γ_1	Γ_{15} (x)	Γ_{12} (x^2-y^2)	Γ_{25}' (xy)		
$a\ 0\ 0$	+	+	+	0		
$0\ a\ 0$	+	0	−	0		
$0\ 0\ a$	+	0	0	0		
$\bar{a}\ 0\ 0$	+	−	+	0		
$0\ \bar{a}\ 0$	+	0	−	0		
$0\ 0\ \bar{a}$	+	0	0	0		
$a\ a\ a$	+	+	0	+		
$\bar{a}\ a\ a$	+	−	0	−		
$a\ \bar{a}\ a$	+	+	0	−		
$a\ a\ \bar{a}$	+	+	0	+		
$\bar{a}\ \bar{a}\ a$	+	−	0	+		
$\bar{a}\ a\ \bar{a}$	+	−	0	−		
$a\ \bar{a}\ \bar{a}$	+	+	0	−		
$\bar{a}\ \bar{a}\ \bar{a}$	+	−	0	+		
$a\ a\ 0$	+	+	0	+		
$a\ 0\ a$	+	+	+	0		
$0\ a\ a$	+	0	−	0		
$\bar{a}\ a\ 0$	+	−	0	−		
$\bar{a}\ 0\ a$	+	−	+	0		
$0\ \bar{a}\ a$	+	0	−	0		
$a\ \bar{a}\ 0$	+	+	0	−		
$a\ 0\ \bar{a}$	+	+	+	0		
$0\ a\ \bar{a}$	+	0	−	0		
$\bar{a}\ \bar{a}\ 0$	+	−	0	+		
$\bar{a}\ 0\ \bar{a}$	+	−	+	0		
$0\ \bar{a}\ \bar{a}$	+	0	−	0		
$b\ a\ a$	+	+	0	+	+	0
$a\ b\ a$	+	0	+	−	+	0
$a\ a\ b$	+	0	+	0	0	+
$\bar{b}\ a\ a$	+	−	0	+	−	0
$a\ \bar{b}\ a$	+	0	+	−	−	0
$a\ a\ \bar{b}$	+	0	+	0	0	+
$b\ \bar{a}\ a$	+	+	0	+	−	0
$\bar{a}\ b\ a$	+	0	−	−	−	0
$\bar{a}\ a\ b$	+	0	−	0	0	−
$b\ a\ \bar{a}$	+	+	0	+	+	0
$a\ b\ \bar{a}$	+	0	+	−	+	0
$a\ \bar{a}\ b$	+	0	+	0	0	−

APPENDIX 1. COMBINATIONS OF PLANE WAVES

Wave vectors	Γ_1	Γ_{15} (x)			Γ_{12} (x^2-y^2)		Γ_{25}' (xy)		
$\bar{b}\,\bar{a}\,a$	+	−	0		+	+	0		
$\bar{a}\,\bar{b}\,a$	+	0	−		−	+	0		
$\bar{a}\,a\,\bar{b}$	+	0	−		0	0	−		
$\bar{b}\,a\,\bar{a}$	+	−	0		+	−	0		
$a\,\bar{b}\,\bar{a}$	+	0	+		−	−	0		
$a\,\bar{a}\,\bar{b}$	+	0	+		0	0	−		
$b\,\bar{a}\,\bar{a}$	+	+	0		+	−	0		
$\bar{a}\,b\,\bar{a}$	+	0	−		−	−	0		
$\bar{a}\,\bar{a}\,b$	+	0	−		0	0	+		
$\bar{b}\,\bar{a}\,\bar{a}$	+	−	0		+	+	0		
$\bar{a}\,\bar{b}\,\bar{a}$	+	0	−		−	+	0		
$\bar{a}\,\bar{a}\,\bar{b}$	+	0	−		0	0	+		
$a\,b\,0$	+	+	0		2	0	+		
$a\,0\,b$	+	+	0		+	+	0		
$0\,a\,b$	+	0	0		−	−	0		
$b\,a\,0$	+	0	+		−2	0	+		
$b\,0\,a$	+	0	+		−	+	0		
$0\,b\,a$	+	0	0		+	−	0		
$\bar{a}\,b\,0$	+	−	0		2	0	−		
$\bar{a}\,0\,b$	+	−	0		+	+	0		
$0\,\bar{a}\,b$	+	0	0		−	−	0		
$b\,\bar{a}\,0$	+	0	+		−2	0	−		
$b\,0\,\bar{a}$	+	0	+		−	+	0		
$0\,b\,\bar{a}$	+	0	0		+	−	0		
$a\,\bar{b}\,0$	+	+	0		2	0	−		
$a\,0\,\bar{b}$	+	+	0		+	+	0		
$0\,a\,\bar{b}$	+	0	0		−	−	0		
$\bar{b}\,a\,0$	+	0	−		−2	0	−		
$\bar{b}\,0\,a$	+	0	−		−	+	0		
$0\,\bar{b}\,a$	+	0	0		+	−	0		
$\bar{a}\,\bar{b}\,0$	+	−	0		2	0	+		
$\bar{a}\,0\,\bar{b}$	+	−	0		+	+	0		
$0\,\bar{a}\,\bar{b}$	+	0	0		−	−	0		
$\bar{b}\,\bar{a}\,0$	+	0	−		−2	0	+		
$\bar{b}\,0\,\bar{a}$	+	0	−		−	+	0		
$0\,\bar{b}\,\bar{a}$	+	0	0		+	−	0		
$a\,b\,c$	+	+	0	0	+	−	+	0	0
$\bar{a}\,b\,c$	+	−	0	0	+	−	+	0	0
$a\,b\,\bar{c}$	+	+	0	0	+	−	−	0	0
$\bar{a}\,b\,\bar{c}$	+	−	0	0	+	−	−	0	0
$b\,\bar{a}\,\bar{c}$	+	0	+	0	−	+	−	0	0
$\bar{b}\,a\,\bar{c}$	+	0	−	0	−	+	−	0	0

Wave vectors	Γ_1	Γ_{15} (x)			Γ_{12} (x^2-y^2)		Γ_{25}' (xy)		
$\bar{a}\ c\ \bar{b}$	+	−	0	0	+	+	0	−	0
$\bar{a}\ \bar{c}\ b$	+	−	0	0	+	+	0	+	0
$\bar{c}\ \bar{b}\ a$	+	0	0	−	0	−2	0	0	+
$c\ \bar{b}\ \bar{a}$	+	0	0	+	0	−2	0	0	−
$\bar{b}\ \bar{a}\ c$	+	0	−	0	−	+	+	0	0
$\bar{c}\ b\ \bar{a}$	+	0	0	−	0	−2	0	0	−
$a\ \bar{c}\ \bar{b}$	+	+	0	0	+	+	0	−	0
$b\ \bar{a}\ c$	+	0	+	0	−	+	+	0	0
$c\ b\ a$	+	0	0	+	0	−2	0	0	+
$a\ c\ b$	+	+	0	0	+	+	0	+	0
$c\ a\ b$	+	0	0	+	−	−	0	+	0
$b\ c\ a$	+	0	+	0	0	+2	0	0	+
$c\ \bar{a}\ \bar{b}$	+	0	0	+	−	−	0	−	0
$\bar{b}\ \bar{c}\ a$	+	0	−	0	0	+2	0	0	+
$\bar{c}\ \bar{a}\ b$	+	0	0	−	−	−	0	+	0
$\bar{b}\ c\ \bar{a}$	+	0	−	0	0	+2	0	0	−
$\bar{c}\ a\ \bar{b}$	+	0	0	−	−	−	0	−	0
$b\ \bar{c}\ \bar{a}$	+	0	+	0	0	+2	0	0	−
$\bar{a}\ b\ \bar{c}$	+	−	0	0	+	−	+	0	0
$a\ b\ \bar{c}$	+	+	0	0	+	−	+	0	0
$\bar{a}\ b\ c$	+	−	0	0	+	−	−	0	0
$a\ \bar{b}\ c$	+	+	0	0	+	−	−	0	0
$\bar{b}\ a\ c$	+	0	−	0	−	+	−	0	0
$b\ \bar{a}\ c$	+	0	+	0	−	+	−	0	0
$a\ \bar{c}\ b$	+	+	0	0	+	+	0	−	0
$a\ c\ \bar{b}$	+	+	0	0	+	+	0	+	0
$c\ b\ \bar{a}$	+	0	0	+	0	−2	0	0	+
$\bar{c}\ b\ a$	+	0	0	−	0	−2	0	0	−
$b\ a\ \bar{c}$	+	0	+	0	−	+	+	0	0
$c\ \bar{b}\ a$	+	0	0	+	0	−2	0	0	−
$\bar{a}\ c\ b$	+	−	0	0	+	+	0	−	0
$\bar{b}\ \bar{a}\ \bar{c}$	+	0	−	0	−	+	+	0	0
$\bar{c}\ b\ \bar{a}$	+	0	0	−	0	−2	0	0	+
$\bar{a}\ \bar{c}\ \bar{b}$	+	−	0	0	+	+	0	+	0
$\bar{c}\ \bar{a}\ b$	+	0	0	−	−	−	0	+	0
$\bar{b}\ \bar{c}\ \bar{a}$	+	0	−	0	0	+2	0	0	+
$\bar{c}\ a\ b$	+	0	0	−	−	−	0	−	0
$b\ c\ \bar{a}$	+	0	+	0	0	+2	0	0	+
$c\ a\ \bar{b}$	+	0	0	+	−	−	0	+	0
$b\ \bar{c}\ a$	+	0	+	0	0	+2	0	0	−
$c\ \bar{a}\ b$	+	0	0	+	−	−	0	−	0
$\bar{b}\ c\ a$	+	0	−	0	0	+2	0	0	−

Appendix 2

Summation Relations

Certain useful summation and closure relations are obtained in this appendix. These concern two expressions which occur frequently in the theory

$$\sum_{\nu}' e^{i(\mathbf{k}-\mathbf{k}')\cdot\mathbf{R}_\nu} \quad \text{and} \quad \sum_{k}' e^{i\mathbf{k}\cdot(\mathbf{R}_\nu-\mathbf{R}_{\nu'})}$$

It is useful to consider first a crystal on which periodic boundary conditions have been imposed, and subsequently to pass to the limit as the periodic volume becomes infinite. Periodic boundary conditions imply that we suppose the crystal repeats itself in all respects after translation through vectors $\mathbf{R}_1 = 2N_1\mathbf{a}$; $\mathbf{R}_2 = 2N_2\mathbf{b}$; $\mathbf{R}_3 = 2N_3\mathbf{c}$, where $\mathbf{a}$, $\mathbf{b}$, and $\mathbf{c}$ are the three primitive translation vectors of the lattice. The concept is made more precise by employing the translation operators $T(\mathbf{R})$. It is then required

$$T(2N_1\mathbf{a}) = [T(\mathbf{a})]^{2N_1} = \mathbf{I} \quad (A2.1)$$

etc., where $\mathbf{I}$ designates the unit operator.

The derivation of the rules is given below for the case of a simple cubic lattice: $|\mathbf{a}| = |\mathbf{b}| = |\mathbf{c}| = a$. An extension to other cubic structures can be made quite simply by considering a cubic cell, which may contain several atoms.

Let $\mathbf{k} - \mathbf{k}' = \mathbf{s}$; $\mathbf{s} = (s_1, s_2, s_3)$; $\mathbf{R}_\nu = a(n_1, n_2, n_3)$. Then consider the summation

$$S_1 = \frac{1}{8N_1 N_2 N_3} \sum_{n_1=-N_1}^{N_1-1} \sum_{n_2=-N_2}^{N_2-1} \sum_{n_3=-N_3}^{N_3-1} e^{i(s_1 n_1 + s_2 n_2 + s_3 n_3)a} \quad (A2.2)$$

The summations are evidently independent. A single one is just a geometric series:

$$\frac{1}{2N_1} \sum_{-N_1}^{N_1-1} e^{is_1 n_1 a} = \frac{1}{2N_1} e^{-is_1 N_1 a} \sum_{0}^{2N_1-1} e^{is_1 n_1 a}$$

$$= \frac{e^{-is_1 N_1 a}}{2N_1} \left[\frac{1 - e^{is_1 2 N_1 a}}{1 - e^{is_1 a}} \right]$$

$$= \frac{\sin s_1 N_1 a}{2N_1 \sin s_1 a/2} \tag{A2.3}$$

In the limit that N_1 is large, the sum is appreciable only when both the numerator and denominator are zero. When this happens, the result is unity. The condition under which it will occur is $s_1 a/2 = l\pi$ or $s_1 = 2\pi l/a$, where l is any integer. Identical results are obtained in the other summations and one observes that for the condition above to hold for all three components, $\mathbf{s}$ must be a reciprocal lattice vector. Further, the quantity $8N_1 N_2 N_3$ equals the number of atoms in the block of atoms on which we have imposed periodic boundary conditions, and we denote this quantity by $\mathcal{N}$. Hence, we have in the limit of large $\mathcal{N}$

$$S_1 = \frac{1}{\mathcal{N}} \sum_{\nu} e^{i(\mathbf{k} - \mathbf{k}') \cdot \mathbf{R}_\nu} = 1 \quad \text{if} \quad \mathbf{k} - \mathbf{k}' = \mathbf{K}_s$$

(a reciprocal lattice vector)

$$= 0 \quad \text{otherwise} \tag{A2.4}$$

(more precisely, instead of 0 we should have a quantity of order $1/\mathcal{N}$).

In the second summation, it is convenient to put $\mathbf{R}_\nu - \mathbf{R}_{\nu'} = \mathbf{R}$, $\mathbf{k} = t(n_1/N_1, n_2/N_2, n_3/N_3)$, $\mathbf{R} = a(m_1, m_2, m_3)$.

$$k \cdot R = at\left(\frac{n_1}{N_1} m_1 + \frac{n_2}{N_2} m_2 + \frac{n_3}{N_3} m_3\right)$$

in which n_1, m_1, etc., are integers. Since the sum includes only values of k lying inside or on the surface of a single Brillouin zone, we have $-N_1 \leqslant n_1 \leqslant N_1 - 1$, etc. So consider

APPENDIX 2. SUMMATION RELATIONS

$$S_2 = \frac{1}{\mathcal{N}} \sum_{n_1=-N_1}^{N_1-1} \sum_{n_2=-N_2}^{N_2-1} \sum_{n_3=-N_3}^{N_3-1} \exp\left[iat\left(\frac{n_1 m_1}{N_1} + \frac{n_2 m_2}{N_2} + \frac{n_3 m_3}{N_3}\right)\right]$$
(A2.5)

As before, a single one of the independent sums gives

$$\sum_{-N_1}^{N_1-1} e^{ian_1 m_1 t/N_1} = \frac{\sin am_1 tN_1/N_1}{\sin am_1 t/2N_1}$$
(A2.6)

The quantity $am_1 t/n_1$ will not be an integral multiple of 2π, so there will be a contribution of order N_1 only when $m_1 = 0$. Hence, for large N_1, N_2, N_3

$$S_2 = \frac{1}{\mathcal{N}} \sum_k e^{i\mathbf{k}\cdot(\mathbf{R}_\nu - \mathbf{R}_{\nu'})} = \delta_{\nu,\nu'}$$
(A2.7)

Equations (A2.4) and (A2.7) are the fundamental summation rules. It is useful to examine these sums in the limit $\mathcal{N}$ becomes infinite. To do this, we write (A2.3) in the form

$$\sum_{-N_1}^{N_1-1} e^{ias_1 n_1} = \frac{2\pi}{a} \frac{as_1/2}{\sin as_1/2} \frac{\sin as_1 N_1}{\pi s_1}$$

Suppose first that $s_1 a/2$ is not an integral multiple of π.

The Dirac delta function can be defined through the relation

$$\delta(x) = \frac{1}{2\pi} \lim_{b\to\infty} \int_{-b}^{b} e^{ixk} \, dk = \lim_{b\to\infty} \frac{\sin xb}{\pi x}$$

Hence

$$\lim_{N_1\to\infty} \sum_{-N_1}^{N_1-1} e^{ias_1 n_1} = \frac{2\pi}{a} \delta(s_1)$$
(A2.8)

If $s_1 a = 2l\pi$, this situation is identical with the case $s_1 = 0$, since the exponentials on the left are unity. Hence we have

$$\sum_\nu e^{i(\mathbf{k}-\mathbf{k}')\cdot\mathbf{R}_\nu} = \frac{(2\pi)^3}{\Omega} \sum_l \delta(\mathbf{k}-\mathbf{k}'-\mathbf{K}_l)$$
(A2.9a)

where the sum over $\mathbf{K}_l$ includes all reciprocal lattice vectors and Ω is the volume of the unit cell. If, however, $\mathbf{k}$ and $\mathbf{k}'$ are constrained to lie *inside* a single zone, then only the zero reciprocal lattice vector contributes, so that we get

$$\sum_{\nu}' e^{i(\mathbf{k} - \mathbf{k}') \cdot \mathbf{R}_\nu} = \frac{(2\pi)^3}{\Omega} \delta(\mathbf{k} - \mathbf{k}') \qquad (A2.9b)$$

The continuous $\mathbf{k}$ limit of (A2.7) is obvious. Since the volume of $\mathbf{k}$-space which is included in the integral is finite and equal to $(2\pi)^3/\Omega$, we have immediately

$$\int d^3k \, e^{i\mathbf{k} \cdot (\mathbf{R}_\nu - \mathbf{R}_{\nu'})} d^3k = \frac{(2\pi)^3}{\Omega} \delta_{\nu, \nu'} \qquad (A2.10)$$

Appendix 3

The Effective Mass Equation in the Many-Body Problem

In Chapter 4, the effects of external fields on the energy levels of electrons were studied. The entire discussion was based on a one particle theory, no account being taken of the electron interactions. In this appendix we will reconsider one aspect of this problem from the point of view of many-body theory. We will show, following the work of Kohn (1957, 1958), that an effective mass equation is valid for the description of the change in the energy levels of an insulator in response to a slowly varying perturbation. Other studies have been presented by Ambegaokar and Kohn (1960), Ambegaokar (1961), and Klein (1959), who have extended the original treatment due to Kohn in the references cited.

We consider a system consisting of an insulator plus one extra electron. It can be proved that the response of this system to a long wavelength, low frequency electric field is the same as that of a free electron of effective mass m^* moving in a medium having the dielectric constant, κ, of the perfect insulator. The point impurity is included as a special case in that the bound states which occur when a small positive charge Q is embedded in the system have a hydrogenlike spectrum given by

$$E_n = -\frac{m^* e^2 Q^2}{2\kappa^2 n^2 \hbar^2} \tag{A3.1}$$

In addition, it has been shown that the low-lying energy levels in a constant, external magnetic field B are characterized by a cyclotron frequency

$$\omega_c = \frac{eB}{m^*} \tag{A3.2}$$

where m^* is the same effective mass as in (A3.1).

We will not attempt to prove these results in general, but will consider only the case in which a weak, slowly varying, external electric potential is applied as a perturbation. We are therefore concerned with the Schrödinger equation.

$$(H_0 + U)\Psi = i\hbar \frac{\partial \Psi}{\partial t}$$

for a system of $(N + 1)$ electrons (the N electron system is the "perfect" insulator) in which U is the perturbation mentioned above and H_0 is given by

$$H_0 = \sum_i \left[\frac{p_i^2}{2m} + V(\mathbf{r}_i) + \sum_{j>i} \frac{2}{r_{ij}} \right] \quad (A3.3)$$

where $V(\mathbf{r}_i)$ represents the periodic potential. It follows from (2.179) that the $(N + 1)$ electron wave function which is an eigenfunction of H_0 can be written in the form

$$\psi_n(\mathbf{k}, \mathbf{r}_1, \ldots, \mathbf{r}_{N+1}) = e^{i\mathbf{k}\cdot\mathbf{R}} u_n(\mathbf{k}, \mathbf{r}_1, \ldots, \mathbf{r}_{N+1}) \quad (A3.4)$$

where

$$\mathbf{R} = \sum_{i=1}^{N+1} \mathbf{r}_i$$

and $\mathbf{k}$ is the total wave vector of the system. The quantity n denotes the set of quantum numbers other than $\mathbf{k}$ required to specify a state. The energy of a state is denoted by $E_n(\mathbf{k})$. We suppose that the ground state of the system has $\mathbf{k} = (0)$, and we put $n = 0$ for this state. Our semiconductor is defined by the following postulated property of the energy spectrum: All excited states of the system with $\mathbf{k} = 0$ have energies which differ from that of the ground state by a finite amount, ΔE.

$$E_n(0) - E_0(0) \geq \Delta E$$

This property is characteristic of the independent particle model of a semiconductor, but not of a metal: it is found to be true experimentally for real semiconductors since electromagnetic radiation, which causes transitions between states of the same $\mathbf{k}$, is not absorbed if the photons

APPENDIX 3. EFFECTIVE MASS EQ. IN MANY-BODY PROBLEM

have less than a definite energy. The energy $E_0(0)$ is assumed to be an absolute minimum, with respect to $\mathbf{k}$.

The perturbation problem will be attacked in the crystal momentum representation. The perturbed wave function for the system is expanded in the Bloch type functions:

$$\Psi(\mathbf{r}_1\ldots\mathbf{r}_{N+1}) = \sum_n \int \varphi_n(\mathbf{k})\psi_n(\mathbf{k},\mathbf{r}_1\ldots\mathbf{r}_{N+1})\, d^3k \qquad (A3.5)$$

in accord with the procedures of Section 4.1. Equation (4.6) for the coefficients $\varphi_n(\mathbf{k})$ is still valid

$$\left[E_n(\mathbf{k}) - i\hbar\frac{\partial}{\partial t}\right]\varphi_n(\mathbf{k}) + \sum_{n'}\int d^3k'\, \langle n\mathbf{k}|U|n'\,\mathbf{k}'\rangle\varphi_{n'}(\mathbf{k}') = 0 \qquad (4.6)$$

in which the matrix element is now given by

$$\langle n\mathbf{k}|U|n'\,\mathbf{k}'\rangle = \int \psi_n{}^*(\mathbf{k},\mathbf{r}_1\ldots\mathbf{r}_{N+1})U\psi_{n'}(\mathbf{k}',\mathbf{r}_1\ldots\mathbf{r}_{N+1})\, d^{3(N+1)}r \qquad (A3.6)$$

We are interested in solutions for a weak perturbation U which describe the states close to the ground state. Hence we set $n=0$ in (4.6). By perturbation theory $\varphi_{n'}(\mathbf{k})$ is of order U for $n' \neq 0$, and the contribution from such states if of order U^2. This we will neglect. Equation (4.6) then has the form

$$\left[E_0(\mathbf{k}) - i\hbar\frac{\partial}{\partial t}\right]\varphi_0(\mathbf{k}) + \int d^3k'\, \langle 0\mathbf{k}|U|0\mathbf{k}'\rangle\varphi_0(\mathbf{k}') = 0 \qquad (A3.7)$$

It is assumed that the perturbation U can be expressed as a sum of perturbing potentials acting on each electron

$$U = \sum_i U(\mathbf{r}_i) \qquad (A3.8)$$

The functions $U(\mathbf{r}_i)$ are identical except for the labeling of the independent variable. Each $U(\mathbf{r}_i)$ may be represented as a Fourier integral

$$U(\mathbf{r}_i) = \frac{1}{(2\pi)^{3/2}}\int V(\mathbf{q})\, e^{i\mathbf{q}\cdot\mathbf{r}_i}\, d^3q \qquad (A3.9)$$

Then the total perturbation can be expressed simply in terms of $V(\mathbf{q})$:

$$U = \frac{1}{(2\pi)^{3/2}} \int V(\mathbf{q}) \rho(\mathbf{q}) \, d^3q \tag{A3.10}$$

where

$$\rho(\mathbf{q}) = \sum_i{}' e^{i\mathbf{q}\cdot\mathbf{r}_i} \tag{A3.11}$$

The quantity $\rho(\mathbf{q})$ may be interpreted as a Fourier coefficient of the electron density; since if we have $(N+1)$ electrons located at points $\mathbf{r}_i$, the electron density is $\sum_i \delta(\mathbf{r} - \mathbf{r}_i)$, and the Fourier transform of this function is $\rho(\mathbf{q})$. Then we have for the matrix element of the potential

$$\langle 0\mathbf{k}|U|0\mathbf{k}'\rangle = \int d^3q\, V(\mathbf{q}) \int \psi_0^*(\mathbf{k}\ldots\mathbf{r}\ldots) \sum_i{}' e^{i\mathbf{q}\cdot\mathbf{r}_i} \psi_0(\mathbf{k}'\ldots\mathbf{r}\ldots)\, d^{3(N+1)}r$$

$$= \int d^3q\, V(\mathbf{q}) \delta(\mathbf{k}' + \mathbf{q} - \mathbf{k}) \langle 0\mathbf{k}|\rho(\mathbf{q})|0\mathbf{k}+\mathbf{q}\rangle \tag{A3.12}$$

The last step follows from the Bloch form of the wave function, since we have in the first line the Fourier transform of a periodic function. The perturbing potential is assumed to be sufficiently slowly varying so that only small values of $\mathbf{q}$ are important. Kohn has shown that

$$\lim_{\mathbf{q}\to 0} \langle n, \mathbf{k}|\rho(\mathbf{q})|n', \mathbf{k}+\mathbf{q}\rangle = \frac{1}{\kappa} \delta_{nn'} \tag{A3.13}$$

where κ is the static dielectric constant of the insulator. Hence we approximate the matrix element of the potential by

$$\langle 0\mathbf{k}|U|0\mathbf{k}'\rangle = \frac{1}{\kappa} V(\mathbf{k}-\mathbf{k}') \tag{A3.14}$$

The contributions from terms of higher order in the expansion of $\rho(\mathbf{q})$ will produce deviations from this simple, static, screening. Equation (A3.7) now has the simple form

$$\left[E_0(\mathbf{k}) - i\hbar\frac{\partial}{\partial t}\right]\varphi_0(\mathbf{k}) + \frac{1}{\kappa}\int d^3k'\, V(\mathbf{k}-\mathbf{k}')\varphi_0(\mathbf{k}') = 0 \tag{A3.15}$$

APPENDIX 3. EFFECTIVE MASS EQ. IN MANY-BODY PROBLEM

For small values of **k**, it is possible to replace $E_0(\mathbf{k})$ by the leading terms of its power series expansion. In accord with our fundamental assumptions, this is

$$E_0(\mathbf{k}) = E_0(0) + \frac{\hbar^2 \mathbf{k}^2}{2m^*} + \cdots \qquad (A3.16)$$

where **k** is the total wave vector previously described, and m^* is the effective mass.[1] When this is substituted into (A3.15), the resulting equation may be transformed back to position space exactly as was done in Section 4.3. The result is just the effective mass equation (4.87) with the one difference that the potential which appears is screened by the dielectric constant of the system.

$$\left[-\frac{\hbar^2}{2m^*} \nabla^2 + \frac{U(\mathbf{r})}{\kappa} \right] F(\mathbf{r}) = i\hbar \frac{\partial F}{\partial t} \qquad (A3.17)$$

in which $F(\mathbf{r})$ is the envelope function given by:

$$F(\mathbf{r}) = \int e^{i\mathbf{k}\cdot\mathbf{r}} \varphi_0(\mathbf{k}) \, d^3k$$

which was introduced in Section 4.3.

---

[1] The generalization to a tensor effective mass is trivial.

Appendix 4

Evaluation of a Tunneling Integral

In Section 4.8 the theory of tunneling between bands in the presence of a uniform electric field was discussed. Let the field be in the x direction. Then the tunnel current is determined as the integral of the probability for transitions between bands, w, over a plane in **k**-space perpendicular to the field. The transition probability is proportional to the square of a matrix element, M_{cv}, for which a formal expression was given in Eq. (4.243):

$$M_{cv} = \frac{F}{\kappa} \int_{-\kappa}^{\kappa} X_{cv}(\mathbf{k}) \exp\left\{\frac{i}{F} \int_0^{k_x} [E_c^{(1)}(\mathbf{k}') - E_v^{(1)}(\mathbf{k}')] \, dk_x'\right\} dk_x \quad (4.243)$$

Here $F = e\mathscr{E}$. In this appendix we will discuss the evaluation of this integral following the procedure of Kane (1959b).

In order to evaluate the integral it is desirable to have an expression for the difference in energy between the bands which is valid for complex **k** as well as for real **k**. To find this, it is necessary to examine the Hamiltonian of which E_c and E_v are eigenvalues. The simplest situation is one in which only two spherical bands E_c and E_v need be considered. In that case we have in atomic units

$$H = \begin{pmatrix} E_g + \mathbf{k}^2 & 2kp \\ 2kp & \mathbf{k}^2 \end{pmatrix} \quad \text{(A4.1)}$$

The zero of energy is taken at the top of the valence band. E_g is the band gap at $\mathbf{k} = 0$. The bands are connected by the momentum matrix element, p, which has been chosen, for simplicity, to be a real scalar quantity. The eigenvalues of H are easily found to be

APPENDIX 4. EVALUATION OF A TUNNELING INTEGRAL

$$E_{cv} = \mathbf{k}^2 + \frac{E_g}{2} \pm \frac{1}{2}\eta \qquad (A4.2)$$

$$\eta = (E_g^2 + 16\mathbf{k}^2 p^2)^{1/2}$$

The $+$ sign corresponds to the conduction band, the $-$ sign to the valence band. Near $\mathbf{k} = 0$, we have

$$E_c = E_g + \mathbf{k}^2 + \frac{4\mathbf{k}^2 p^2}{E_g} \qquad (A4.3)$$

$$E_v = \mathbf{k}^2 - \frac{4\mathbf{k}^2 p^2}{E_g}$$

The reciprocal effective masses are:

$$\frac{1}{m_c} = 1 + \frac{4p^2}{E_g} \; ; \qquad \frac{1}{m_v} = -1 + \frac{4p^2}{E_g} \qquad (A4.4)$$

(We write the valence band energy as $E_v = -\mathbf{k}^2/m_v$.) We define a reduced effective mass, μ, by

$$\mu^{-1} = m_c^{-1} + m_v^{-1}$$

The momentum matrix element is related to μ by

$$p^2 = E_g/8\mu \qquad (A4.5)$$

The difference in energy between the conduction and valence bands is

$$E_c - E_v = \eta = \left(E_g^2 + \frac{2E_g \mathbf{k}^2}{\mu}\right)^{1/2} \qquad (A4.6)$$

It is seen that the conduction and valence bands join for certain complex values of $\mathbf{k}$. It suffices in this calculation to allow only k_x to be complex. The bands then join if $k_x = iq$, where

$$q = \left(\frac{\mu E_g}{2} + k_y^2 + k_z^2\right)^{1/2} \qquad (A4.7)$$

The integral in the exponential of (4.243) is elementary:

$$\int_0^{k_x} (E_c - E_v) \, dk_x' = \int_0^{k_x} \eta \, dk_x' = \frac{k_x \eta}{2} + \quad (A4.8)$$

$$\frac{1}{2}\left(\frac{\mu E_g}{2}\right)^{1/2} \left[E_g + \frac{2}{\mu}(k_y^2 + k_z^2)\right] \ln\left\{\frac{\eta + k_x(2E_g/\mu)^{1/2}}{[E_g^2 + (2E_g/\mu)(k_y^2 + k_z^2)]^{1/2}}\right\}$$

It is also necessary to compute the matrix element X_{cv}

$$X_{cv} = \frac{(2\pi)^3}{\Omega_0} i \int U_c \frac{\partial}{\partial k_x} U_v \, d^3r \quad (A4.9)$$

where U_c and U_v are the eigenvectors of the matrix H. Let u_c and u_v be the vectors at $\mathbf{k} = 0$. A straightforward calculation yields

$$U_c = \frac{1}{\sqrt{2\eta}} [(\eta + E_g)^{1/2} u_c + (\eta - E_g)^{1/2} u_v]$$

$$U_v = \frac{1}{\sqrt{2\eta}} [(\eta - E_g)^{1/2} u_c - (\eta + E_g)^{1/2} u_v] \quad (A4.10)$$

It is now possible to evaluate X_{cv}, making use of the orthonormality of u_c and u_v since the k_x dependence of the solutions are entirely contained in η. A straightforward calculation yields

$$X_{cv} = \frac{iE_g^2 k_x}{\mu \eta^2 (\eta^2 - E_g^2)^{1/2}} = \frac{iE_g^{3/2}}{(2\mu)^{1/2} \eta^2}\left(\frac{k_x}{k}\right) \quad (A4.11)$$

We must now calculate M_{cv}. On substitution into (4.243) we obtain

$$M_{cv} = \frac{iE_g^{3/2} F}{(2\mu)^{1/2} \kappa} \int_{-\kappa}^{\kappa} \frac{dk_x}{\eta^2} \left[\exp \frac{i}{F} \int_0^{k_x} \eta \, dk_x'\right]$$

This integral is approximated by deforming the path into the complex k_x-plane to pass close to the branch point located at $k_x = iq$ where the bands join. When the field, F, is small the dominant contributions to the integral consists of two portions: a contribution from a small semicircle about the branch point, and a contribution from the horizontal portion of the contour parallel to the real k_x-axis. In the latter case, the limits may be extended to $\pm \infty$ without serious error. We find

APPENDIX 4. EVALUATION OF A TUNNELING INTEGRAL

$$\int_{-\kappa}^{\kappa} \frac{dk_x}{\eta^2}\left[\exp\frac{i}{F}\int_0^{k_x}\eta\,dk_x'\right] = \frac{\pi\mu}{3E_g}\exp\left[\frac{i}{F}\int_0^{iq}\eta\,dk_x'\right] \quad (A4.12)$$

$$= \frac{\pi\mu}{3E_g q}\exp\left[-\frac{\pi}{2}\left(\frac{E_g}{2\mu}\right)^{1/2}\frac{q^2}{F}\right]$$

It is now possible to compute the rate of tunneling from (4.245):

$$n = \frac{FE_g\mu}{72\pi}\int\frac{1}{q^2}\exp\left[-\pi\left(\frac{E_g}{2\mu}\right)^{1/2}\frac{q^2}{F}\right]dk_y\,dk_z \quad (A4.13)$$

For small fields, the integral may be done by transformation to polar coordinates and extension of the limits to infinity. Under these circumstances, the exponential dominates, and we find:

$$n = \frac{F^2}{36\pi}\left(\frac{2\mu}{E_g}\right)^{1/2}\exp\left[-\frac{\pi}{F}\frac{\mu^{1/2}E_g^{3/2}}{2\sqrt{2}}\right] \quad (A4.14)$$

This result agrees with that of Kane (his $m_r = \mu/2$).

Appendix 5

Spin Density Waves

We will examine briefly the spin density waves discussed by Overhauser (1962). The Hartree-Fock equations (2.183) are approximated by

$$[H_0 + A]\psi = E\psi \tag{A5.1}$$

where A represents the exchange term, which is treated as a potential with a possible explicit spin dependence. The one-electron wave function ψ is not assumed to be an eigenfunction of σ_z and A may have off-diagonal elements connecting the different spin components. Let us assume with Overhauser that A is given by

$$A = A_0 + A_1$$

where

$$A_1 = -g\boldsymbol{\sigma} \cdot (\hat{\mathbf{i}} \cos qz + \hat{\mathbf{j}} \sin qz) \tag{A5.2}$$

where $\hat{\mathbf{i}}$ and $\hat{\mathbf{j}}$ are the usual unit vectors in the x and y directions. Such an A would occur for a system possessing a fractional spin polarization at every point, but with direction of the polarization a continuous function of position. Such an alternating polarization is referred to as a spin density wave by Overhauser. (It is not necessary for the waves to be transverse, as in the example.) If the explicit expressions for the spin operators are used, A_1 has the form

$$A_1 = -g \begin{pmatrix} 0 & e^{-iqz} \\ e^{iqz} & 0 \end{pmatrix} \tag{A5.3}$$

We suppose for simplicity that the eigenfunctions of the operator $H_0 + A_0$ are plane waves $e^{i\mathbf{k}\cdot\mathbf{r}}$, and the eigenvalues of this operator are $\varepsilon_\mathbf{k}$. This

APPENDIX 5. SPIN DENSITY WAVES

is correct for a free electron gas. Then it is easy to see that the operator A' connects plane waves of different wave vectors: the two components of the spinor ψ have wave vectors $\mathbf{k}$ and $\mathbf{k}+\mathbf{q}$. The eigenvalues of H may be found by diagonalizing the simple matrix

$$\begin{pmatrix} \varepsilon_{\mathbf{k}} & -g \\ -g & \varepsilon_{\mathbf{k}+\mathbf{q}} \end{pmatrix}$$

to be

$$E_{\mathbf{k}} = \tfrac{1}{2}(\varepsilon_{\mathbf{k}} + \varepsilon_{\mathbf{k}+\mathbf{q}}) \pm [\tfrac{1}{4}(\varepsilon_{\mathbf{k}} - \varepsilon_{\mathbf{k}+\mathbf{q}})^2 + g^2]^{1/2} \qquad (A5.4)$$

and the eigenfunctions are, for the lower branch of the spectrum

$$\frac{1}{\Omega^{1/2}} \begin{pmatrix} \cos\theta & e^{i\mathbf{k}\cdot\mathbf{r}} \\ \sin\theta & e^{i(\mathbf{k}+\mathbf{q})\cdot\mathbf{r}} \end{pmatrix} \qquad (A5.5)$$

in which $\mathbf{q}$ points in the z direction, and

$$\cos\theta(\mathbf{k}) = g/[g^2 + (\varepsilon_{\mathbf{k}} - E_{\mathbf{k}})^2]^{1/2} \qquad (A5.6)$$

The square modulus of ψ is constant, as is the case for a single plane wave. Hence, no fluctuation in the charge density is associated with the wave, and no alternating term need be included in H_0.

The interaction (A5.2) lowers the energy of some of the states and raises that of others. The total energy of the system can be reduced. The surfaces of constant energy no longer have spherical symmetry, so that the existence of spin density waves in a material would lead to modifications of the Fermi surface.

It does not suffice, however, the show that the total energy of the system is reduced with an exchange potential of the form (A5.2) to prove that the ground state of the system will contain spin density waves. It is necessary, in addition, to show that such a solution is self-consistent and this is a more difficult problem. Kohn and Nettel (1960) have shown that for sufficiently weak- and short-range particle interactions, a self-consistent solution of the Hartree-Fock equations containing a spin density wave is impossible. Their discussion does not, however, apply to the Coulomb interaction for which Overhauser has proved that the ground state in the Hartree-Fock approximation must contain spin density waves.

The question then arises as to whether spin density waves are present in real materials. Consideration of electron correlation effects not included in the Hartree-Fock approximation tends to reduce the strength of the effective electron-electron interaction. The effective interaction is frequently approximated by a potential $e^{-\alpha r}/r$. For interactions of the strength and range actually encountered in most metals, it is probable that spin density waves are not contained in the ground state. Chromium seems, however, to be an exception in that the weak antiferromagnetism probably can be best described in terms of spin density waves.

Bibliography

Adams, E. N. (1952). Motion of an electron in a perturbed periodic potential. *Phys. Rev.* **85**, 41.

Adams, E. N. (1953). The crystal momentum as a quantum mechanical operator. *J. Chem. Phys.* **21**, 2013.

Adams, E. N. (1957). Definition of energy bands in the presence of an external force field. *Phys. Rev.* **107**, 698.

Adams, E. N., and Argyres, P. N. (1956). Acceleration of electrons by an external force field. *Phys. Rev.* **102**, 605.

Adams, W. H. (1962). Stability of Hartree-Fock states. *Phys. Rev.* **127**, 1650.

Altmann, S. L. (1956). Spherical harmonics with the symmetry of the closed packed hexagonal lattice. *Proc. Phys. Soc. (London)* **A69**, 184.

Altmann, S. L. (1958a). The cellular method for a close packed hexagonal lattice. I. Theory. *Proc. Roy. Soc.* **A244**, 141.

Altmann, S. L. (1958b). The cellular method for a close packed hexagonal lattice. II. Application to zirconium. *Proc. Roy. Soc.* **A244**, 153.

Altmann, S. L., and Cohan, N. V. (1958). Cellular eigenvalues for titanium. *Proc. Phys. Soc. (London)* **71**, 383.

Ambegaokar, V., and Kohn, W. (1960). Electromagnetic properties of insulators. I. *Phys. Rev.* **117**, 423.

Ambegaokar, V. (1961). Electromagnetic properties of insulators. II. *Phys. Rev.* **121**, 91.

Antoncik, E. (1953). The electron theory of metallic aluminium. *Czechoslov. J. Phys.* **2**, 18.

Argyres, P. N. (1962). Theory of tunneling and its dependence on a longitudinal magnetic field. *Phys. Rev.* **126**, 1386.

Asdente, M., and Friedel, J. (1961). $3d$ band structure of Cr. *Phys. Rev.* **124**, 384.

Asdente, M. (1962). Fermi surface for the $3d$ band of chromium. *Phys. Rev.* **127**, 1949.

Bader, F., Ganzhorn, K., and Dehlinger, U. (1954). Ferromagnetism and band structure of the transition metals. *Z. Physik* **137**, 190.

Bardeen, J. (1938a). An improved calculation of the energies of metallic Li and Na. *J. Chem. Phys.* **6**, 367.

Bardeen, J. (1938b). Compressibilities of the alkali metals. *J. Chem. Phys.* **6**, 372.

Bardeen, J., Blatt, F. J., and Hall, L. J. (1957). Indirect transitions from the valence to the conduction bands. *In* "Photoconductivity Conference" (R. G. Breckenridge *et al.*, eds.), p. 146. Wiley, New York.

Barrett, S. (1956). X-ray study of the alkali metals at low temperatures. *Acta Cryst.* **9**, 671.

Bassani, F. (1957). Energy band structure in silicon crystals by the orthogonalized plane wave method. *Phys. Rev.* **108**, 263.

Bassani, F., and Celli, V. (1961). Energy band structure of solids from a perturbation on the empty lattice. *J. Phys. Chem. Solids* **20**, 64.

Bassani, F., and Yoshimine, M. (1963). Electronic band structure of group IV elements and of III–V compounds. *Phys. Rev.* **130**, 20.

Bassani, F., and Brust, D. (1963). Effect of alloying and pressure on the band structure of germanium and silicon. To be published.

Batterman, B. W., Chipman, D. R., and De Marco, J. J. (1961). Absolute measurement of the atomic scattering factor of iron, copper, and aluminum. *Phys. Rev.* **122**, 68.

Behringer, R. E. (1958). Note on: The band structure of aluminium: A self-consistent calculation. *J. Phys. Chem. Solids* **5**, 145.

Belding, E. F. (1959). $3d$ band structure of some transition elements. *Phil. Mag.* [8] **4**, 1145.

Bell, D. G. (1953). Group theory and crystal lattices. *Revs. Modern Phys.* **26**, 311.

Bell, D. G., Hensman, R., Jenkins, D. P., and Pincherle, L. (1954). A note on the band structure of silicon. *Proc. Phys. Soc. (London)* **A67**, 562.

Benedek, G. B., and Kushida, T. (1958). The pressure dependence of the Knight shift in the alkali metals and copper. *J. Phys. Chem. Solids* **5**, 241.

Bernal, M. J. M., and Boys, S. F. (1952). Electronic wave functions. VIII. A calculation of the ground states Na^+, Ne and F^-. *Phil. Trans. Roy. Soc.* **A245**, 139.

Bethe, H., Schweber, S., and de Hoffman, F. (1955). ,,Mesons and Fields,'' Vol. 1, Row, Peterson, New York.

Biondi, M. A., and Rayne, J. A. (1959). Band structure of noble metal alloys: optical absorption in α-Brasses at $4.2°K$. *Phys. Rev.* **115**, 1522.

Birman, J. L. (1958). Electronic energy bands in ZnS: potential in zincblende and wurtzite. *Phys. Rev.* **109**, 610.

Birman, J. L. (1959a). Calculation of electronic energy bands in ZnS. *J. Phys. Chem. Solids* **8**, 35.

Birman, J. L. (1959b). Simplified LCAO method for zincblende, wurtzite, and mixed crystal structures. *Phys. Rev.* **115**, 1493.

Blackman, M. (1938). The diamagnetic susceptibility of bismuth. *Proc. Roy. Soc.* **A166**, 1.

Bloch, F. (1928). Über die Quantenmechanik der Elektronen in Kristallgittern. *Z. Physik* **52**, 555.

Blount, E. I. (1962a). Formalisms of band theory. *Solid State Phys.* **13**, 305.

Blount, E. I. (1962b). Bloch electrons in a magnetic field. *Phys. Rev.* **126**, 1636.

Blume, M., Briggs, N., and Brooks, H. (1959). "Tables of Coulomb Wave Functions,'' Cruft Laboratory, Harvard, Massachusetts.

Bohm, D. (1951). "Quantum Theory.'' Prentice-Hall, Englewood Cliffs, New Jersey.

Bohr, A., and Weisskopf, V. (1950). The influence of nuclear structure on the hyperfine structure of heavy elements. *Phys. Rev.* **77**, 94.

Bouckaert, L. P., Smoluchowski, R., and Wigner, E. (1936). Theory of Brillouin zones and symmetry properties of wave functions in crystals. *Phys. Rev.* **50**, 58.
Bowlden, H. J. (1957). Radiative transitions in semiconductors. *Phys. Rev.* **106**, 427.
Brandt, N. B. (1960). On the hole component of the Fermi surface in bismuth. *Soviet Phys. JETP* **11**, 975.
Breit, G. (1930). Possible effects of nuclear spin on X-ray terms. *Phys. Rev.* **35**, 1447.
Brillouin, L. (1931). "Quantenstatistik." Springer, Berlin.
Brooks, H. (1953). Cohesive energy of alkali metals. *Phys. Rev.* **91**, 1027.
Brooks, H. (1958). Quantum theory of cohesion. *Nuovo. cimento* **7**, Suppl., 165.
Brooks, H., and Ham, F. S. (1958). Energy bands in solids: The quantum defect method. *Phys. Rev.* **112**, 344.
Brown, E. (1962). Role of orthogonalization in the determination of valence states in crystals. *Phys. Rev.* **126**, 421.
Brown, R. N., Mavroides, J. G., Dresselhaus, M. S., and Lax, B. (1960). Interband magnetoreflection experiments in bismuth. *In* "The Fermi Surface" (W. A. Harrison and M. B. Webb, eds.), p. 203. Wiley, New York.
Brown, R. N., Mavroides, J. G., and Lax, B. (1963). Magnetoreflection in bismuth. *Phys. Rev.* **129**, 2055.
Brust, D., Phillips, J. C., and Bassani, F. (1962). Critical points and ultraviolet reflectivity of semiconductors. *Phys. Rev. Letters* **9**, 94.
Brust, D., Cohen, M. L., and Phillips, J. C. (1962). Reflectance and photoemission from Si. *Phys. Rev. Letters* **9**, 389.
Burdick, G. A. (1963). Energy band structure of copper. *Phys. Rev.* **129**, 138.
Burstein, E. (1954). Anomalous optical absorption limit in InSb. *Phys. Rev.* **93**, 632.
Burstein, E., Picus, G. S., Wallis, R. F., and Blatt, F. (1959). Zeeman type magneto-optical studies of interband transitions in semiconductors. *Phys. Rev.* **112**, 15.
Butcher, P. N. (1951). The absorption of light by alkali metals. *Proc. Phys. Soc.* (*London*) **A64**, 765.
Callaway, J. (1955). Electronic energy bands in iron. *Phys. Rev.* **99**, 500.
Callaway, J. (1957a). Contribution of core polarization to the cohesive energies of the alkali metals. *Phys. Rev.* **106**, 868.
Callaway, J. (1957b). Energy bands in gallium arsenide. *J. Electronics* **2**, 330.
Callaway, J. (1958a). Electron energy bands in solids. *Solid State Phys.* **7**, 99.
Callaway, J. (1958b). Electron energy bands in sodium. *Phys. Rev.* **112**, 322.
Callaway, J. (1958c). Electron wave functions in metallic cesium. *Phys. Rev.* **112**, 1061.
Callaway, J. (1959). *d* Bands in the body centered cubic lattice. *Phys. Rev.* **115**, 346.
Callaway, J. (1960a). Electron wave functions in metallic potassium. *Phys. Rev.* **119**, 1012.
Callaway, J. (1960b). *d* Bands in cubic lattices, II. *Phys. Rev.* **120**, 731.
Callaway, J. (1961a). *d* Bands in cubic lattices, III. *Phys. Rev.* **121**, 1351.
Callaway, J. (1961b). Electron wave functions in metallic sodium. *Phys. Rev.* **123**, 1255.

Callaway, J. (1961c). Energy bands in lithium. *Phys. Rev.* **124**, 1824.
Callaway, J. (1963). Optical absorption in a electric field. *Phys. Rev.* **130**, 549.
Callaway, J., and Edwards, D. M. (1960). Cubic field splitting of D levels in metals. *Phys. Rev.* **118**, 923.
Callaway, J., and Glasser, M. L. (1958). Fourier coefficients of crystal potentials. *Phys. Rev.* **112**, 73.
Callaway, J., and Hughes, A. J. (1962). Moment singularity expansion for the density of states. *Phys. Rev.* **128**, 134.
Callaway, J., and Kohn, W. (1962). Electron wave functions in metallic lithium. *Phys. Rev.* **127**, 1913.
Callaway, J., and Morgan, D. F. Jr. (1958). Cohesive energy and wave functions for rubidium. *Phys. Rev.* **112**, 334.
Callaway, J., Woods, R. D., and Sirounian, V. (1957). Relativistic effects in cohesive energies of alkali metals. *Phys. Rev.* **107**, 934.
Cardona, M., and Sommers, H. S., Jr. (1961). Effect of temperature and doping on the reflectivity of germanium in the fundamental absorption region. *Phys. Rev.* **122**, 1382.
Carr, W. J. (1961). Energy, specific heat, and magnetic properties of a low density electron gas. *Phys. Rev.* **122**, 1437.
Carr, W. J., Coldwell-Horsfall, R. A., and Fein, A. E. (1961). Anharmonic contribution to the energy of a dilute electron gas — Interpolation for the correlation energy. *Phys. Rev.* **124**, 747.
Carter, J. L. (1953). Thesis, Cornell University (unpublished).
Chambers, R. G. (1960). Magneto-resistance. *In* "The Fermi Surface" (W. A. Harrison and M. B. Webb, eds.), p. 100. Wiley, New York.
Cheng, C. H., Wei, C. T., and Beck, P. A. (1960). Low-temperature specific heat of bcc alloys of $3d$ transition elements. *Phys. Rev.* **120**, 426.
Chodorow, M. I. (1939). The band structure of metallic copper. *Phys. Rev.* **55**, 675.
Clogston, A. M. (1962). Impurity states in metals. *Phys. Rev.* **125**, 439.
Cohen, M. H. (1958). Optical constants, heat capacity, and the Fermi surface. *Phil. Mag.* **3**, 762.
Cohen, M. H. (1960). *In* "The Fermi Surface" (W. A. Harrison and M. B. Webb, eds.), p. 176. Wiley, New York.
Cohen, M. H. (1961). Energy bands in the bismuth structure. I. A non-ellipsoidal model for electrons in Bi. *Phys. Rev.* **121**, 387.
Cohen, M. H., and Blount, E. I. (1960). The g factor and de Haas-van Alphen effect of electrons in bismuth. *Phil. Mag.* **5**, 115.
Cohen, M. H., and Falicov, L. M. (1961). Magnetic breakdown in crystals. *Phys. Rev. Letters* **7**, 231.
Cohen, M. H., and Ham, F. S. (1960). Electron effective mass in solids — a generalization of Bardeen's formula. *J. Phys. Chem. Solids* **16**, 177.
Cohen, M. H., and Heine, V. (1958). Electronic band structures of the alkali metals and of the noble metals and their α phase alloys. *Advances in Phys.* **7**, 395.

Cohen, M. H., and Heine, V. (1961). Calcellation of kinetic and potential energy in atoms, molecules, and solids. *Phys. Rev.* **122**, 1821.

Cohen, M. H., Goodings, D. A., and Heine, V. A. (1959). Contribution of core polarization to the atomic hyperfine structure and Knight shift of Li and Na. *Proc. Phys. Soc. (London)* **73**, 811.

Coldwell-Horsfall, R., and Maradudin, A. A. (1960). Zero-point energy of an electron lattice. *J. Math. Phys.* **1**, 395.

Cooper, M. J. (1962). The electron distribution in chromium. *Phil. Mag.* **7**, 2059.

Corak, W. S., Garfunkel, M. P., Satterthwaite, C. B., and Wexler, A. (1955). Atomic heats of copper, silver, and gold from 1°K to 5°K. *Phys. Rev.* **98**, 1699.

Corbato, F. J. (1956). A calculation of the energy bands of the graphite crystal by means of the tight-binding method. Ph. D. Thesis, Mass. Inst. Technol. (unpublished).

Cornwell, J. (1961a). The electronic energy bands of the alkali metals metallic beryllium. *Proc. Roy. Soc.* **261**, 551.

Cornwell, J. (1961b). The Fermi surfaces of the noble metals. *Phil. Mag.* **6**, 727.

Crawford, M., and Schawlow, A. (1949). Electron-nuclear potential fields from hyperfine structure. *Phys. Rev.* **76**, 1310.

Crisp, R. S., and Williams, S. E. (1960). The K emission spectrum of metallic lithium. *Phil. Mag.* **5**, 525.

Dash, W. C., and Newman, R. (1955). Intrinsic optical absorption in single-crystal Ge and Si at 77°K and 300°K. *Phys. Rev.* **99**, 1151.

des Cloizeau, J. (1963). Orthogonal orbitals and generalized Wannier functions. *Phys. Rev.* **129**, 554.

Dresselhaus, G. (1954). Ph. D. Thesis, Univ. of California, Berkeley (unpublished).

Dresselhaus, G. (1955). Spin-orbit coupling effects in Zn blende structures. *Phys. Rev.* **100**, 580.

Dresselhaus, G., Kip, A. F., and Kittel, C. (1955). Cyclotron resonance of electrons and holes in Si and Ge crystals. *Phys. Rev.* **98**, 368.

Dresselhaus, G., Kip, A. F., Ku, Han-Ying, Wagoner, G., and Christian, S. M. (1955b). Cyclotron resonance in Ge-Si alloys. *Phys. Rev.* **100**, 1218.

Drude, P. (1902). "The Theory of Optics." Longmans, Green, New York.

DuBois, D. F. (1959). Electron interactions, II. Properties of a dense electron gas. *Ann. Phys. (N.Y.)* **8**, 24.

Easterling, V. J., and Bohm, H. V. (1962). Magneto-acoustic measurements in silver at 230 Mc/sec and 4.2°K. *Phys. Rev.* **125**, 812.

Ehrenreich, H. (1960). Band structure and electron transport of GaAs. *Phys. Rev.* **120**, 1951.

Ehrenreich, H. (1961). Band structure and transport properties of some 3–5 compounds. *J. Appl. Phys.* **32**, 2155.

Ehrenreich, H., and Philipp, H. R. (1962). Optical properties of Ag and Cu. *Phys. Rev.* **128**, 1622.

Ehrenreich, H., Philipp, H. R., and Phillips, J. C. (1962). Interband transitions in groups 4, 3–5, and 2–6 semiconductors. *Phys. Rev. Letters* **8**, 59.

Elliott, R. J. (1954). Spin orbit coupling in band theory: Character tables for some double space groups. *Phys. Rev.* **96**, 280.

Enz, C. P. (1960). Theory of the magnetic susceptibility of crystals. *Helv. Phys. Acta* **33**, 89.

Erdelyi, A. (1954). "Table of Integral Transforms," Vol. I. McGraw-Hill, New York.

Everett, G. E. (1962). Cyclotron resonance measurements in bismuth. *Phys. Rev.* **128**, 2564.

Evtuhov, V. (1962). Valence bands of germanium and silicon in an external magnetic field. *Phys. Rev.* **125**, 1869.

Eyges, L. (1961). Solution of Schrödinger equation for a periodic lattice, I. *Phys. Rev.* **123**, 1673.

Eyges, L. (1962). Solution of Schrödinger equation for a periodic lattice, II. *Phys. Rev.* **126**, 93.

Fawcett, E. (1960). Cyclotron resonance in aluminum. Anomalous skin effect. *In* "The Fermi Surface" (W. A. Harrison and M. B. Webb, eds.), pp. 166 and 197. Wiley, New York.

Fawcett, E., and Reed, W. A. (1962). Multiple connectivity of the Fermi surface of nickel from its magnetoresistance anisotropy. *Phys. Rev. Letters* **9**, 336.

Feher, G. (1959). Electron spin resonance experiments on donors in Si. I. Electronic structure of donors by electron nuclear double resonance. *Phys. Rev.* **114**, 1219.

Fermi, E. (1930). Magnetic moments of atomic nuclei. *Z. Physik* **60**, 320.

Fletcher, G. C. (1952). Density of states curve for the $3d$ electrons in Nickel. *Proc. Phys. Soc. (London)* **A65**, 192.

Fletcher, G. C., and Wohlfarth, E. P. (1951). Calculation of the density of states curve for $3d$ electrons in nickel. *Phil. Mag.* [7] **42**, 106.

Flower, M., March, N. H., and Murray, A. M. (1960). Metallic transitions in ionic crystals: Some group theoretical results. *Phys. Rev.* **119**, 1885.

Fock, V. (1930a). Näherungsmethode zur Lösung des quantenmechanischen Mehrkörperproblems. *Z. Physik* **61**, 126.

Fock, V. (1930b). "Selfconsistent field" mit Austausch für Natrium. *Z. Physik* **62**, 795.

Foner, S., and Pugh, E. M. (1953). Hall effects of the cobalt nickel alloys and of Armco iron. *Phys. Rev.* **91**, 20.

Foner, S. (1957). Hall effect in titanium, vanadium, chromium, and manganese. *Phys. Rev.* **107**, 1513.

Franz, W. (1958a). The Zener effect and impact ionization, *In* "International Conference on Semiconductors," Garmisch-Partenkirchen, 1956, p. 317. New York. Interscience.

Franz, W. (1958b). Influence of an electric field on an optical absorption edge. *Z. Naturforsch.* **13**, 484.

Fredkin, D. R., and Wannier, G. H. (1962). Theory of electron tunneling in semiconductor junctions. *Phys. Rev.* **128**, 2054.

Freeman, A. J., and Watson, R. E. (1960). Contribution of the Fermi contact term to the magnetic field at the nucleus in ferromagnets. *Phys. Rev. Letters* **5**, 498.

Friedel, J. (1952). The Absorption of light by noble metals and its relation to the van der Waals contribution to the cohesive energy. *Proc. Phys. Soc. (London)* **B65**, 769.

Friedel, J. (1958). Metallic alloys. *Nuovo cimento* [10] **7**, Suppl., 287.

Froese, C. (1957). The limiting behavior of atomic wave functions for large atomic number. *Proc. Roy. Soc.* **A239**, 311.

Fröhlich, H. (1937). Quantum mechanical discussion of the cohesive forces and thermal expansion coefficients of the alkali metals. *Proc. Roy. Soc.* **A158**, 97.

Fuchs, K. (1935). Quantum mechanical investigation of the cohesive forces of metallic copper. *Proc. Roy. Soc.* **A151**, 585.

Fuchs, K. (1936). Quantum mechanical calculation of the elastic constants of monovalent metals. *Proc. Roy. Soc.* **A153**, 622.

Fukuchi, M. (1956). The energy band structure of metallic copper. *Progr. Theoret. Phys. (Kyoto)* **16**, 222.

Galt, J. K., Yager, W. A., Merritt, F. R., Celtin, B. B., and Brailsford, D. A. (1959). Cyclotron absorption in metallic Bi and its alloys. *Phys. Rev.* **114**, 1396.

Garcia-Moliner, F. (1958). On the Fermi surface of copper. *Phil. Mag.* [8] **3**, 207.

Gashimzade, F. M., and Khartsiev, V. E. (1961). Energy structure of complex semiconductors: calculation of the band structures of Si, Ge, and GaAs by a simplified orthogonalized plane wave method. *Soviet Physics, Solid State* **3**, 1054.

Gaspar, R. (1952). The binding of metallic aluminum. *Acta Phys. Acad. Sci. Hung.* **2**, 31.

Gell-Mann, M., and Bruckner, K. (1957). Correlation of electron gas at high density. *Phys. Rev.* **106**, 364.

Glicksman, M., and Christian, S. M. (1956). Conduction band structure of Ge-Si alloys. *Phys. Rev.* **104**, 1278.

Goertzel, G., and Tralli, N. (1960). "Some Mathematical Methods of Physics," p. 165. McGraw-Hill, New York.

Goldberger, M. L., and Adams, E. N. (1952). The configurational distribution function in quantum-statistical mechanics. *J. Chem. Phys.* **20**, 240.

Goodenough, J. B. (1960). Band structure of transition metals and their alloys. *Phys. Rev.* **120**, 67.

Goodings, D. A., and Heine, V. (1960). Contribution of exchange polarization of core electrons to the magnetic field at the nucleus of Fe. *Phys. Rev. Letters* **5**, 370.

Goodman, R. R. (1961). Cyclotron resonance in germanium. *Phys. Rev.* **122**, 397.

Gorin, E. (1936). The theoretical constitution of metallic potassium. *Physik. Z. Sowjetunion* **9**, 328.

Greene, J. B., and Manning, M. F. (1943). Electronic energy bands in face centered iron. *Phys. Rev.* **63**, 203.

Grimes, C. C. (1962). Cyclotron resonance in sodium and potassium. Ph. D. Thesis, University of California, Berkeley (unpublished).

Gubanov, A. I., and Nran'yan, A. A. (1960). Application of the equivalent orbital method to the study of the band structure in A(III)B(V) compounds. *Soviet Physics, Solid State* **1**, 956.

Gunnersen, E. M. (1957). The de Haas-van Alphen effect in aluminium. *Phil. Trans. Roy. Soc.* **A249**, 299.

Hagstrum, H. D. (1961). Theory of Auger neutralization of ions at the surface of a diamond type semiconductor. *Phys. Rev.* **122**, 83.

Hall, G. G. (1952). The electronic structure of diamond. *Phil. Mag.* [7] **43**, 338.

Ham, F. S. (1954). Electronic energy bands in metals. Ph. D. Thesis, Harvard Univ. (unpublished).

Ham, F. S. (1955). The quantum defect method. *Solid State Phys.* **1**, 127.

Ham, F. S. (1957). Expansions of the irregular Coulomb function. *Quart. Appl. Math.* **15**, 31.

Ham, F. S. (1960). *In* "The Fermi Surface" (W. A. Harrison and M. B. Webb, eds.), p. 9. Wiley, New York.

Ham, F. S. (1962a). Energy bands of alkali metals. I. Calculated bands. *Phys. Rev.* **128**, 82.

Ham, F. S. (1962b). Energy bands of alkali metals. II. Fermi surface. *Phys. Rev.* **128**, 2524.

Ham, F. S., and Segall, B. (1961). Energy bands in periodic lattices, Green's function method. *Phys. Rev.* **124**, 1786.

Hanna, S. S., Heberle, J., Perlow, G. J., Preston, R. S., and Vincent, D. H. (1960). Direction of the effective magnetic field at the nucleus in ferromagnetic iron. *Phys. Rev. Letters* **4**, 513.

Harrison, W. A. (1959). Fermi surface in Al. *Phys. Rev.* **116**, 555.

Harrison, W. (1960a). Band structure of aluminum. *Phys. Rev.* **118**, 1182.

Harrison, W. (1960b). Electronic structure of polyvalent metals. *Phys. Rev.* **118**, 1190.

Harrison, W. (1960c). Bismuth Fermi surface. *J. Phys. Chem. Solids* **17**, 171.

Hartree, D. R. (1957). "The Calculation of Atomic Structures." Wiley, New York.

Hartree, D. R. (1958). Representation of exchange terms in Fock's equations by a quasi-potential. *Phys. Rev.* **109**, 840.

Hartree, D. R., and Hartree, W. (1938). Wave functions for negative ions of sodium and potassium. *Proc. Cam. Phil. Soc.* **34**, 550.

Hebborn, J. E. and Sondheimer, E. H. (1960). The diamagnetism of conduction electrons in metals. *J. Phys. Chem. Solids* **13**, 105.

Heine, V. (1957a). The band structure of aluminum, I. *Proc. Roy. Soc.* **A240**, 340.

Heine, V. (1957b). The band structure of aluminum. II. *Proc. Roy. Soc.* **A240**, 354.

Heine, V. (1957c). The band structure of aluminum. III. *Proc. Roy. Soc.* **A240**, 363.

Heine, V. (1957d). Hfs of paramagnetic ions. *Phys. Rev.* **107**, 1002.

Hellmann, H., and Kassatotschkin, W. (1936a). Metallic binding according to the combined approximation procedure. *J. Chem. Phys.* **4**, 324.

Hellmann, H., and Kassatotschkin, W. (1936b). Metallic bond, according to the combined approximation method. *Acta Physicochim.* **5**, 23.

Hensel, J. C., and Feher, G. (1963). Cyclotron resonance experiments in uniaxially stressed silicon: valence band inverse mass parameters and deformation potentials. *Phys. Rev.* **129**, 1041.

Herman, F. (1952). Electronic structure of the diamond crystal. *Phys. Rev.* **88**, 1210.

Herman, F. (1954a). Speculations on energy band structure of Ge-Si alloys. *Phys. Rev.* **95**, 847.

Herman, F. (1954b). Some recent developments in the calculation of crystal energy bands — New results for the germanium crystal. *Physica* **20**, 801.

Herman, F. (1954c). Calculation of energy band structures of diamond and Ge crystals by method of orthogonalized plane waves. *Phys. Rev.* **93**, 1214.

Herman, F. (1955). Speculations on the energy band structure of zinc-blende type crystals. *J. Electronics* **1**, 103.

Herman, F. (1958). Energy band structure of solids. *Revs. Modern Phys.* **30**, 102.

Herman, F., and Callaway, J. (1953). Electronic structure of the germanium crystal. *Phys. Rev.* **89**, 518.

Herman, F., and Skillman, S. (1960). Theoretical investigation of the energy band structure of semiconductors. *Proc. Intern. Conf. on Semiconductor Phys., Prague, 1960* (Publishing House of the Czechoslovak Acad. Sci.).

Herman, F., Callaway, J., and Acton, F. S. (1954). Comparison of various approximate exchange potentials. *Phys. Rev.* **95**, 371.

Herring, C. (1937a). Effect of time reversal symmetry on the energy bands of crystals. *Phys. Rev.* **52**, 361.

Herring, C. (1937b). Accidental degeneracy in the energy bands of crystals. *Phys. Rev.* **52**, 365.

Herring, C. (1940). A new method for calculating wave functions in crystals. *Phys. Rev.* **57**, 1169.

Herring, C. (1942). Character tables for two space groups. *J. Franklin Inst.* **233**, 525.

Holmes, D. K. (1952). An application of the cellular method to silicon. *Phys. Rev.* **87**, 782.

Houston, M. V. (1940). Acceleration of electrons in a crystal lattice. *Phys. Rev.* **57**, 184.

Howarth, D. J. (1953). Electronic eigenvalues of copper. *Proc. Roy. Soc.* **A220**, 513.

Howarth, D. J. (1955). Application of the augmented plane wave method to copper. *Phys. Rev.* **99**, 469.

Howarth, D. J., and Jones, H. (1952). The cellular method of determining electronic wave functions and eigenvalues in crystals with applications to sodium. *Proc. Phys. Soc. (London)* **A65**, 355.

Howling, D. H., Mendoza, E., and Zimmerman, J. E. (1955). Preliminary experiments on the temperature wave method of measuring specific heats of metals at low temperatures. *Proc. Roy. Soc.* **A229**, 86.

Hsu, Y. C. (1951). Relation of antiferromagnetic structure of the binding energies of some bcc transition metals. *Phys. Rev.* **83**, 975.

Hubbard, J. (1958). The description of collective motion in terms of many body perturbation theory, II. The correlation energy of a free electron gas. *Proc. Roy. Soc.* **A243**, 336.

Hund, F., and Mrowka, B. (1935a). Electronic states in a crystal lattice, particularly Diamond. *Ber. Verhandl. sachs. Akad. Wiss. (Leipzig), Math.-naturwiss. Kl.* **87**, 185.

Hund, F., and Mrowka, B. (1935b). State of electrons in crystal lattices. *Physik. Z.* **36**, 888.

Ives, H. E., and Briggs, H. B. (1936). Optical constants of potassium. *J. Opt. Soc. Am.* **26**, 238.

Ives, H. E., and Briggs, H. B. (1937a). Optical constants of sodium. *J. Opt. Soc. Am.* **27**, 181.

Ives, H. E., and Briggs, H. B. (1937b). Optical constants of rubidium and cesium. *J. Opt. Soc. Am.* **27**, 395.

Jain, A. L., and Koenig, S. H. (1962). On electrons and holes in bismuth. *Phys. Rev.* **127**, 442.

Jeffreys, H., and Jeffreys, B. S. (1950). "Methods of Mathematical Physics." Cambridge Univ. Press, London and New York.

Jenkins, D. P. (1956). Calculations on the band structure of silicon. *Proc. Phys. Soc.* (*London*) **A69**, 548.

Johnson, E., and Christian, S. (1954). Properties of Ge-Si alloys. *Phys. Rev.* **95**, 560.

Johnston, D. F. (1955). The structure of the π-band of graphite. *Proc. Roy. Soc.* **A227**, 349.

Johnston, D. F. (1956). The effect of the mixing of π- and σ-orbitals on the Fermi surface in the l.c.a.o. model for graphite. *Proc. Roy. Soc.* **A237**, 48.

Jones, H. (1960). "The Theory of Brillouin Zones and Electronic States in Crystals." North-Holland Publ. Co., Amsterdam.

Jones, W. H., Graham, T. P., and Barnes, R. G. (1960). Knight shifts in potassium, indium, and yttrium metals. *Acta Metallurgica* **8**, 663.

Kahn, A. H. (1955). Theory of infrared absorption of carriers in Ge and Si. *Phys. Rev.* **97**, 1647.

Kalinkina, I. N., and Strelkov, P. G. (1958). Specific heat of bismuth between 0.3 and 4.4°K. *Soviet Phys. JETP* **7**, 426.

Kambe, K. (1955). Cohesive energy of noble metals. *Phys. Rev.* **99**, 419.

Kane, E. O. (1956a). Energy band structure in p-type germanium and silicon. *J. Phys. Chem. Solids* **1**, 83.

Kane, E. O. (1956b). Band structure of indium antimonide. *J. Phys. Chem. Solids* **1**, 249.

Kane, E. O. (1959a). The semi-empirical approach to band structure. *J. Phys. Chem. Solids* **8**, 38.

Kane, E. O. (1959b). Zener tunnelling in semiconductors. *J. Phys. Chem. Solids* **12**, 181.

Keldysh, L. V. (1958). Behavior of nonmetallic crystals in strong electric fields. *Soviet Phys. JETP* **6**, 763.

Keldysh, L. V. (1958b). The effect of a strong electric field on the optical properties of insulating crystals. *Soviet Phys. JETP* **7**, 788.

Kern, B. (1960). Die SiK β-Banden der Röntgenemissionsspektren von elementarem Silicium, Silicium Carbid, und Siliciumdioxyd. *Z. Physik* **159**, 178.

Kimball, G. E. (1935). The electronic structure of diamond. *J. Chem. Phys.* **3**, 560.

Kip, A. F. (1960). Cyclotron resonance in metals — Experimental. *In* "The Fermi Surface" (W. A. Harrison and M. B. Webb, eds.), p. 146. Wiley, New York.

Kip, A. F., Langenberg, D. N., and Moore, T. W. (1961). Cyclotron resonance in copper. *Phys. Rev.* **124**, 359.

Kittel, C., and Mitchell, A. H. (1954). Theory of donor and acceptor states in Si and Ge. *Phys. Rev.* **96**, 1488.

Kjeldaas, T., and Kohn, W. (1956). Interaction of conduction electrons and nuclear magnetic moments in metallic Na. *Phys. Rev.* **101**, 66.

Kjeldaas, T., and Kohn, W. (1957). Theory of the diamagnetism of Bloch electrons. *Phys. Rev.* **105**, 806.

Klauder, J. R., and Kunzler, J. E. (1960). Higher order open orbits and the interpretation of magnetoresistance and Hall effect data for copper. *In* "The Fermi Surface" (W. A. Harrison and M. B. Webb, eds.), p. 125. Wiley, New York.

Klein, A. (1959). Many-particle approach to one-electron problem in insulators and semiconductors. *Phys. Rev.* **115**, 1136.

Kleinman, L. (1962). Deformation potentials in silicon. I. Uniaxial strain. *Phys. Rev.* **128**, 2614.

Kleinman, L., and Phillips, J. C. (1959). Crystal potential and energy bands of semiconductors. I. Self-consistent calculations for diamond. *Phys. Rev.* **116**, 880.

Kleinman, L., and Phillips, J. C. (1960a). Crystal potential and energy bands of semiconductors. II. Self-consistent calculations for cubic BN. *Phys. Rev.* **117**, 460.

Kleinman, L., and Phillips, J. C. (1960b). Crystal potential and energy bands of semiconductors. III. Self-consistent calculations for Si. *Phys. Rev.* **118**, 1153.

Kleinman, L., and Phillips, J. C. (1962). Covalent bonding and charge density in diamond. *Phys. Rev.* **125**, 819.

Knight, W. D. (1956). Electron paramagnetism and nuclear magnetic resonance in Metals. *Solid State Phys.* **2**, 93.

Kobayasi, S. (1958). Structure of β-SiC by the orthogonalized plane wave method. *J. Phys. Soc. Japan* **13**, 261.

Kohn, W. (1952). Variational methods for periodic lattices. *Phys. Rev.* **87**, 472.

Kohn, W. (1954). Interaction of conduction electrons and nuclear magnetic moments in metallic Li. *Phys. Rev.* **96**, 590.

Kohn, W. (1957). Effective-mass theory in solids from many-particle standpoint. *Phys. Rev.* **105**, 509.

Kohn, W. (1958). Interaction of charged particles in dielectric. *Phys. Rev.* **110**, 857.

Kohn, W. (1959). Analytic properties of Bloch waves and Wannier functions. *Phys. Rev.* **115**, 809.

Kohn, W., and Nettel, S. J. (1960). Giant fluctuations in a degenerate Fermi gas. *Phys. Rev. Letters* **5**, 8.

Kohn, W., and Rostoker, N. (1954). Solution of the Schrödinger equation in periodic lattices with application to metallic lithium. *Phys. Rev.* **94**, 1411.

Kohn, W., and Schechter, D. (1955). Theory of acceptor levels in germanium. *Phys. Rev.* **99**, 1903.

Koopmans, T. (1934). Über die Zuordnung von Wellenfunktionen und Eigenwerten zu den einzelnen Elektronen eines Atoms. *Physica* **1**, 104.

Korringa, J. (1947). On the calculation of the energy of a Bloch wave in a metal. *Physica* **13**, 392.

Koster, G. F. (1953). Localized functions in molecules and crystals. *Phys. Rev.* **89**, 67.

Koster, G. F. (1954). Theory of scattering in solids. *Phys. Rev.* **95**, 1436.

Koster, G. F. (1955). Density of states curve for Ni. *Phys. Rev.* **98**, 901.

Koster, G. F. (1957). Space groups and their representations. *Solid State Phys.* **5**, 173.

Koster, G. F., and Slater, J. C. (1954a). Wave functions for impurity levels. *Phys. Rev.* **95**, 1167.

Koster, G. F., and Slater, J. C. (1954b). Simplified impurity calculation. *Phys. Rev.* **96**, 1208.

Krutter, H. M. (1935). The energy bands of copper. *Phys. Rev.* **48**, 664.

Kuhn, T. S. (1951). A general solution of the confluent hypergeometric equation: Analytic and numerical development. *Quart. Appl. Math.* **9**, 1.

Kuhn, T. S., and Van Vleck, J. H. (1950). A simplified method of computing the cohesive energies of monovalent metals. *Phys. Rev.* **79**, 382.

Lampert, M. (1955). Ground state of impurity atoms in semiconductors having anisotropic energy surfaces. *Phys. Rev.* **97**, 352.

Landau, L. (1930). The diamagnetism of metals. *Z. Physik* **64**, 629.

Langenberg, D. N., and Moore, T. W. (1959). Cyclotron resonance in aluminum. *Phys. Rev. Letters* **3**, 137.

Lax, B. (1958). Experimental investigations of the electronic band structure of solids. *Revs. Modern Phys.* **30**, 122.

Lax, M. (1954). Localized perturbations. *Phys. Rev.* **94**, 1391.

Leigh, R. S. (1956). The augmented plane wave and related methods for crystal eigenvalue problems. *Proc. Phys. Soc. (London)* **A69**, 388.

Lerner, L. S. (1962). The Shubnikov-de Haas effect in bismuth. *Phys. Rev.* **127**, 1480.

Levinger, B., and Frankl, D. (1961). Cyclotron resonance measurement of the energy band parameters of germanium. *J. Phys. Chem. Solids* **20**, 281.

Lien, W. H., and Phillips, N. E. (1960). The electronic heat capacities of potassium, rubidium, and cesium. *In* "Proceedings of the 7th International Conference on Low Temperature Physics," p. 675. Univ. of Toronto Press, Toronto.

Lifshitz, I. M. (1956). Some problems of the dynamic theory of nonideal crystal lattices. *Nuovo Cimento* **3**, *Suppl.*, 716.

Lifshitz, I. M., and Kosevich, A. M. (1956). Theory of magnetic susceptibility in metals at low temperatures. *Soviet Phys. JETP* **2**, 636.

Liu, L. (1961). Valence spin-orbit splitting and conduction g tensor in silicon. *Phys. Rev. Letters* **6**, 683.

Liu, L. (1962). Effect of spin orbit coupling in Si and Ge. *Phys. Rev.* **126**, 1317.

Löwdin, P. (1955). Quantum theory of many-particle Systems. III. Extension of Hartree-Fock scheme to include degenerate systems and correlation effects. *Phys. Rev.* **97**, 1509.

Lomer, W. M. (1955). The valence bands in two-dimensional graphite. *Proc. Roy. Soc.* **A227**, 330.

Lomer, W. M. (1962). Electronic structure of chromium group metals. *Proc. Phys. Soc. (London)* **80**, 489.

Lomer, W. M., and Marshall, W. (1958). The electronic structure of the metals of the first transition period. *Phil. Mag.* [8] **3**, 185.

Luttinger, J. M. (1956). Quantum theory of cyclotron resonance in semi-conductors: General theory. *Phys. Rev.* **102**, 1030.

Luttinger, J. M., and Goodman, R. R. (1955). Classical theory of cyclotron resonance for holes in Ge. *Phys. Rev.* **100**, 673.

Luttinger, J. M., and Kohn, W. (1955). Motion of electrons and holes in perturbed periodic fields. *Phys. Rev.* **97**, 869.

Luttinger, J. M., and Ward, J. C. (1960). Ground state energy of a many fermion system. *Phys. Rev.* **118**, 1417.

McAfee, K. B., Ryder, E. J., Shockley, W., and Sparks, M. (1951). Observations of Zener current in germanium $p - n$ junctions. *Phys. Rev.* **83**, 650.

McClure, J. W. (1957). Band structure of graphite and de Haas-van Alphen effect. *Phys. Rev.* **108**, 612.

McClure, J. W. (1960). Theory of diamagnetism of graphite. *Phys. Rev.* **119**, 606.

McDougall, J. (1932). Calculation of the terms of the optical spectrum of an atom with one series electron. *Proc. Roy. Soc.* **A138**, 550.

Macfarlane, G. G., and Roberts, V. (1955). Infrared absorption of germanium near the lattice edge. *Phys. Rev.* **97**, 1714.

Macfarlane, G. G., McLean, T. P., Quarrington, J. E., and Roberts, V. (1957). Fine structure in absorption-edge spectrum of Ge. *Phys. Rev.* **108**, 1377.

Macfarlane, G. G., McLean, T. P., Quarrington, J. E., and Roberts, V. (1958). Fine structure in absorption edge spectrum of Si. *Phys. Rev.* **111**, 1245.

Manning, M. F. (1943). Electronic energy bands in body centered iron. *Phys. Rev.* **63**, 190.

Manning, M. F., and Goldberg, L. (1938). Self-consistent field for iron. *Phys. Rev.* **53**, 662.

Marshall, W. (1958). Orientation of nuclei in ferromagnets. *Phys. Rev.* **110**, 1280.

Martin, D. L. (1961a). Electronic specific heat of lithium isotopes. *Proc. Roy. Soc.* **A263**, 378.

Martin, D. L. (1961b). Specific heat of sodium at low temperatures. *Phys. Rev.* **124**, 438.

Mase, S. (1958). Electronic structure of bismuth type crystals. *J. Phys. Soc. Japan* **13**, 434.

Maslen, V. (1956). The Fermi, or exchange, hole in atoms. *Proc. Phys. Soc. (London)* **A69**, 734.

Matyas, Z. (1948). The energies of electrons in aluminium. *Phil. Mag.* [7] **39**, 429.

Miasek, M. (1957). Tight-binding method for hexagonal close-packed structure. *Phys. Rev.* **107**, 92.

Milford, F. J., and Gager, W. B. (1961). Knight shift in K. *Phys. Rev.* **121**, 716.

Moffitt, W., and Ballhausen, C. J. (1956). Quantum theory. *Ann. Rev. Phys. Chem.* **7**, 107.

Montroll, E. W. (1954). Frequency spectrum of vibrations of a crystal lattice. *Am. Math. Monthly* **61**, 46.

Moore, T. W., and Spong, F. W. (1962). Cyclotron resonance in aluminum. *Phys. Rev.* **125**, 846.

Morita, A. (1949a). Electronic structure of diamond crystal. *Sci. Repts. Tohoku Univ., First Ser.* **33**, 92.

Morita, A. (1949b). Electronic structure of bismuth crystal. *Sci. Repts. Tohoku Univ., First Ser.* **33**, 144.

Morita, A. (1958). Theory of cohesive energies and structures of diamond type valence crystals. *Progr. Theoret. Phys. (Kyoto)* **19**, 534.

Morse, M. (1938). "Functional Topology and Abstract Variational Theory," Mem. sci. math., Fasc. 92. Gauthier-Villars, Paris.

Morse, R. W. (1960). The Fermi surfaces of the noble metals by ultrasonics. *In* "The Fermi Surface" (W. A. Harrison and M. B. Webb, eds.), p. 214. Wiley, New York.

Morse, R. W., Myers, A., and Walker, C. T. (1961). On the Fermi surface shapes of the noble metals by ultrasonics. *J. Acoust. Soc. Am.* **33**, 699.

Moss, T. S. (1961). Optical absorption edge in GaAs and its dependence on electric field. *J. Appl. Phys.* **32**, 2136.

Moss, T. S., and Walton, A. K. (1959). Measurement of effective mass of electrons in InP by infra-red Faraday effect. *Physica* **25**, 1142.

Mott, N. F. (1935). Discussion of the transition metals on the basis of quantum mechanics. *Proc. Phys. Soc. (London)* **47**, 571.

Mott, N. F. (1949). The basis of the electron theory of metals, with special reference to the transition metals. *Proc. Phys. Soc. (London)* **A62**, 416.

Mott, N. F. (1953). Note on the electronic structure of the transition metals. *Phil. Mag.* [7] **44**, 187.

Mott, N. F., and Stevens, K. W. H. (1957). The band structure of the transition metals. *Phil. Mag.* [8] **2**, 1364.

Mullaney, J. F. (1944). Optical properties and electronic structure of solid silicon. *Phys. Rev.* **66**, 326.

Murnaghan, F. D. (1938). "Theory of Group Representations." Johns Hopkins Press, Baltimore, Maryland.

Nozieres, P. (1958). Cyclotron resonance in graphite. *Phys. Rev.* **109**, 1510.

Nozieres, P., and Pines, D. (1958). Correlation energy of a free electron gas. *Phys. Rev.* **111**, 442.

Nran'yan, A. A. (1960). Computing the integrals in the method of equivalent orbitals and evaluating the valence band parameters for semiconductors of the A(III)B(V) type. *Soviet Physics, Solid State* **2**, 439.

Nran'yan, A. A. (1961). Concerning the zone structure of diamond type crystals. *Soviet Physics, Solid State* **2**, 1494.

Onsager, L. (1952). Interpretation of the de Haas-van Alphen effect. *Phil. Mag.* [7] **43**, 1006.

Overhauser, A. (1962). Spin-density waves in an electron gas. *Phys. Rev.* **128**, 1437.
Parmenter, R. H. (1955). Energy levels of crystal modified by alloying or by pressure. *Phys. Rev.* **100**, 573.
Parzen, G. (1953). Electronic energy bands in metals. *Phys. Rev.* **89**, 237.
Paul, W. (1959). The effect of pressure on the electrical properties of germanium and silicon. *J. Phys. Chem. Solids* **8**, 196.
Pauling, L. (1938). The nature of the interatomic forces in metals. *Phys. Rev.* **54**, 899.
Pauling, L. (1953). A theory of ferromagnetism. *Proc. Natl. Acad. Sci. U.S.* **39**, 551.
Peierls, R. (1933a). The diamagnetism of conduction electrons. *Z. Physik* **80**, 763.
Peierls, R. (1933b). The diamagnetism of conduction electrons; Strong fields. *Z. Physik* **81**, 186.
Philipp, H. R., and Taft, E. A. (1959). Optical constants of Ge in region 1 to 10 ev. *Phys. Rev.* **113**, 1002.
Philipp, H. R., and Taft, E. A. (1960). Optical constants of Si in region 1 to 10 ev. *Phys. Rev.* **120**, 37.
Phillips, J. C. (1956). Critical points and lattice vibration spectra. *Phys. Rev.* **104**, 1263.
Phillips, J. C. (1958). Energy-band interpolation scheme based on pseudopotential. *Phys. Rev.* **112**, 685.
Phillips, J. C. (1961). Generalized Koopmans' theorem. *Phys. Rev.* **123**, 420.
Phillips, J. C. (1962). Band structure of silicon, germanium, and related semiconductors. *Phys. Rev.* **125**, 1931.
Phillips, J. C., and Kleinman, L. (1959). New method for calculating wave functions in crystals and molecules. *Phys. Rev.* **116**, 287.
Phillips, J. C., and Kleinman, L. (1962). Crystal potential and energy bands of semiconductors. IV. Exchange and correlation. *Phys. Rev.* **128**, 2098.
Phillips, N. E. (1960). Nuclear quadrupole and electronic heat capacities of bismuth. *Phys. Rev.* **118**, 644.
Pincherle, L. (1960). Band structure calculations in solids. *Repts. Progr. in Phys.* **23**, 355.
Pines, D. (1955). Electron interaction in metals. *Solid State Phys.* **1**, 367.
Pines, D. (1958). Electrons and plasmons. *Nuovo cimento* **7**, Suppl., 329.
Pippard, A. B. (1957). An experimental determination of the Fermi surface in copper. *Phil. Trans. Roy. Soc.* **A250**, 325.
Pippard, A. B. (1960). Experimental analysis of the electronic structure of metals. *Repts. Progr. in Phys.* **23**, 176.
Pippard, A. B. (1962). Quantization of coupled orbits in metals. *Proc. Roy. Soc. (London)* **270**, 1.
Pratt, G. W., (1956). Unrestricted Hartree-Fock method. *Phys. Rev.* **102**, 1303.
Pratt, G. W. (1957). Exchange and superexchange coupling between conduction electrons and d-electrons in magnetic materials. *Phys. Rev.* **106**, 53.
Price, P. J., and Radcliffe, J. M. (1959). Eskai tunnelling. *IBM J. Research Develop.* **3**, 364.

Prokofjew, W. K. (1929). Berechnung der Zahlen der Dispersionszentren des Natriums. *Z. Physik* **58**, 255.

Quelle, F. W. (1962). Energy band calculations in solids. *Bull. Am. Phys. Soc.* [2] **7**, 214.

Quinn, J. (1960). Electron-phonon interaction in normal metals. *In* "The Fermi Surface" (W. A. Harrison and M. B. Webb, eds.), p. 58. Wiley, New York.

Racah, G. (1931). Theory of hyperfine structure. *Z. Physik* **7**, 431.

Raimes, S. (1952). A calculation of the cohesive energies and pressure/volume relations of the divalent metals. *Phil. Mag.* [7] **43**, 327.

Raimes, S. (1953). Compressibility of metallic aluminum. *Proc. Phys. Soc. (London)* **A66**, 949.

Rauch, C. J., Stickler, J. J., Zeiger, H. J., and Heller, G. S. (1960). Millimeter cyclotron resonance in silicon. *Phys. Rev. Letters* **4**, 64.

Raynor, G. V. (1952). The band structure of metals. *Repts. Progr. in Phys.* **15**, 173.

Reeh, H. (1960). The polarization potential of electrons in the field of an atom or atom core. *Z. Naturforsch.* **15a**, 377.

Reitz, J. R. (1955). Methods of the one electron theory of solids. *Solid State Phys.* **1**, 1.

Roaf, D. J. (1962). Unpublished work quoted by Shoenberg (1962).

Rosenthal, J., and Breit, G. (1932). The isotope shift in hyperfine structure. *Phys. Rev.* **41**, 459.

Roth, L. M. (1960). g Factor and donor spin-lattice relaxation for electrons in germanium and silicon. *Phys. Rev.* **118**, 1534.

Roth, L. M. (1962). Theory of Bloch electrons in a magnetic field. *J. Phys. Chem. Solids* **23**, 433.

Roth, L. M., Lax, B., and Zwerdling, S. (1959). Theory of optical magneto-absorption effects in semiconductors. *Phys. Rev.* **114**, 90.

Ryter, C. (1960). Direct measurement of the electron density at the nucleus in metallic lithium at liquid helium temperatures. *Phys. Rev. Letters* **5**, 10.

Saffern, M. M., and Slater, J. C. (1953). An augmented plane wave method for the periodic potential problem. *Phys. Rev.* **92**, 1126.

Schiff, B. (1955). The cellular method for a close packed hexagonal lattice with applications to titanium. *Proc. Phys. Soc. (London)* **A68**, 686.

Schiff, B. (1956). The cellular method for close packed hexagonal titanium. *Proc. Phys. Soc. (London)* **A69**, 185.

Schiff, L. I. (1955). "Quantum Mechanics." McGraw-Hill, New York.

Schlosser, H. (1960). A new method for calculation of electronic energy bands with application to lithium and sodium. Ph. D. Thesis, Carnegie Inst. Technol. (unpublished).

Schlosser, H. (1962). Symmetrized combinations of plane waves and matrix elements of the Hamiltonian for cubic lattices. *J. Phys. Chem. Solids* **23**, 963.

Schmid, L. A. (1953). Calculation of the cohesive energy of diamond. *Phys. Rev.* **92**, 1373.

Schulz, L. G. (1957). The experimental study of the optical properties of metals and the relation of the results to the Drude free electron theory. *Advances in Phys.* **6**, 102.

Segall, B. (1958). Band structure calculations of semiconductors by the Kohn-Rostoker-Korringa method: Application to germanium. *J. Phys. Chem. Solids* **8**, 371.

Segall, B. (1961a). Calculation of the band structure of silver. *Bull. Am. Phys. Soc.* [2] **6**, 145.

Segall, B. (1961b). Energy bands of aluminum. *Phys. Rev.* **124**, 1797.

Segall, B. (1962). Fermi surface and energy bands of copper. *Phys. Rev.* **125**, 109.

Seitz, F. (1935). The theoretical constitution of metallic lithium. *Phys. Rev.* **47**, 400.

Seitz, F. (1940). "The Modern Theory of Solids." McGraw-Hill, New York.

Shakin, C., and Birman, J. L. (1958). Electronic energy bands in ZnS. Preliminary results. *Phys. Rev.* **109**, 818.

Shockley, W. (1937). The empty lattice test of the cellular method in solids. *Phys. Rev.* **52**, 866.

Shockley, W. (1950). Energy band structures in semiconductors. *Phys. Rev.* **78**, 173.

Shoenberg, D. (1960). The de Haas-van Alphen effect in copper, silver and gold. *Phil. Mag.* [8] **5**, 105.

Shoenberg, D. (1962). The de Haas-van Alphen effect and the electronic structure of metals. *Proc. Phys. Soc. (London)* **79**, 1.

Silverman, R. A. (1952). Fermi energy of metallic lithium. *Phys. Rev.* **85**, 227.

Silverman, R. A., and Kohn, W. (1950). On the cohesive energy of metallic lithium. *Phys. Rev.* **80**, 912.

Silverman, R. A., and Kohn, W. (1951). Erratum: On the cohesive energy of metallic lithium. *Phys. Rev.* **82**, 283.

Slater, J. C. (1934). Electronic energy bands in metals. *Phys. Rev.* **45**, 794.

Slater, J. C. (1936). The ferromagnetism of nickel. *Phys. Rev.* **49**, 537.

Slater, J. C. (1937). Wave functions in a periodic potential. *Phys. Rev.* **51**, 846.

Slater, J. C. (1949). Electrons in perturbed periodic lattices. *Phys. Rev.* **76**, 1592.

Slater, J. C. (1951a). A simplification of the Hartree-Fock method. *Phys. Rev.* **81**, 385.

Slater, J. C. (1951b). Magnetic effects and the Hartree-Fock equation. *Phys. Rev.* **82**, 538.

Slater, J. C. (1953a). An augmented plane wave method for the periodic potential problem. *Phys. Rev.* **92**, 603.

Slater, J. C. (1953b). Ferromagnetism and the band theory. *Revs. Modern Phys.* **25**, 199.

Slater, J. C. (1956). The electronic structure of solids. *In* "Handbuch der Physik," Vol. 19, p. 1. Springer, Berlin.

Slater, J. C., and Koster, G. F. (1954). Simplified LCAO method for the periodic potential problem. *Phys. Rev.* **58**, 1498.

Slonczewski, J. C., and Weiss, P. R. (1958). Band structure of graphite. *Phys. Rev.* **109**, 272.

Smith, G. E., Galt, J. K., and Merritt, F. R. (1960). Electron spin resonance in bismuth and antimony. *Phys. Rev. Letters* **4**, 276.

Sondheimer, E. H., and Wilson, A. H. (1951). The diamagnetism of free electrons. *Proc. Roy. Soc.* **A210**, 173.

Spicer, W. E., and Simon, R. E. (1962). Photoemissive studies of the band structure of silicon. *Phys. Rev. Letters* **9**, 385.

Spitzer, W. G., and Fan, H. Y. (1957). Determination of optical constants and carrier effective mass of semiconductors. *Phys. Rev.* **106**, 882.

Steinberger, J., and Wick, G. C. (1949). On the polarization of slow neutrons. *Phys. Rev.* **76**, 994.

Stern, F. (1956). Self-consistent field calculation for three configurations of atomic iron. *Phys. Rev.* **104**, 684.

Stern, F. (1959). Calculation of the cohesive energy of metallic iron. *Phys. Rev.* **116**, 1399.

Stern, F. (1961). A spherical $3d$ electron distribution in body centered cubic metals. *Phys. Rev. Letters* **6**, 675.

Stern, F., and Talley, R. M. (1957). Optical absorption in p-type InAs. *Phys. Rev.* **108**, 158.

Sternheimer, R. M. (1954). Electronic polarizabilities of ions from the Hartree-Fock wave functions. *Phys. Rev.* **96**, 951.

Stickler, J. J., Zeiger, H. D., and Heller, G. S. (1962). Quantum effects in Ge and Si, I. *Phys. Rev.* **127**, 1077.

Sturge, M. D. (1962). Optical absorption of gallium arsenide between 0.6 and 2.75 ev. *Phys. Rev.* **127**, 768.

Suffczynski, M. (1955). Energy integrals for $3d$ electrons in body-centered iron. *Acta Physica Polon.* **14**, 493.

Suffczynski, M. (1956a). Two-center integrals for body-centered iron using atomic functions with exchange. *Acta Physica Polon.* **15**, 111.

Suffczynski, M. (1956b). Two-center integrals in solids. *Acta Physica Polon.* **15**, 287.

Suffczynski, M. (1957). On the width of the $3d$ electron band in iron. *Acta Physica Polon.* **16**, 161.

Suffczynski, M. (1960). Optical constants of metals. *Phys. Rev.* **117**, 663.

Swenson, C. (1955). Compression of the alkali metals to 10,000 atmospheres at low temperatures. *Phys. Rev.* **99**, 423.

Tauc, J., and Abraham, A. (1961). Optical investigation of the band structure of Ge-Si alloys. *J. Phys. Chem. Solids* **20**, 192.

Tessman, J. R., Kahn, A. H., and Shockley, W. (1953). Electronic polarizabilities of ions in crystals. *Phys. Rev.* **92**, 890.

Thompson, E. D. (1960). Low lying energy levels in ferromagnetic metals. Ph. D. Thesis, Mass. Inst. Technol. (unpublished).

Thorsen, A. C., and Berlincourt, T. G. (1961a). De Haas–van Alphen effect in potassium. *Phys. Rev. Letters* **6**, 617.

Thorsen, A. C., and Berlincourt, T. G. (1961b). De Haas–van Alphen effect in rubidium. *Bull. Am. Phys. Soc.* [2] **6**, 511.

Tibbs, S. R. (1938). Electronic energy bands in metallic copper and silver. *Proc. Cambridge Phil. Soc.* **34**, 89.

Tomboulian, D. H., and Bedo, D. E. (1956). Absorption and emission spectra of Si and Ge in soft X-ray region. *Phys. Rev.* **104**, 590.
Van Hove, L. (1953). The occurrence of singularities in the elastic frequency distribution of a crystal. *Phys. Rev.* **89**, 1189.
Van Vleck, J. H. (1934). "Theory of Electric and Magnetic Susceptibilities." Oxford Univ. Press, London and New York.
Von der Lage, F. C., and Bethe, H. A. (1947). A method for obtaining electronic eigenfunctions and eigenvalues in solids with an application to sodium. *Phys. Rev.* **71**, 612.
Wainwright, T., and Parzen, G. (1953). Electronic energy bands in crystals. *Phys. Rev.* **92**, 1129.
Wallace, P. R. (1947a). The band theory of graphite. *Phys. Rev.* **71**, 622.
Wallace, P. R. (1947b). Erratum: The band theory of graphite. *Phys. Rev.* **72**, 258.
Walmsley, R. H. (1962). Linear shift of the Fermi level in iron with applied magnetic field. *Phys. Rev. Letters* **8**, 242.
Wannier, G. (1937). Structure of electronic excitation levels in insulating crystals. *Phys. Rev.* **52**, 191.
Wannier, G. (1943). Energy eigenvalues for the Coulomb potential with cut-off. I. *Phys. Rev.* **64**, 358.
Wannier, G. (1959). "Elements of Solid State Theory," p. 66. Cambridge Univ. Press, London and New York.
Wannier, G. (1960). Wave function and effective Hamiltonian for Bloch electrons in an electric field. *Phys. Rev.* **117**, 432.
Wannier, G., and Fredkin, D. R. (1962). Decoupling of Bloch bands in the presence of homogenous fields. *Phys. Rev.* **125**, 1910.
Watson, R. E. (1960a). Iron series Hartree-Fock calculations. I. *Phys. Rev.* **118**, 1036.
Watson, R. E. (1960b). Iron series Hartree-Fock calculations. II. *Phys. Rev.* **119**, 1934.
Weiner, D. (1962). De Haas–van Alphen effect in bismuth-tellurium alloys. *Phys. Rev.* **125**, 1227.
Weiss, R. J., and De Marco, J. (1958). X-ray determination of the number of $3d$ electrons in Cu, Ni, Co, Fe, and Cr. *Revs. Modern Phys.* **30**, 59.
Weiss, R. J., and Freeman, A. J. (1959). X-ray and neutron scattering from electrons in a crystalline field and the determination of the outer electron configuration in iron and nickel. *J. Phys. Chem. Solids* **10**, 147.
Whittaker, E. T., and Watson, G. N. (1952). "A Course in Modern Analysis," 4th ed, Chapter 16. Cambridge Univ. Press, London and New York.
Wigner, E. P. (1934). On the interaction of electrons in metals. *Phys. Rev.* **46**, 1002.
Wigner, E. P. (1938). Effects of electron interaction on energy levels of electrons in metals. *Trans. Faraday Soc.* **34**, 678.
Wigner, E. P. (1959). "Group Theory and Its Application to the Quantum Mechanics of Atomic Spectra" (translated by J. Griffin). Academic Press, New York.
Wigner, E., and Seitz, F. (1933). On the constitution of metallic sodium, I. *Phys. Rev.* **43**, 804.

Wigner, E., and Seitz, F. (1934). On the constitution of metallic sodium, II. *Phys. Rev.* **46**, 509.

Wigner, E., and Seitz, F. (1955). Qualitative analysis of the cohesion in metals. *Solid State Phys.* **1**, 97.

Wilson, A. H. (1936). "The Theory of Metals," 1st ed., p. 126. Cambridge Univ. Press, London and New York.

Wilson, A. H. (1953). "The Theory of Metals," 2nd ed. Cambridge Univ. Press, London and New York.

Wohlfarth, E. P. (1953). The energy band structure of a linear metal. *Proc. Phys. Soc. (London)* **A66**, 889.

Wohlfarth, E. P., and Cornwell, J. F. (1961). Critical points and ferromagnetism. *Phys. Rev. Letters* **7**, 342.

Wolff, P. A. (1961a). Localized moments in metals. *Phys. Rev.* **124**, 1030.

Wolff, P. A. (1961b). Theory of conduction band structure in bismuth (to be published).

Wolfram, T., and Callaway, J. (1962). Exchange narrowing of d-bands in ferromagnets. *Phys. Rev.* **127**, 1605.

Wollan, E. O. (1960). Magnetic coupling in crystalline compounds. A phenomenological theory of magnetism in $3d$ metals. *Phys. Rev.* **117**, 387.

Wood, J. H. (1960). Wave functions for Fe d band. *Phys. Rev.* **117**, 714.

Wood, J. H. (1962). Energy bands in iron via the augmented plane wave method. *Phys. Rev.* **126**, 517.

Woodruff, T. O. (1955). Solution of Hartree-Fock-Slater equations for Si crystal by method of orthogonalized plane waves. *Phys. Rev.* **98**, 1741.

Woodruff, T. O. (1956). Application of orthogonalized plane-wave method to Si crystal. *Phys. Rev.* **103**, 1159.

Woodruff, T. O. (1957). The orthogonalized plane-wave method. *Solid State Phys.* **4**, 367.

Yafet, Y. (1952). Calculation of the g factor of metallic sodium. *Phys. Rev.* **85**, 478.

Yafet, Y. (1957). The g value in conduction electron spin resonance. *Phys. Rev.* **106**, 679.

Yamaka, E., and Sugita, T. (1953). Energy band structure in silicon crystal. *Phys. Rev.* **90**, 992.

Zehler, V. (1953). The calculation of the energy bands in diamond. *Ann. Physik* [6] **13**, 229.

Zener, C. (1933). Remarkable optical properties of the alkali metals. *Nature* **132**, 968.

Zener, C. (1934). Theory of the electrical breakdown of solid dielectrics. *Proc. Roy. Soc.* **145**, 523.

Zener, C., and Heikes, R. R. (1953). Exchange interaction. *Revs. Modern Phys.* **25**, 191.

Ziman, J. M. (1961). The ordinary energy bands of partially disordered alloys. *Advances in Phys.* **10**, 1.

Zitter, R. N. (1962). Small field galvanomagnetic tensor of bismuth at 4.2°K. *Phys. Rev.* **127**, 1471.

Zwerdling, S., Lax, B., and Roth, L. M. (1957). Oscillatory magneto-absorption in semiconductors. *Phys. Rev.* **108**, 1402.

Zwerdling, S. Lax, B., Roth, L. M., and Button, K. J. (1959). Exciton and magneto-absorption of direct and indirect transitions in Ge. *Phys. Rev.* **114**, 80.

Zwerdling, S., Kleiner, W. H., and Theriault, W. H. (1961). Oscillatory magneto-absorption in InSb under high resolution. *J. Appl. Phys.* **32**, 2119.

Author Index

Numbers in italic show the page on which the full reference is listed.

A

Abraham, A., 165, *344*
Action, F. S., 121, *335*
Adams, E. N., 216, 220, 270, 271, 283, *327*, *333*
Adams, W. H., *327*
Altmann, S. L., 78, 86, 196, *327*
Ambegaokar, V., 315, *327*
Antoncik, E., 177, *327*
Argyres, P. N., 281, 282, *327*
Asdente, M., 196, *327*

B

Bader, F., 191, 193, *327*
Ballhausen, C. J., 193, *339*
Bardeen, I., 80, 141, 143, 287, *327*
Barnes, R. G., 155, *336*
Barrett, C., 143, *328*
Bassani, F., 76, 158, 166, 167, 168, 174, 177, *328*, *329*
Batterman, B. W., 193, *328*
Beck, P. A., 201, *330*
Bedo, D. E., 162, 164, *345*
Behringer, R. E., 179, *328*
Belding, E. F., 196, 200, *328*
Bell, D. G., 78, 158, 167, *328*
Benedek, G. B., 155, *328*
Berlincourt, T. G., 150, *344*
Bernal, M. I. M., 179, *328*
Bethe, H., 23, 77, 285, *328*, *345*
Biondi, M. A., 187, *328*
Birman, J. L., 174, *328*, *343*
Blackman, M., 265, *328*
Blatt, F., 287, 302, 306, *327*, *329*

Bloch, F., 4, *328*
Blount, E. I., 34, 109, 110, 111, 205, 206, 257, 269, 270, 278, 305, *328*, *330*
Blume, M., 128, *328*
Bohm, D., 277, *328*
Bohm, H. V., *184*, *331*
Bohr, A., 156, *329*
Bouckaert, L. P., 8, 21, 23, *329*
Bowlden, H. J., 306, *329*
Boys, S. F., 179, *328*
Brailsford, D. A., 204, *333*
Brandt, N. B., 204, *329*
Breit, G., 156, *329*, *342*
Briggs, H. B., 299, *336*
Briggs, N., 128, *328*
Brillouin, L., 6, *329*
Brooks, H., 85, 125, 126, 128, 129, 130, 140, 143, 145, 154, 155, *328*, *329*
Brown, C., 76, *329*
Brown, R. N., 205, *329*
Bruckner, K., 136, 152, *333*
Brust, D., 158, 166, 167, 168, *329*
Burdick, G. A., 181, 183, 188, *329*
Burstein, E., 175, 302, 306, *329*
Butcher, P. N., *329*
Button, K. J., 306, *347*

C

Callaway, J., 43, 64, 76, 85, 121, 124, 130, 131, 133, 139, 140, 143, 146, 147, 148, 151, 152, 154, 155, 158, 168, 169, 176, 193, 194, 195, 196, 202, 306, *329*, *330*, *335*, *346*
Cardona, M., 165, *330*
Carr, W. J., 137, 138, *330*

Carter, J. L., 210, *330*
Celli, V., 76, 158, 167, 168, 174, 177, *328*
Celtin, B. B., 204, *333*
Chambers, R. F., 268, *330*
Cheng, C. H., 201, *330*
Chipman, D. R., 193, *328*
Chodorow, M. I., 101, 181, 189, *330*
Christian, S., 163, 164, *331*, *333*, *336*
Clogston, A. M., 227, *330*
Cohan, N. V., 196, *327*
Cohen, M. H., 76, 80, 124, 146, 149, 157, 188, 203, 204, 205, 208, 257, 269, 297, 298, 299, *330*, *331*
Cohen, M. L., 158, 166, 167, *329*
Coldwell-Horsfall, R., 137, *331*
Coldwell-Horsfall, R. A., 137, 138, *330*,
Cooper, M. J., 193, *331*
Corak, W. S., 187, *331*
Corbato, F. J., 210, *331*
Cornwell, J., 148, 181, 188, 200, *331*, *346*
Crawford, M., 156, *331*
Crisp, R. S., 149, *331*

D

Dash, W. C., *331*
Dehlinger, U., 191, 193, *327*
de Hoffman, F., 285, *328*
De Marco, J., 193, *328*, *345*
des Cloizeau, J., 220, *331*
Dresselhaus, G., 161, 164, 172, *331*
Dresselhaus, M. S., 174, 205, *329*
Drude, P., 301, *331*
Du Bois, D. F., 152, *331*

E

Easterling, V. J., 184, *331*
Edwards, D. M., 195, *330*
Ehrenreich, H., 166, 174, 176, 188, 299, *331*
Elliott, R. J., 51, 54, 159, 166, *332*
Enz, C. P., 270, *332*
Erdelyi, A., 263, *332*
Everett, G. E., 208, *332*

Evtuhov, V., 257, *332*
Eyges, L., 57, *332*

F

Falicov, L. M., 269, *330*
Fan, H. Y., 299, *344*
Fawcett, E., 177, 201, *332*
Feher, G., 160, 161, *332*, *334*
Fein, A. E., 137, 138, *330*
Fermi, E., 156, *332*
Fletcher, G. C., 196, *332*
Flower, M., 61, *332*
Fock, V., 115, *332*
Foner, S., 193, *332*
Frankl, D., 163, *338*
Franz, W., 282, 306, *332*
Fredkin, D. R., 246, 282, *332*, *345*
Freeman, A. J., 200, 203, *332*, *345*
Friedel, J., 183, 231, *327*, *333*
Fröhlich, H., 141, *333*
Froese, C., 179, *333*
Fuchs, K., 136, 181, *333*
Fukuchi, M., 181, *333*

G

Gager, W. B., 129, 154, 155, *339*
Galt, J. K., 204, 208, *333*, *343*
Ganzhorn, K., 191, 193, *327*
Garcia-Moliner, F., 184, *333*
Garfunkel, M. P., 187, *331*
Gashimzade, F. M., 158, 174, *333*
Gaspar, R., 177, *333*
Gell-Mann, M., 136, 152, *333*
Glasser, M. L., 124, *330*
Glicksman, M., 163, *333*
Goertzel, G., 87, *333*
Goldberg, L., 196, *339*
Goldberger, M. L., 271, *333*
Goodenough, J. B., 191, 193, *333*
Goodings, D. A., 157, 202, *331*, *333*
Goodman, R. R., 257, *333*, *339*
Gorin, E., 124, 125, *333*
Graham, T. P., 155, *336*
Greene, J. B., 196, *333*

Grimes, C. C., 150, *333*
Gubanov, A. I., 174, *333*
Gunnersen, E. M., 177, *334*

H

Hagstrum, H. D., 162, 164, *334*
Hall, L. J., 287, *327*
Hall, G. G., 158, *334*
Ham, F. S., 80, 86, 95, 125, 126, 128, 129, 130, 143, 147, 148, 151, 152, 154, 155, *330*, *334*
Han-Ying, 164, *331*
Hanna, S. S., 157, 202, *334*
Harrison, W., 44, 45, 177, 179, 181, 208, *334*
Hartree, D. R., 118, 121, 143, *334*
Hartree, W., 143, *334*
Hebborn, J. E., 269, 273, *334*
Heberle, J., 157, 202, *334*
Heikes, R. R., 192, *346*
Heine, V., 76, 146, 149, 157, 171, 177, 188, 202, 297, *330*, *331*, *333*, *334*
Heller, G. S., 160, 163, *342*, *344*
Hellmann, H., 131, *334*
Hensel, J. C., 161, *334*
Hensman, R., 158, 167, *328*
Herman, F., 121, 133, 158, 164, 168, 169, 176, *334*, *335*
Herring, C., 36, 54, 68, 159, *335*
Holmes, D. K., 158, *335*
Houston, M. V., 282, *335*
Howarth, D. J., 80, 86, 95, 101, 181, *335*
Howling, D. H., 177, *335*
Hsu, Y. C., 183, *335*
Hubbard, J., 137, 138, *335*
Hughes, A. J., 43, *330*
Hund, F., 158, *335*

I

Ives, H. E., 299, *336*

J

Jain, A. L., 204, *336*
Jeffreys, B. S., 225, 262, *336*
Jeffreys, H., 225, 262, *336*
Jenkins, D. P., 158, 167, *328*, *336*
Johnson, E., 164, *336*
Johnston, D. F., 210, 211, *336*
Jones, H., 12, 55, 80, 86, *335*, *336*
Jones, W. H., 155, *336*

K

Kahn, A. H., 164, 299, *336*, *344*
Kalinkina, I. N., 205, *336*
Kambe, K., 126, 181, 182, *336*
Kane, E. O., 34, 158, 172, 173, 174, 205, 281, 282, 320, *336*
Kassatdschkin, W., 131, *334*
Keldysh, L. V., 282, 306, *336*
Kern, B., 162, *336*
Khartsiev, V. E., 158, 174, *333*
Kimball, G. E., 158, *336*
Kip, A. F., 161, 164, 172, 184, *331*, *336*, *337*
Kittel, C., 161, 172, 242, *331*, *337*
Kjeldaas, T., 154, 155, 246, 270, *337*
Klauder, J. R., 184, *337*
Klein, A., 315, *337*
Kleiner, W. H., 174, *347*
Kleinman, L., 74, 123, 158, 169, 171, 172, 174, *337*, *341*
Knight, W. D., 154, 155, *337*
Kobayasi, S., 174, *337*
Koenig, S. H., 204, *336*
Kohn, W., 86, 111, 143, 154, 155, 216, 241, 242, 244, 246, 270, 315, 325, 327, 330, *337*, *339*, *343*
Koopmans, T., 117, *337*
Korringa, J., 86, *338*
Kosevich, A. M., 265, *338*
Koster, G. F., 12, 18, 19, 107, 108, 112, 158, 168, 184, 196, 220, 221, 225, 229, 232, *338*, *343*
Krutter, H. M., 181, 182, *338*
Ku, 164, *331*
Kuhn, T. S., 125, 126, 127, 128, 145, *338*
Kunzler, J. E., 184, *337*
Kushida, T., 155, *328*

AUTHOR INDEX

L

Lampert, M., 242, *338*
Landau, L., 253, *338*
Langenberg, D. N., 184, *337, 338*
Lax, B., 133, 163, 205, 257, 302, 306, *329, 338, 342, 347*
Lax, M., 227, *338*
Leigh, R. S., 95, 101, *338*
Lerner, L. S., 205, *338*
Levinger, B., 163, *338*
Lien, W. H., 151, *338*
Lifshitz, I. M., 229, 265, *338*
Liu, L., 158, 173, 257, *338*
Löwdin, P., 114, *338*
Lomer, W. M., 191, 196, 209, *338, 339*
Luttinger, J. M., 43, 150, 216, 242, 246, 257, *339*

M

McAfee, K. B., 282, *339*
McClure, J. W., 211, 212, *339*
McDougall, J., 171, *339*
Macfarlane, G. G., 160, 289, *339*
McLean, T. P., 160, *339*
Manning, M. F., 196, 199, *333, 339*
Maradudin, A. A., 137, *331*
March, N. H., 61, *332*
Marshall, W., 191, 202, *339*
Martin, D. L., 151, *339*
Mase, S., 208, *339*
Maslen, V., 121, *339*
Matyas, Z., 177, *339*
Mavroides, J. G., 205, *329*
Mendoza, E., 177, *335*
Merritt, F. R., 204, 208, *333, 343*
Miasek, M., 107, *339*
Milford, F. J., 129, 154, *339*
Mitchell, A. H., 242, *337*
Moffitt, W., 193, *339*
Montroll, E. W., 39, *340*
Moore, T. W., 177, 184, *337, 338, 340*
Morgan, D. F. Jr., 143, 154, 155, *330*
Morita, A., 158, 208, *340*
Morse, M., 39, 184, 189, *340*

Moss, T. S., 176, *340*
Mott, N. F., 183, 189, 191, 193, *34*
Mrowka, B., 158, *335*
Mullaney, J. F., 158, *340*
Murnaghan, F. D., 50, *340*
Murray, A. M., 61, *332*
Myers, A., 184, 189, *340*

N

Nettel, S. J., 325, *337*
Newman, R., *331*
Nozieres, P., 138, 211, 212, *340*
Nran'yan, A. A., 158, 174, *333, 340*

O

Onsager, L., 265, *340*
Overhauser, A., 118, 324, *340*

P

Parmenter, R. H., 174, *341*
Parzen, G., 112, *341, 345*
Paul, W., 163, *341*
Pauling, L., 191, *341*
Peierls, R., 270, *341*
Perlow, G. J., 157, 202, *334*
Philipp, H. R., 165, 166, 176, 188, 299, *331, 341*
Phillips, J., *337*
Phillips, J. C., 39, 74, 117, 123, 158, 160, 161, 162, 166, 167, 168, 169, 171, 172, 174, 176, *329, 331, 337, 341*
Phillips, N. E., 151, 205, *338, 341*
Picus, G. S., 302, 306, *329*
Pincherle, L., 55, 158, 167, *341*
Pines, D., 136, 137, 138, 154, 155, 179, *340, 341*
Pippard, A. B., 133, 184, 266, 269, *341*
Pratt, G. W., 116, 202, *341*
Preston, R. S., 157, 202, *334*
Price, P. J., 282, *341*
Prokofjew, W. K., 124, *342*
Pugh, E. M., 193, *332*

AUTHOR INDEX

Q
Quarrington, J. E., 160, *339*
Quelle, F. W., 158, *342*
Quinn, J., 152, *342*

R
Racah, G., 156, *342*
Radcliffe, J. M., 282, *341*
Raimes, S., 145, 146, 177, *342*
Rauch, C. J., 160, *342*
Rayne, J. A., 187, *328*
Raynor, G. V., 133, *342*
Reed, W. A., 201, *332*
Reeh, H., 139, *342*
Reitz, J. R., 55, *342*
Roaf, D. J., 184, 185, *342*
Roberts, V., 160, 289, *339*
Rosenthal, J., 156, *342*
Rostoker, N., 86, *337*
Roth, L. M., 163, 246, 256, 257, 270, 302, 306, *347*, *347*
Ryder, E. J., 282, *339*
Ryter, C., 155, *342*

S
Saffern, M. M., 95, 96, *342*
Satterhwaite, C. B., 187, *331*
Schawlow, A., 156, *331*
Schecter, D., 244, *337*
Schiff, B., 196, *342*
Schiff, L. I., 46, 230, 288, *342*
Schlosser, H., 57, 95, 101, 147, 148, *342*
Schmid, L. A., 158, *342*
Schulz, L. G., 188, *342*
Schweber, S., 285, *328*
Segall, B., 86, 95, 158, 167, 177, 180, 181, 183, 185, 187, 188, 190, *334*, *343*
Seitz, F., 77, 115, 120, 124, 133, 134, 135, *343*, *345*, *346*
Shakin, C., 174, *343*
Shockley, W., 33, 77, 282, 299, *339*, *343*, *344*
Shoenberg, D., 184, *343*

Silverman, R. A., 80, 143, *343*
Simon, R. E., 165, *344*
Sirounean, V., 85, 130, *330*
Slater, J. C., 77, 95, 96, 107, 108, 115, 116, 120, 133, 158, 168, 169, 181, 184, 191, 196, 200, 220, 221, 225, 232, *338*, *342*, *343*
Skillman, S., 168, 169, *335*
Slonczewski, J. C., 209, 211, *343*
Smith, G. E., 208, *343*
Smoluchowski, R., 8, 21, 23, *329*
Sommers, H. S., Jr., 165, *330*
Sondheimer, E. H., 257, 269, 273, *334*, *344*
Sparks, M., 282, *339*
Spicer, W. E., 165, *344*
Spitzer, W. G., 299, *344*
Spong, F. W., 177, *340*
Steinberger, J., 196, *344*
Stern, F., 175, 196, 197, 198, 200, *344*
Sternheimer, R. M., 139, 140, *344*
Stevens, K. W. H., 191, 193, *340*
Stickler, J. J., 160, 163, *342*, *344*
Strelkov, P. G., 205, *336*
Sturge, M. D., 175, *344*
Suffczynski, M., 189, 195, 196, *344*
Sugita, T., 158, *346*
Swenson, C. A., 143, 145, *344*

T
Taft, E. A., 165, *341*
Talley, R. M., 175, *344*
Tauc, J., 165, *344*
Tessman, J. R., 299, *344*
Theriault, W. H., 174, *347*
Thompson, E. D., 115, 118, *344*
Thorsen, A. C., 149, 150, *344*
Tibbs, S. R., 181, *344*
Tomboulian, D. H., 162, 164, *345*
Tralli, N., 87, *333*

V
Van Hove, L., 39, *345*
Van Vleck, J. H., 125, 140, 145, *338*, *345*

Vincent, D. H., 157, 202, *334*
Von der Lage, F. C., 23, 77, *345*

W

Wagoner, G., 164, *331*
Wainwright, T., 112, *345*
Walker, C. T., 184, 189, *340*
Wallace, P. R., 209, 210, *345*
Wallis, R. F., 302, 306, *329*
Walmsley, R. H., 201, *345*
Walton, A. K., 176, *340*
Wannier, G., 41, 109, 126, 128, 220, 282, *345*
Wannier, G. H., 246, 282, 283, *332*
Ward, J. C., 43, 150, *339*
Watson, G. N., 128, *345*
Watson, R. E., 196, 203, *332, 345*
Wei, C. T., 201, *330*
Weiner, D., 205, *345*
Weiss, P. R., 209, *343*
Weiss, R. J., 193, 200, 211, *345*
Weisskopf, V., 156, *329*
Wexler, A., 187, *331*
Wick, G. C., 196, *344*
Wigner, E., 8, 12, 13, 14, 16, 19, 21, 23, 52, 54, 57, 58, 77, 124, 133, 134, 135 137, *329, 345, 346*
Williams, S. E., 149, *331*
Wilson, A. H., 257, 258, 270, 290, *344, 346*
Whittaker, E. T., 128, *345*
Wohlfarth, E. P., 107, 196, 200, *332, 346*
Wolff, P. A., 205, 227, *346*
Wolfram, T., 193, *346*
Wollan, E. O., 191, 193, *346*
Wood, J. H., 101, 196, 199, *346*
Woodruff, T. O., 158, *346*
Woods, R. D., 85, 130, *330*

Y

Yafet, Y., 257, *346*
Yager, W. A., 204, *333*
Yamaka, E., 158, *346*
Yoshimine, M., 158, 174, *328*

Z

Zehler, V., 158, *346*
Zeiger, H. D., 163, *344*
Zeiger, H. J., 160, *342*
Zener, C., 192, 282, 301, *346*
Ziman, J. M., 186, *346*
Zimmerman, J. E., 177, *335*
Zitter, R. N., 204, *346*
Zwerdling, S., 163, 174, 257, 302, 306, *342, 347*

Subject Index

A

Absorption constant, 286, *see also* Optical properties
Acceleration, 279
Accidental degeneracy, 36
Alkali metals, 133–157
    band structure, 146–154
    cohesive energy, 133, 140
    Fermi surfaces, 147
    lattice constant and compressibility, 142
    optical properties, 299
    wave functions, 154
Aluminum, 146, 177–181
Atomic units, 55
Augmented plane waves, 95–102

B

Bismuth, 203–208
Bloch functions, *see* Tight binding approximation
Bloch's Theorem, 2–4, 46, 56, 69, 78, 87
Boundary correction, 134
Brillouin zone, 5–9
    body centered cubic lattice, 7
    face centered cubic lattice, 8

C

Cellular method, 77–85
    boundary conditions, 78
    spherical approximation, 80
Cesium, 142, 148, 151, 155
Chemical potential, *see* Fermi energy
Chromium, 196, 326
Cohesive energy, 133, 145
    alkali metals, 133, 140
    iron, 197
    noble metals, 182
Complex wave vector, 34, 321
Compressibility, 141
Conductivity, electrical, 290
Copper, 181, 182, 184, 185, 188
Correlation energy, 135–138
Critical points, 39
Crystal field theory, 193
Crystal momentum representation, 216, 277, 317
Crystal potential, 120–124
Cyclotron frequency, 253, 265–269
Cyclotron resonance, 304, *see also* cyclotron frequency
    alkali metals, 150
    aluminum, 180
    bismuth, 204
    noble metals, 184
d Bands, 191–203
de Haas-van Alphen effect, 262
    alkali metals, 149
    aluminum, 180
    bismuth, 204
    noble metals, 184

D

Density of states, 37–43, 229
Determinantal wave function, 114, 119
Diamond, 158–173
Dielectric constant, 296, 318
Dirac equation, 46, 85
Double group, 48

E

Effective mass, 29–37, 80–85
    alkali metals, 151
    degenerate bands, 33

equation, impurities, 233, 315–319
equation, magnetic effects, 250
noble metals, 188
optical, 188, 297
thermal, 151, 300
valence crystals, 160, 163, 174, 175
Electric fields, 276–283, 320–323
Electron lattice, 136
Energy bands, 6
connections, 35
degeneracies, 37, 53
free electrons, 9
methods of calculation, 55–131
Exchange energy, 115, 121, 135
Exchange potential, 120

F

"f" sum rule, 31
Fermi energy, 38, 150
in a magnetic field, 264
Fermi surface, 43–46
alkali metals, 147
aluminum, 177
bismuth, 204
nickel, 201
noble metals, 183
Ferromagnetism, 192
Fourier coefficient of potential, 63, 65
Free electrons, 9–11
exchange energy, 121
Fermi surface, 43
magnetic susceptibility, 257–265
Free energy, 258
Friedel sum rule, 231

G

g-factor, 252, 256
Gallium arsenide, 175
Germanium, 34, 158–173, 242
band parameters, 163
Germanium-silicon alloys, 164
Gold, 182, 185, 188
Graphite, 209
Green's function, 86, 227, *see also* variational methods

H

Hartree-Fock equations, 112–120

I

Indium antimonide, 34, 174
Iron, 196–201
Irreducible representations, 16–28,
see also space groups
characters, 17
character tables, 22, 23, 25–28
compatibility tables, 36
direct product, 32
double group, 48

K

Knight shift, 154
Koopmans' theorem, 117
$K \cdot p$ method, 173
Kramers' theorem, 53
Kubic harmonics, 77

L

Landau levels, 253, 265
Lattice constant, equilibrium, 141
Lithium, 142, 147, 148, 151, 155
Luttinger-Kohn representation, 216–219, 234, 246

M

Magnetic breakdown, 269
Magnetic fields, 244–276
Magnetic susceptibility, 245
Diamagnetic, steady, 269–276
Free electrons, 257–265
Magneto-optical effect, 304
Many body problem
effective mass equation, 315
Momentum matrix element, 30, 32, 287
Mossbauer effect, 202

N

Nickel, 182, 196, 201
Noble metals, 181–190
cohesive energies, 182

SUBJECT INDEX

effective masses, 188, 299
Fermi surface, 183
Normal level order, 152

O

Optical properties, 283–306
  semi-conductors, 283–289
    direct transitions, 287
    indirect transitions, 289
  free carriers, 289–302
    relaxation processes 301
    in a magnetic field, 302–306
Orbits, open, 186, 269
Orthogonalized plane waves, 68–74
  orthogonality coefficients, 70

P

Partition function, 258, 270
Plane wave expansions, 56–68
  symmetrized linear combinations, 57, 307
Point impurities, 221–244, *see also* Effective mass
  bound states, 222
  Koster-Slater model, 221
  resonant (virtual) states, 233
  scattering amplitude, 230
Polarization, 130, 139, 140
  exchange, 157, 202
  potential, 139
Potassium, 140, 142, 148, 150, 151, 155
Pseudo potential, 74–76, 130–131, 167, 169, 178, 186

Q

Quantum defect method, 124–131

R

Reciprocal lattice, 4
Rubidium, 142, 148, 151, 155

S

Self consistent field, 118, 123
Silicon, 34, 158–173
  band parameters, 160
Silver, 182, 185, 188
Sodium, 134, 135, 142, 148, 150, 151, 155
Space groups, 11–16, *see also* irreducible representations; double group
  class, 13
  point group, 13
  full cubic group, 14
  symmorphic, 14
Specific heat electrons
  alkali metals, 150
  noble metals, 187
  transition elements, 201
Spin density waves, 324
Spin orbit coupling, 46–51, 163, 172, 255
Stark levels, 281, 306
Summation relations, 311

T

Tight binding method, 102–108, 197
  overlap integral, 104
  two center approximation, 104
Time reversal symmetry, 52–54, 292
Titanium, 196
Transition elements, 191–203
Translation operators, 2, 113, *see also* space groups
Tunneling, 276, 282, 320–323

U

Unitary transformation, 235, 248

V

Variational methods, 86–95, *see also* Green's function
  integral equation, 88

W

Wannier functions, 108–112, 119, 215, 220, 223, 227
Wigner Seitz approximation, 124
Wigner Seitz cell, 6

Z

Zinc blende structures, 174